ANIMAL MANURE RECYCLING: TREATMENT AND MANAGEMENT

Animal Manure Recycling: Treatment and Management

Editor:

Vikas Mishra, Ph.D.

Delve Publishing
www.delvepublishing.com

Animal Manure Recycling: Treatment and Management
Editor: **Vikas Mishra, Ph.D.**

708 3rd Avenue, 6th Floor
New York, NY 10017
United States of America

www.delvepublishing.com

ISBN: 978-1-68095-773-0

Library of Congress Control Number: 2016951529

About the Editor

Vikas Mishra, Ph.D.

Vikas obtained his Ph.D. from King George's Medical University, India in 2010. His interests span from bio-resources, fermentation technology to medical virology and neuroscience research. He is currently working as a research associate at the University of British Columbia.

Contents

List of Contributors

A. Nasiru

Environmental Technology Division, School of Industrial Technology, Universiti Sains Malaysia, 11800 Penang, Malaysia

N. Ismail

Environmental Technology Division, School of Industrial Technology, Universiti Sains Malaysia, 11800 Penang, Malaysia

M. H. Ibrahim

Environmental Technology Division, School of Industrial Technology, Universiti Sains Malaysia, 11800 Penang, Malaysia

Hargopal Singh
Department of Soil Science, PAU, Ludhiana-141 004

Pritpal Singh
Department of Soil Science, PAU, Ludhiana-141 004

S.S. Hundal
Department of Zoology, PAU, Ludhiana-140 004, India

Andrew Sharpley
University of Arkansas/Division of Agriculture University of Arkansas System - Dept. of Crop, Soil and Environmental Sciences, 115 Plant Sciences Building – 72701 – Fayetteville, AR - USA

Ke Wang
School of Municipal and Environmental Engineering, State Key Laboratory of Urban Water Resource and Environment (SKLUWER), Harbin Institute of Technology, 73 Huanghe road, Harbin, Heilongjiang 150090, China

Weiguang Li
School of Municipal and Environmental Engineering, State Key Laboratory of Urban Water Resource and Environment (SKLUWER), Harbin Institute of Technology, 73 Huanghe road, Harbin, Heilongjiang 150090, China

Xiangkun Li
School of Municipal and Environmental Engineering, State Key Laboratory of Urban

Water Resource and Environment (SKLUWER), Harbin Institute of Technology, 73 Huanghe road, Harbin, Heilongjiang 150090, China

Nanqi Ren
School of Municipal and Environmental Engineering, State Key Laboratory of Urban Water Resource and Environment (SKLUWER), Harbin Institute of Technology, 73 Huanghe road, Harbin, Heilongjiang 150090, China

Francesca Nardin
Faculty of Science and Technology, Free University of Bolzano-Bozen, piazza Università 5, Bozen I-39100, Italy

Fabrizio Mazzetto
Faculty of Science and Technology, Free University of Bolzano-Bozen, piazza Università 5, Bozen I-39100, Italy

Kwang-Hwa Jeong
Animal Environment Division, National Institute of Animal Science, R.D.A., Suwon 441-706, Korea

Modabber Ahmed Khan
Animal Environment Division, National Institute of Animal Science, R.D.A., Suwon 441-706, Korea

Ho Kang
Department of Environmental Engineering, Chungnam National University, Daejeon 305-764, Korea

Jung-Kon Kim
Animal Environment Division, National Institute of Animal Science, R.D.A., Suwon 441-706, Korea

Jong-Hoon Kwag
Animal Environment Division, National Institute of Animal Science, R.D.A., Suwon 441-706, Korea

S. M. Thamsborg
Organic Animal Health and Production, Dept. of Animal Science and Animal Health, The Royal Veterinary and Agricultural University

Jayakumar Pathma
Department of Biotechnology School of Life Sciences, Pondicherry University

Natarajan Sakthivel
Department of Biotechnology School of Life Sciences, Pondicherry University

Tânia M. B. Dos Santos
Zootechnician, Prof. Doctor, Department of Animal Science, State University of Mato Grosso do Sul-UEMS / Aquidauana-MS

Paula S. F. Trevizan
Zootechnician, Master of Animal Science - Animal Production in the Cerrado Pantanal, State University of Mato Grosso do Sul-UEMS / Aquidauana-MS

Cristiane A. N. Xavier
Zootechnician Doctor Fellow PNPD / CAPES, Department of Animal Science, Federal University of Mato Grosso South UFMS / Campo Grande-MS

Charles Kiefer
Zootechnician, Prof. Doctor, Department of Animal Science, Federal University of Mato Grosso do Sul-UFMS / Campo Grande-MS

André L. J. Ferraz
Zootechnician, Prof. Doctor, Department of Animal Science, State University of Mato Grosso do Sul-UEMS / Aquidauana-MS

M. García
Total Area Microbiology, Biomedical Sciences Institute, Faculty of Health Sciences, University CEU Cardenal Herrera, Moncada (Valencia), Spain

S. Fernández-Barredo
Veterinary Diagnostic Center (CEDIVET), Valencia, Spain

M. T. Pérez-Gracia
Total Area Microbiology, Biomedical Sciences Institute, Faculty of Health Sciences, University CEU Cardenal Herrera, Moncada (Valencia), Spain
Veterinary Diagnostic Center (CEDIVET), Valencia, Spain

Budiyono
Department of Chemical Engineering, University of Diponegoro, Jl. Prof. Sudharto SH, Semarang, Indonesia

I.N. Widiasa
Department of Chemical Engineering, University of Diponegoro, Jl. Prof. Sudharto SH, Semarang, Indonesia

Seno Johari
Department of Animal Agriculture, University of Diponegoro, Jl. Prof. Sudharto SH Semarang, Indonesia

Sunarso
Department of Animal Agriculture, University of Diponegoro, Jl. Prof. Sudharto SH Semarang, Indonesia

Paul J. A. Withers
School of Environment, Natural Resources and Geography, Bangor University, Bangor, Gwynedd LL57 2UW, UK

James J. Elser
School of Life Sciences, Arizona State University, Tempe, Arizona 85287-4501, USA

Julian Hilton
Aleff Group, Cross Keys Centre, 36 Erith High Street, Erith, London, DA8 1QY, UK

Hisao Ohtake
Department of Biotechnology, Osaka University, Yamada-oka 2-1, Suita-shi, Osaka, Japan

Willem J. Schipper
Willem Schipper Consulting, Oude Vlissingseweg 4, 4336 AD Middelburg, The Netherlands

Kimo C. van Dijk
Department of Soil Quality, Wageningen University, 6700 AA Wageningen, The Netherlands

Fengsong Zhang
State Key Laboratory of Water Environment Simulation, School of Environment, Beijing Normal University, Beijing 100875, China
State Key Laboratory of Environmental Criteria and Risk Assessment, Chinese Research Academy of Environmental Sciences, Beijing 100012, China

Yanxia Li
State Key Laboratory of Water Environment Simulation, School of Environment, Beijing Normal University, Beijing 100875, China

Xiong Xiong
State Key Laboratory of Water Environment Simulation, School of Environment, Beijing Normal University, Beijing 100875, China

Ming Yang
State Key Laboratory of Water Environment Simulation, School of Environment, Beijing Normal University, Beijing 100875, China

Wei Li
State Key Laboratory of Water Environment Simulation, School of Environment, Beijing Normal University, Beijing 100875, China

Anita Maienza
National Research Centre for Agriculture, Research Center for the Soil-Plant System (CRA-RPS), Via della Navicella 2-4, 00184 Roma, Italy
National Research Council, Institute of Biometeorology (IBIMET-CNR), Via Caproni 8, 50145 Firenze, Italy

Giovanni Mughini
National Research Centre for Agriculture, Research Unit for Intensive Wood Production, Via Valle della Quistione 27, 00166 Roma, Italy

Luca Salvati
National Research Centre for Agriculture, Research Center for the Soil-Plant System (CRA-RPS), Via della Navicella 2-4, 00184 Roma, Italy

Anna Benedetti
National Research Centre for Agriculture, Research Center for the Soil-Plant System (CRA-RPS), Via della Navicella 2-4, 00184 Roma, Italy

Maria Teresa Dell'Abate
National Research Centre for Agriculture, Research Center for the Soil-Plant System (CRA-RPS), Via della Navicella 2-4, 00184 Roma, Italy

Amira Oun
Department of Civil and Environmental Engineering, Michigan State University, East Lansing, 48824, USA

Arun Kumar
Department of Civil Engineering, Indian Institute of Technology Delhi, New Delhi 110016, India

Timothy Harrigan
Department of Biosystems and Agricultural Engineering, Michigan State University, East Lansing, 48824, USA

Andreas Angelakis
National Agricultural Research Foundation, Iraklion 71307, Hellas

Irene Xagoraraki
Department of Civil and Environmental Engineering, Michigan State University, East Lansing, 48824, USA

Sasikumar Elumalai
Department of Biological Systems Engineering, 460 Henry Mall, University of Wisconsin-Madison, Madison, WI 53706, USA

Aicardo Roa- Espinosa
Department of Biological Systems Engineering, 460 Henry Mall, University of Wisconsin-Madison, Madison, WI 53706, USA
Soil Net LLC, 560 Enterprise Ave, Belleville, WI 53508, USA

John L Markley
Department of Biochemistry, 433 Babcock Drive, University of Wisconsin-Madison, Madison, WI 53706, USA.

Troy M Runge
Department of Biological Systems Engineering, 460 Henry Mall, University of Wisconsin-Madison, Madison, WI 53706, USA

Preface

Many environmental problems of current concern are due to the high production and local accumulations of organic wastes that are too great for the basic degradation processes inherent in nature. With adequate application rates, animal manure constitutes a valuable resource as a soil fertilizer, as it provides a high content of macro- and micronutrients for crop growth and represents a low-cost, environmental-friendly alternative to mineral fertilizers. Improper disposal of animal manure poses a threat to the environment. Recycling manure cannot only protect the local environment but can also save you money. Spreading animal manure can contribute to harmful urban run-off that contains bacteria and nutrients that pollute our local groundwater. Throwing it away in the trash takes up valuable landfill space and wastes natural resources. A rapidly changing and expanding livestock and poultry production sector, in addition to increasing human waste, due to ever increasing human population, means there is also an increase in livestock manure, which if not utilized and managed can lead to a range of environmental problems. Historically, manure was used as a fertilizer, but after overuse of chemical fertilizers for decades, manure is again being used for organic farming for crop fertilization and soil amelioration, and biofuel production. This book, "Animal Manure Recycling Treatment and Management" presents an overview of environment friendly principles and practices used for managing animal manure more efficiently and in a sustainable manner for creating a consistent fertilizer product. In addition to the manure management, this book also describes technologies for converting manure to biogas and management of gaseous emissions and run off of manure derived greenhouse gases. Livestock production is experiencing dynamic growth and specialisation, and is facing the challenge of ensuring that changes in production systems include the development of environmentally friendly manure management systems. This book provides an overview of manure management systems and technologies for removing solid and liquid manure from animal houses, storage, transport and application in the field. The overview includes specific descriptions and applicability of potential technologies in terms of capability, costs, environmental impact and so on. Methods needed to manage the manure are also presented. The introduction of appropriate management technologies could thus mitigate the health and environmental risks associated with the overproduction of organic wastes derived from the livestock industry by stabilizing them before their use or disposal. Stabilisation involves the decomposition of an organic material to the extent of eliminating the hazards and is normally reflected by decreases in microbial biomass and its activity and in concentrations.

Editor

Vikas Mishra, Ph.D.

INTRODUCTION

Manure, organic material that is used to fertilize land, usually consisting of the feces and urine of domestic livestock, with or without accompanying litter such as straw, hay, or bedding. Farm animals void most of the nitrogen, phosphorus, and potassium that is present in the food they eat, and this constitutes an enormous fertility resource. In some countries, human excrement is also used. Livestock manure is less rich in nitrogen, phosphorus, and potash than synthetic fertilizers and hence must be applied in much greater quantities than the latter. A ton of manure from cattle, hogs, or horses usually contains only 10 pounds of nitrogen, 5 pounds of phosphorus pentoxide, and 10 pounds of potash. But manure is rich in organic matter, or humus, and thus improves the soil's capacity to absorb and store water, thus preventing erosion. Much of the potassium and nitrogen in manure can be lost through leaching if the material is exposed to rainfall before being applied to the field. These nutrient losses may be prevented by such methods as stacking manure under cover or in pits to prevent leaching, spreading it on fields as soon as it is feasible, and spreading preservative materials in the stable. A green manure is a cover crop of some kind, such as rye, that is plowed under while still green to add fertility and conditioning to the soil.

The use of manure as fertilizer dates to the beginnings of agriculture. On modern farms manure is usually applied with a manure spreader, a four-wheeled self-propelled or two-wheeled tractor-drawn wagon. As the spreader moves, a drag-chain conveyor located at the bottom of the box sweeps the manure to the rear, where it is successively shredded by a pair of beaters before being spread by rotating spiral fins. Home gardeners like to use well-rotted manure, since it is less odorous, more easily spread, and less likely to "burn" plants. See also fertilizer. Animal manures have been handled as a solid and were either deposited directly to pasture by the animals or collected along with the bedding used in the animal shelter and applied to land as a crop nutrient. As the number of animals on farms has increased the need for more efficient methods of manure management has also developed. The emphasis of the resources on these web pages are to provide an overview of manure treatment technologies, research being done in this field, and highlight the considerations involved in selecting candidates for a particular farm.

Why Treat Manure

Manure from animals is a very wet, bulky material, containing as much as 90% water. Of the 10% that is dry matter, much of it is carbon. The manure handling system can impact the characteristics of the animal waste stream. Water-based waste management systems may have a liquid stream with greater than 98% moisture while manure scraped from open lots can contain a very high percentage of soil. As a result, the plant nutrient content of the manure is a very small portion of the weight/volume. Hauling nutrients to cropland in the form of manure can get very expensive compared to hauling the same amount of nitrogen or phosphorus in the form of commercial fertilizer.

Correcting the imbalance in manure nutrient distribution will involve finding ways to economically move excess nutrients off the livestock or poultry farm. Many alternatives to land application of raw manure are being examined. Many of these "manure treatment technologies" have been used in one form or another for many years, while others are recent solutions. Often, technologies are used in combination to create a system that can be tailored to the species, management of the farm, climate, or other factors. Technology that may be very successful on one farm may not be appropriate for another.

To evaluate whether you need to treat manure produced on your farm, questions to consider include:

- Do you have enough land to apply all the manure produced to meet the agronomic needs of crops?
- Is salt accumulating in the soil where manure is being applied?
- Is nitrate or salt in groundwater or nutrient enrichment of surface waters a local concern?
- Do community members complain about your manure application activities?
- Are odors and/or gaseous emissions from your farm a concern to citizens and/or a regulatory authority?

If the answer to any of these questions is yes, then you are a potential candidate for a manure-treatment technology. The objectives of manure treatment depend on the needs of a particular farm and applicable regulations. However, some general objectives of manure treatment are:

- stabilization of the manure,
- odor reduction,
- nutrient management,
- proper storage and utilization,
- pathogen reduction, and

- reduction of gaseous emissions such as ammonia, hydrogen sulfide, and greenhouse gases.

Treating manure can result in:

- Reduced amounts of nutrients; thus, less land is required for manure application. Alternatively, a treatment technology can be used to produce "designer" manure in which nutrients are in the right agronomic proportions for use as a fertilizer. Producing designer manure minimizes the risk of nutrient pollution of the environment.
- Reduced manure mass. Treatment results in less material to transport and apply to cropland.
- Controlled pathogens and antibiotics. Reduced concentrations of pathogens and antibiotics in animal manure lessen biosecurity and health risks.
- Production of value-added byproducts. Treatment processes that produce energy (heat, electricity), fertilizer, and organic materials can be used on the farm or sold to provide an extra income source to the farm.
- Increased costs and labor requirement to manage manure.

Principles of Manure-Treatment Processes

The basic principles underlying manure-treatment systems or technologies can be grouped into: physical, chemical, and biological processes. Manure-treatment technologies typically include one or a combination of these basic processes.

Physical Processes

The treatment of manure or wastewater can happen by or through application of physical forces. Physical processes include those technologies that involve liquid-solids separation and/or the use of heat and pressure. Liquid-solids separation can be achieved through settling (sedimentation) or by using mechanical methods (e.g., use of screens, centrifuges, or belt presses).

Advantages of liquid-solids separation include concentration of solids for separate treatment or reuse, reduction of solids that can settle before treating and storing manure in lagoons, and prevention of clogging of manure pipes. In combination with chemicals, liquid -solids separation can be used to remove nutrients from manure. Other physical methods include drying, incineration, pyrolysis, combustion, and gasification.

Chemical Processes

Principal chemical processes for wastewater include chemical coagulation, chemical precipitation, chemical disinfection, chemical oxidation, chemical neutralization, chemical stabilization, and ion exchange. Chemicals, in conjunction with physical processes, can be used to remove nitrogen, phosphorus, and heavy metals. Chemicals have also been used to control odors and pH and to remove organic compounds. It should be noted that adding chemicals means adding external material into the manure-treatment system that is not generated on the farm, thereby increasing the volume of the waste stream to be handled after treatment.

Biological Processes

Biological-treatment systems or processes are designed to oxidize dissolved and particulate biodegradable constituents of manure into acceptable end products, and to transform or remove nitrogen and phosphorus. With proper environmental controls, all wastewater containing biodegradable constituents can be treated biologically. Biological processes make use of naturally occurring microorganisms to degrade manure in the presence of oxygen (aerobic) or absence of oxygen (anaerobic). It is important that the characteristics of each biological process be understood to operate biological-treatment-based technologies efficiently.

In aerobic processes-based technologies, oxygen is provided to the manure so that microorganisms can:

- Remove organic matter, often referred to as biochemical oxygen demand (BOD) or chemical oxygen demand (COD).
- Remove nitrogen and phosphorus through biological uptake.
- Manage nitrogen by converting ammonia in the manure to nitrogen gas (through a process called nitrification and denitrification).

Depending on the technology, air can be supplied using bubble or surface aerators. Thus, the success of aeration-based technologies depends on the survival of aerobic microorganisms. Anaerobic-treatment processes occur naturally in soils or in engineered systems, such as reactors or landfills, in environments where there is no oxygen. The major benefits of anaerobic digestion are reduced COD or BOD and solids; production of methane gas – a potential energy source; less energy required compared to aerobic processes; and less biological-sludge production compared to aerobic processes. Anaerobic digestion does not reduce manure's nitrogen and phosphorus content, and has the potential of producing odor and corrosive gases.

Manure Management Principles

The following principles are accepted by the three Departments of Agriculture across the Prairie Provinces as the basis for sustainable manure management.

- Sound manure management requires annual planning and proper record keeping.
- Manure application rates will depend on the type of manure and nutrient availability, and should be calculated using results from annual soil and manure analysis to meet crop nutrient requirements.
- Manure application should account for the season, weather conditions and site-specific conditions relating to soil, topography and water.
- Manure application equipment should be calibrated to ensure consistent and appropriate delivery rates of manure.
- Manure should be managed in a way that maximizes crop nutrient utilization and minimizes negative impacts to soil, water and air resources.

Benefits of Manure

Livestock manure is a valuable resource. It is an effective source of plant nutrients and organic matter, which can be used to improve crop production and soil quality.

Using manure as a fertilizer returns nutrients to the land for crop production.

Applying manure correctly:

- increases crop productivity and yield
- supplies macronutrients and micronutrients required for crops
- provides nitrogen (N) throughout the season, potentially increasing the protein content in the crop
- increases microbial activity, which can increase the availability of nutrients
- increases the organic matter content of the soil, which provides the following benefits:
    - improved soil structure
    - increased water infiltration and water holding capacity, resulting in reduced
    - water loss from runoff and increasing water availability to the crop
    - increased cation exchange capacity (the ability to hold nutrients)
    - reduced wind and water erosion

Manure Management Challenges

All nutrient sources, whether commercial chemical fertilizer or livestock manure, must be managed to ensure that soil, water and air quality are not degraded. However, there are additional challenges to using manure effectively as a fertilizer.

These challenges include:

- The moisture and nutrient contents of manure are highly variable, making it difficult to estimate nutrient availability without a representative manure analysis.
- The nutrients in manure are rarely "balanced" to meet all of the crop's nutrient requirements. For example, when manure is applied to meet the crop's N requirement, other nutrients such as phosphorus (P) and potassium (K) are simultaneously applied at rates that, in many cases, do not match the crop's requirements.
- The low nutrient content per unit weight or volume of manure limits the distance that manure can be transported economically.
- Manure can be a considerable source of odor during land application.
- Because manure may contain pathogens, added care must be taken to ensure it does not enter surface and ground water. Manure storage, handling and application practices will be affected by weather.

1
CHAPTER

VERMICOMPOSTING: TOOL FOR SUSTAINABLE RUMINANT MANURE MANAGEMENT

A. Nasiru, N. Ismail, and M. H. Ibrahim

INTRODUCTION

Nutrients losses from animal production in the form of manure are inevitable. Excessive animal waste results from an intensive ruminant production (management), high stocking density, or from feeding nutrients more than required by the animals [1]. The nutrients loss and waste products may exceed the carrying capacity of an area and become detrimental to the environment. Poor manure management contributes to pollution and eutrophication of surfaces water, ground water, and coastal marine ecosystem. It contributes to air pollution through emissions of odour, ammonia, methane and nitrous oxide, and it also contributes to soil pollution through the accumulation of heavy metals. These pollution and eutrophication effects subsequently lead to loss of human health, biodiversity, climate change, acidification, and ecosystem degradation [2]. Ruminant livestock has the highest contribution to these GHG emissions among livestock [3]. The sources of these GHG emissions and other pollutants from ruminants are respiration, enteric fermentation and manure. Nutrients losses due to ruminant production are either from enteric fermentation or manure that is influence by nature and composition of feed [4]. Factors such as the type and profile of nutrients in the feed, level of feed intake, rate of digesta passage, feed preservation methods, physical processing, chemical, and or biological treatment have direct effect on animal productivity as well as faecal and urine output [5, 6].

Ruminant manure is a valuable resource as a soil fertiliser, providing both macro, and micronutrients required for the plant growth, and is a low cost alternative to mineral

Citation: Nasiru, N. Ismail, and M. H. Ibrahim, "Vermicomposting: Tool for Sustainable Ruminant Manure Management," Journal of Waste Management, vol. 2013, Article ID 732759, 7 pages, 2013. doi:10.1155/2013/732759.

fertiliser [7]. Nevertheless, only a fraction of the nutrients excreted by livestock manure is properly collected and managed as manure [2] and difficulties are encountered in disposing excreta from livestock produced in large feed lots [8]. However, unmanaged and over-production of the manure (from intensive production) have led to inappropriate and indiscriminate application, resulting in overfertilisation, soil toxicity, dispersal of pathogens, odour, water pollution, and increase in greenhouse gas emission [9]. Therefore, livestock farms will need an improved manure management strategy [2].

Composting and vermicomposting are two of the best known processes for the biological stabilisation of solid waste [7]. However, composting reduces agronomic value of compost and contributes to greenhouse gas emission due to nutrients losses during compost making. Moreover, composting requires human labour or fuel in order to turn the compost heap to ensure aeration [10]. On the other hand, earthworm in vermicomposting process serve as an agent for turning, fragmentation, and aeration of the manure; therefore, it drastically increase the rate of microbial activities [9].

Ruminant Global Distribution and Manure Production

In 2010, world's ruminant population was about 3.6 billion, of which 5.38%, 39.59%, 25.19%, and 29.84% were for buffaloes, cattle, goats and sheep, respectively [11]. The relative distribution of the number of ruminant animals in different parts of the world according to FAOSTAT [11] was shown in Table 1. However, there are large differences between regions in the share of animal numbers and production. For example, Europe has only 8.71% of the total cattle but supplies 17.10% and 34.70% of global cow meat and milk respectively. On the other hand, Africa has 20% of the global cattle but produces 9.74% and 5.30% of world's cow meat, and milk, respectively, indicating different production systems.

Table 1: Distribution of ruminant livestock (% of global total) by region 2010

	Buffaloes	Cattle	Goat	Sheep
Africa	2.06	19.78	34.17	27.75
Asia	97.14	33.06	59.29	42.18
Europe	0.20	8.71	1.88	12.14
Latin America	0.60	28.37	3.78	7.99
North America	—	7.47	0.34	0.60
Oceania	—	2.61	0.54	9.34
World total	194, 152, 560	1, 430, 101, 625	909, 691, 076	1, 077, 762, 456

Source: FAOSTAT, 2013 [11].

Livestock excreta (ruminant inclusive) have not been treated with the importance they deserve; hence, estimates of their global amount excreted are highly uncertain

[2, 8, 12]. However, estimates of manure dry matter (DM) excretion are useful for designing manure treatment and handling technologies or for the development of future manure treatment technologies and for nutrient management [13]. Attempts had been made to predict an estimate of nutrient excretion from ruminant and most of the works were done on dairy and beef cattle [8, 13–16]. It had been assumed that under free access to feed, ruminants excrete a constant amount of faeces per unit live or metabolic weight [17]. An equation to estimate potential faecal dry matter (DM) output (, g/day) of cattle had been developed by Konandreas and Anderson [18] as follows:

$$F = fW^{0.75},$$

where (g DM/kg$W^{0.75}$) is a constant that depends on the physiological state of the animals.

The values of 42, 45, and 49 were estimated for dry, pregnant, and lactating cows, respectively. It is also suggested that these estimates could be used for feeds in a range of digestibilities from 42 to 65%. Fernández-Rivera et al. [17] considered wide range of digestibilities from 24.3 to 84.1% and an average daily faecal excretion/DM was determine as 2.383 Kg for cattle, 0.345 Kg for sheep, and 0.197 Kg for goat. From these, it can be assumed that an average of 3,407,932; 371,828, and 179,209 metric tonnes of faecal output/DM are excreted from cattle, sheep, and goat daily in the world. Strong relationship exists between apparent digestibility, feed intake, and faecal output. An increase in ruminant nutrients digestibility result in a decrease to faecal output while, an increase in feed intake increases faecal output [17]. Hoffman et al. [19], reported that heifers limit fed at 80 and 90% of ad libitum intake excreted 0.86 and 0.36 Kg/daily less dry matter, respectively, as compared with control diet with improved feed efficiency.

Based on production system, Sheldrick et al. [8] recognised three types of excreta as (i) from livestock grazing on pasture or rangeland, (ii) from livestock grazing on crop residues in the field, and (iii) from housed livestock. It's excreta from housed livestock can be collected, stored, and used at appropriate time and location. Ruminant manure from animal house consists of faeces, urine, bedding materials, split feed, split drinking water, and water used for washing the pen and it will be collected below slatted floor as slurries. When livestock are tied, the excretions are separated into solid manure, known as farm yard manure [12].

Ruminant Manure Characteristics

The amount and nutrient content of faeces excreted are highly variable and depend on types of animals, animal weight, diets, livestock production systems, and apparent digestibility [20, 21]. Other factors include dietary concentrations of crude protein, neutral detergent fibre, and nutrients intake [13]. Nutrients found in the manure or in compound emitted to the air and water originate from fraction of the feed which is not retained by the animals [22]. The nutrients of most concern are nitrogen (N), phosphorous (P), and potassium (K). Ruminant manure contributes 75, 66.4, and 83.6% N, P, and K, respectively, from the world total livestock excreta [8]. Manure N is partitioned between organic and ammonium N. Organic N is assumed to come primarily from faeces

[15]. The type and amount of crude protein (CP) consumed by ruminant affect total N excretions and the relative amount of N excreted in faeces and urine [6]. Lowering the CP content of diet will decrease the amount of N excreted by animal and vice versa. Somda et al. [20] reported that total amount and proportion of nutrients excreted in faeces and urine varied with the lignin : neutral detergent fibre (NDF), lignin : N, and polyphenol : N ratios of the diet. An increase in dietary fermentable energy content at a similar N intake level increases N excretion in the faeces with decline of N excreted in urine [23]. Tannins and polyphenols also shift N excretion from urine to faeces and from soluble to insoluble N form in faeces [6].

Ruminants excrete P mainly in faeces, which consist of unabsorbed dietary P and endogenous (from saliva, digestive juices, Intestinal) and the remaining source is through urine. Faecal endogenous P is the main pathway of P excretion (up to 85%) and the remaining is mainly unabsorbed dietary P [24]. Total P excretion in faeces depends largely on P intake. The ratio of endogenous faecal P to total faecal P is highly variable depending on the age, diet, and physiological condition of the animal [25]. Endogenous P losses becomes the predominant source of faecal P (70–80%) on diet low in P, as P intakes increase the ability of faecal endogenous excretion become saturated and most of any excess P is excreted in urine [26]. More than 90% of the heavy metals consumed by ruminant in the feed are excreted in the faeces or urine [2, 22].

Impact of Ruminant Manure and Its Management on the Environment

Losses from ruminant manure management systems roughly decrease in order of C, . This order is related to the reactivity, speciation, solubility, and fugacity of the nutrient element species. The double mobility of C, N, and S in soluble waterborne compounds as well as in gases makes their cycling and loss pathways much faster and more complex than those of the mineral elements [27]. Current environmental concerns relate mainly to gaseous emissions of ammonia (NH_3), nitrous oxide (N_2O), and methane (CH_4) from manure management system to atmosphere, to the leaching of nitrate, to groundwater, and to nitrogen (N) both organic and inorganically bound N species and P particulate losses from manure management to surface water [2]. Nitrogen is released in gaseous forms (mainly NH_3, N_2, N_2O, and NO), in dissolved forms as inorganic and organic N (, , DON) and as particulate matter via run off. Carbon is released from manure in gaseous form forms mainly CO_2 and CH_4, in dissolved forms as organic and inorganic C (ΣHCO_3, DOC) and as particulate matter (Via run-off). On the other hand sulphur is lost via volatilisation of sulphides (H_2S) and sulphur dioxides (SO_2) and the leaching of sulphate () and particulate matter [27] (Figure 1).

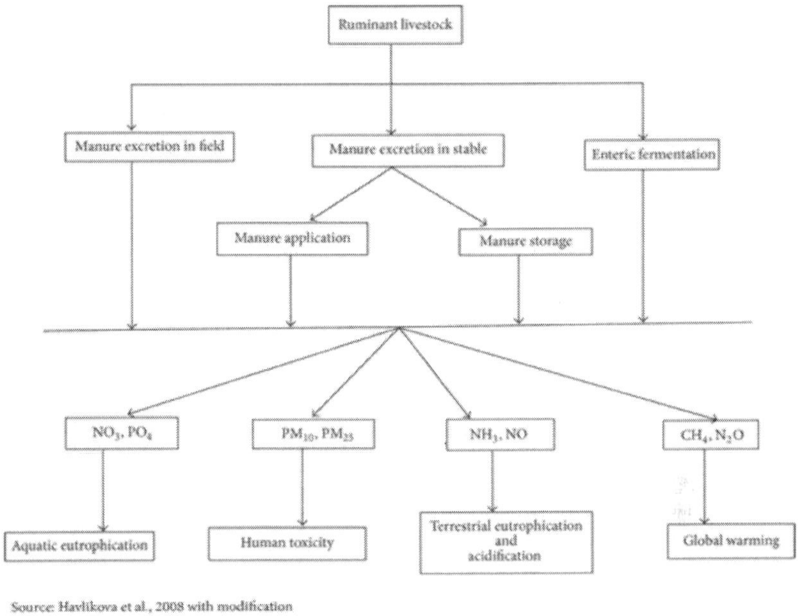

Figure 1: It shows associated nutrients loss in ruminant production and manure management.

Manure Management

Ruminant manure collected in housing systems has to be stored for some time inside or outside the housing system until timely spread of the manure on the field, that is, during the growing season when the crop will be able to utilize the plant nutrients [28]. Traditionally ruminant manure is normally spread in the farm without any treatment [10]. Farmers applied these organic fertilisers for the long-term benefit of their soils. Manure helps in stabilization of soil aggregates preventing erosion; it improves soil structure promoting moisture retention and it even may correct drainage problems in wet areas [29]. Losses or emissions in the form of volatilization, leakage, run-off, and dumping occur during storage and application of the manure [30]. When applied in excess to the land requirement, it can lead to environmental pollution. These include overfertilisation of soil, soil toxicity, dispersal of pathogens and weed seeds, odour, water pollution, and increase in greenhouse gas emission and may pose health risk [9]. However, ruminant manure is stored for a month or more before being applied into the soil [12]. Apart from being less effective compared with mineral fertiliser, slurry poses several problems during storage, likewise during and after its application. As conditions in the slurry are anaerobic, this means that CH_4 and CO_2 are produced as end products. After application, NH_3 emissions increase with an increase in slurry dry matter content. Environmentally friendly slurry application requires the slurry to be evenly applied near or under the surface [31]. Generally, manure storage contributes to the atmospheric pool of gases such as ammonia, nitrous oxide, and methane [32].

$NH_3$and greenhouse gases losses result from microbiological, chemical, and physical processes. The environmental and health risk imposed by ruminant manure mentioned earlier is due to its nonstabilisation. Stabilisation is degree of decomposition of a waste substance, which is reflected by decrease in level of microbial biomass activity and concentrations of labile compounds [7, 33]. Stabilisation reduces the environmental problems associated with manure management by transforming it into a safer and more stabilised material suitable for application to soil [33].

Composting is a widely used method for disposal of animal waste [34]. Farmers used it for various objectives which includes reducing the mass of manure before spreading, homogenizing the manure, destroying pathogens and weed seeds, deodorizing the manure, or making it into a saleable product with the aim of returning it to agricultural land [30, 35]. Composting is continuous aerobic degradation of organic materials by microorganism into humus-like substances. Composting is one of the best known processes for the biological stabilisation of solid waste [7]. Composting is a microbial aerobic decomposition process with the formation of stabilised and matured organic materials. Efficient composting requires the control of several factors in order to obtain a quality agricultural product. It have been grouped into two, those depending on the composting mix such as nutrients balance, pH, particle size, porosity, and moisture, and those depend on the process management such as oxygen concentration, temperature, and water content. Therefore, composting animal manure should be seen as technology which adds value and produces a high quality product for multiple agricultural uses. According to Peigné and Girardin [36], studies have demonstrated that nutrients are lost during the composting operation and may induce environmental problems. Losses are generated in many ways: as ammonia volatilisation, as nitrous oxide and methane emissions, or as nutrients leached in drainage water. Nutrient losses can be up to 62% (C), 42% (N), and 6.5% (K and Na) and less than 6% (C and Mg) and 2% (P), and these are related to the initial manure content [37] and compost management employed. The product of composting sometime is heterogeneous and the process require long period of time to be completed [38]. Air and water are the main environmental components that are affected by composting pollution. Nutrient and salt loss during composting resulted in reduced electrical conductivity of the composted manure. In addition to energy loss due to labour or fuel required to turn the compost heap, these reduce agronomic value of compost and contribute to greenhouse gas emission [10, 37].

Vermicomposting as an Effective Tool in Ruminant Manure Management

Vermicomposting is a mesophilic biooxidation and stabilisation process of organic materials that involve the joint action of earthworm and microorganism. This increases the rate of the decomposition process by accelerating the stabilisation of organic matter and greatly modifying its physical and biochemical properties [39]. Microorganisms produce the enzymes that cause biochemical decomposition of organic matter, but earthworms are the crucial drivers as they stimulate and increase biological activity by fragmentation and ingestion of organic matter and this will increase the surface area to be exposed to microorganism [39]. Earthworms act as mechanical blenders and by comminuting the organic matter they modify its physical and chemical status by gradually reducing the C : N ratio and increasing the surface area exposed to microorganism [40]. They also serve as agent of turning and aeration [38].

A Vermicomposting process has two distinguished phases and is (i) an active phase, where the earthworms process the biomass, modifying its physical state and microbial composition. The effect of earthworm on the decomposition of organic matter during vermicomposting processes is due to gut associated processes (GAPs), and it includes the modification that organic waste and microbes undergo during their passage through the gut of earthworm. (ii) A maturation like phase, also known as cast associated processes (CAPs), is marked by the displacement of the earthworm towards fresher layers of undigested waste, where the microbes take over in the decomposition of waste and the effects of earthworm are mainly indirect and derived from GAPs [7, 9, 39]. Vermicompost is a finely divided, peat-like material with a low C : N ratio, excellent structure, porosity, aeration, drainage, and moisture-holding capacity, and it supplies a suitable mineral balance, improves plant nutrient availability, and could act as complex-nutrient-source granules [9]. Earthworm plays a significant role in processing ruminant manure as it reduces the moisture content, pH, and electrical conductivity compared with composting [41, 42]. These might be attributed to high rate of mineralisation, this can be up to 60% [43], and it brings accumulation of organic acids from microbial metabolism and enhances production of fulvic and humic acids during decomposition [44]. According to Atiyeh et al. [41] carbon dioxide evolution decreases rapidly (44%) one week after the introduction of earthworms, and continued at a lower rate throughout the 17 weeks, 51% reduction as compared to 22% without earthworms, indicating increasing organic matter stability. Vermicomposting enhanced nitrogen mineralisation and increase the rates of conversion of ammonium-nitrogen into nitrate. This will increase the concentration of nitrate-nitrogen to 28% after 17 weeks, while in compost the nitrate-nitrogen concentration will increase by 3%. This suggests that earthworms produced conditions in the manure that favoured nitrification, resulting in rapid conversion of ammonium-nitrogen into nitrates [41]. Comparison of compost and vermicompost showed that vermicompost had significantly lower C : N ratios, and this was due to loss of carbon as CO_2 during bioxidation and production of mucus and nitrogen excrements increase the level of nitrogen which lower the C : N ratio [42, 45]. With regard to lignolysis, Vinceslas-Akpa and Loquet [46] find out that at the beginning of vermicomposting, lignolysis was more efficient compared with composting; however, at the end fraction identified as cellulose increased particularly in the vermicompost, apparently the rate of cellulolysis and lignolysis was slightly faster in the compost. Vermicomposting increases the ash content and accelerates the rate of mineralisation which is essential to make the nutrients available to plant [43]. Ruminant manure vermicompost was found to have the highest total phosphorous compared to other livestock manure vermicomposts [42]. Among the effects of different microorganisms and enzymes contributing to such increased availability of phosphorus, major emphasis may be given to the presence of very high concentration of phosphate-solubilising bacteria in the vermicast [47]. Addition of vermicompost to the soil adds to its mineralogical nutrients and contributes to its biological fertility by adding beneficial microbes to the soil. It favourably affects soil pH, microbial population, and enzyme activities. It also reduces the proportion of water soluble chemicals, which causes possible environmental contamination. All these help in increased production of healthier crops [43, 47].

CONCLUSION

Manure production is an inevitable aspect of ruminant production. Under free access to feed, ruminants excrete a constant amount of faeces per unit liveweight. Millions tonnes of ruminant manure are produced daily; without proper handling management, it results in increase in GHG emission and environmental pollutions. These called for efficient and sustainable way of treating the waste. Composting and vermicomposting are considered as the best option for biomass management. However composting reduces the agronomic value of the products and contributes to environmental pollution, in addition to energy loss due to labour or fuel required to turn the compost heap compared with vermicomposting. Earthworms can break down organic matter very rapidly, resulting in stable, nontoxic vermicomposts with a better structure, microbial content, and available nutrient content than composts. Vermicomposting through the activities of earthworm associated microbes accelerated the process of ruminant manure decomposition and stabilisation and promoted biochemical characteristics that were favourable to plant growth. Vermicomposting is a cost effective and ecofriendly waste management technology and has many advantages over traditional composting.

Conflict of Interests

The authors declared that they have no conflict of interest.

ACKNOWLEDGMENTS

A. Nasiru acknowledges USM-TWAS PG fellowship 2010 for fellowship award. The study was funded through Universiti Sains Malaysia (USM) grant (Grant no. 304/PTEKIND/6730067). The authors acknowledge USM for providing research facilities.

REFERENCES

1. S. Tamminga, "Nutrition management of dairy cows as a contribution to pollution control," Journal of Dairy Science, vol. 75, no. 1, pp. 345–357, 1992. ·
2. H. Menzi, O. Oenema, C. Burton et al., "Impacts of intensive livestock production and manure management on the environment," in Livestock in a Changing Landscape Drivers, Consequences and Responses, Henning Steinfeld, H. A. Mooney, F. Schneider, and L. E. Neville, Eds., vol. 1, pp. 139–163, Island Press, 2010.
3. G. Zervas and E. Tsiplakou, "An assessment of GHG emissions from small ruminants in comparison with GHG emissions from large ruminants and monogastric livestock," Atmospheric Environment, vol. 49, pp. 13–23, 2012. · ·
4. X. Hao, M. B. Benke, C. Li, F. J. Larney, K. A. Beauchemin, and T. A. McAllister, "Nitrogen transformations and greenhouse gas emissions during composting of manure from cattle fed diets containing corn dried distillers grains with solubles

and condensed tannins," Animal Feed Science and Technology, vol. 166-167, pp. 539–549, 2011. · ·

5. J. M. Powell, M. A. Wattiaux, G. A. Broderick, V. R. Moreira, and M. D. Casler, "Dairy diet impacts on fecal chemical properties and nitrogen cycling in soils," Soil Science Society of America Journal, vol. 70, no. 3, pp. 786–794, 2006. · ·
6. J. M. Powell, G. A. Broderick, J. H. Grabber, and U. C. Hymes-Fecht, "Technical note: effects of forage protein-binding polyphenols on chemistry of dairy excreta," Journal of Dairy Science, vol. 92, no. 4, pp. 1765–1769, 2009. · ·
7. C. Lazcano, M. Gómez-Brandón, and J. Domínguez, "Comparison of the effectiveness of composting and vermicomposting for the biological stabilization of cattle manure," Chemosphere, vol. 72, no. 7, pp. 1013–1019, 2008. ·
8. W. Sheldrick, J. K. Syers, and J. Lingard, "Contribution of livestock excreta to nutrient balances," Nutrient Cycling in Agroecosystems, vol. 66, no. 2, pp. 119–131, 2003. · ·
9. Dominguez and C. A. Edwards, "Relationships between composting and vermicomposting," in Vermiculture Technology Earthworms, Organic Wastes, and Environmental Management, C. A. Edwards, N. Q. Arancon, and R. Sherman, Eds., pp. 11–26, Taylor & Francis, New York, NY, USA, 2011.
10. X. Hao, C. Chang, F. J. Larney, and G. R. Travis, "Greenhouse gas emissions during cattle feedlot manure composting," Journal of Environmental Quality, vol. 30, no. 2, pp. 376–386, 2001. · ·
11. "FAO statistical database," Food and Agricultural Organisation, 2013, http://faostat3.fao.org/home/index.html#download.
12. S. G. Sommer and N. J. Hutchings, "Ammonia emission from field applied manure and its reduction—invited paper," European Journal of Agronomy, vol. 15, no. 1, pp. 1–15, 2001. · ·
13. T. D. Nennich, J. H. Harrison, L. M. VanWieringen et al., "Prediction of manure and nutrient excretion from dairy cattle," Journal of Dairy Science, vol. 88, no. 10, pp. 3721–3733, 2005.
14. Standard, D384.2, Manure Production and Characteristics, ASABE, St. Joseph, Mich, USA, 2005.
15. A. Rotz, D. R. Buckmaster, and J. W. Comerford, "A beef herd model for simulating feed intake, animal performance, and manure excretion in farm systems," Journal of Animal Science, vol. 83, no. 1, pp. 231–242, 2005. ·
16. V. A. Wilkerson, D. R. Mertens, and D. P. Casper, "Prediction of excretion of manure and nitrogen by Holstein dairy cattle," Journal of Dairy Science, vol. 80, no. 12, pp. 3193–3204, 1997. ·
17. S. Fernández-Rivera, T. Williams, P. Hiernaux, and J. Powell, "Faecal excretion by ruminants and manure availability for crop production in semi-arid West Africa," 1995.
18. A. Konandreas and F. M. Anderson, Cattle Herd Dynamics: An Integer and Stochastic Model for Evaluating Production Alternatives, ILRI (aka ILCA and

ILRAD), 1982.

19. C. Hoffman, C. R. Simson, and M. Wattiaux, "Limit feeding of gravid holstein heifers: effect on growth, manure nutrient excretion, and subsequent early lactation performance," Journal of Dairy Science, vol. 90, no. 2, pp. 946–954, 2007. · ·
20. Z. C. Somda, J. M. Powell, S. Fernandez-Rivera, and J. Reed, "Feed factors affecting nutrient excretion by ruminants and the fate of nutrients when applied to soil," in Livestock and Sustainable Nutrient Cycling in Mixed Farming Systems of Sub-Saharan Africa, J. M. Powell, S. Fernández-Rivera, T. O. Williams, and C. Renard, Eds., ILRI (aka ILCA and ILRAD), Addis Ababa, Ethiopia, 1995.
21. M. Mathot, V. Decruyenaere, D. Stilmant, and R. Lambert, "Effect of cattle diet and manure storage conditions on carbon dioxide, methane and nitrous oxide emissions from tie-stall barns and stored solid manure," Agriculture, Ecosystems and Environment, vol. 148, pp. 134–144, 2012. · ·
22. O. Petersen, S. G. Sommer, F. Béline et al., "Recycling of livestock manure in a whole-farm perspective," Livestock Science, vol. 112, no. 3, pp. 180–191, 2007. · ·
23. E. Kebreab, J. France, J. A. Mills, R. Allison, and J. Dijkstra, "A dynamic model of N metabolism in the lactating dairy cow and an assessment of impact of N excretion on the environment," Journal of Animal Science, vol. 80, no. 1, pp. 248–259, 2002. ·
24. Bravo, D. Sauvant, C. Bogaert, and F. Meschy, "III. Quantitative aspects of phosphorus excretion in ruminants," Reproduction Nutrition Development, vol. 43, no. 3, pp. 285–300, 2003. · ·
25. M. Vitti and E. Kebreab, Phosphorus and Calcium Utilization and Requirements in Farm Animals, CABI, New York, NY, USA, 2010.
26. J. Underwood and N. F. Suttle, The Mineral Nutrition of Livestock, CABI, New York, NY, USA, 1999.
27. O. Oenema, D. Oudendag, and G. L. Velthof, "Nutrient losses from manure management in the European Union," Livestock Science, vol. 112, no. 3, pp. 261–272, 2007. · ·
28. O. Oenema, A. Bannink, S. G. Sommer, J. W. van Groenigen, and G. L. Velthof, "Gaseous nitrogen emissions from livestock farming systems," in Nitrogen in the Environment, J. L. Hatfield and R. F. Follett, Eds., chapter 12, pp. 395–441, Academic Press, San Diego, Calif, USA, 2nd edition, 2008. ·
29. J. Martinez, P. Dabert, S. Barrington, and C. Burton, "Livestock waste treatment systems for environmental quality, food safety, and sustainability," Bioresource Technology, vol. 100, no. 22, pp. 5527–5536, 2009. · ·
30. M. P. Bernal, J. A. Alburquerque, and R. Moral, "Composting of animal manures and chemical criteria for compost maturity assessment. A review," Bioresource Technology, vol. 100, no. 22, pp. 5444–5453, 2009. · ·
31. B. Amon, V. Kryvoruchko, T. Amon, and S. Zechmeister-Boltenstern, "Methane,

nitrous oxide and ammonia emissions during storage and after application of dairy cattle slurry and influence of slurry treatment," Agriculture, Ecosystems and Environment, vol. 112, no. 2-3, pp. 153–162, 2006. · ·

32. D. R. Külling, H. Menzi, T. F. Kröber et al., "Emissions of ammonia, nitrous oxide and methane from different types of dairy manure during storage as affected by dietary protein content," Journal of Agricultural Science, vol. 137, no. 2, pp. 235–250, 2001. · ·

33. M. Benito, A. Masaguer, A. Moliner, N. Arrigo, and R. M. Palma, "Chemical and microbiological parameters for the characterisation of the stability and maturity of pruning waste compost," Biology and Fertility of Soils, vol. 37, no. 3, pp. 184–189, 2003. · ·

34. S. Goyal, S. K. Dhull, and K. K. Kapoor, "Chemical and biological changes during composting of different organic wastes and assessment of compost maturity," Bioresource Technology, vol. 96, no. 14, pp. 1584–1591, 2005. · ·

35. N. A. El Kader, P. Robin, J. M. Paillat, and P. Leterme, "Turning, compacting and the addition of water as factors affecting gaseous emissions in farm manure composting," Bioresource Technology, vol. 98, no. 14, pp. 2619–2628, 2007. · ·

36. J. Peigné and P. Girardin, "Environmental impacts of farm-scale composting practices," Water, Air, and Soil Pollution, vol. 153, no. 1–4, pp. 45–68, 2004. · ·

37. B. Eghball, J. F. Power, J. E. Gilley, and J. W. Doran, "Nutrient, carbon, and mass loss during composting of beef cattle feedlot manure," Journal of Environmental Quality, vol. 26, no. 1, pp. 189–193, 1997. · ·

38. P. M. Ndegwa and S. A. Thompson, "Integrating composting and vermicomposting in the treatment and bioconversion of biosolids," Bioresource Technology, vol. 76, no. 2, pp. 107–112, 2001. · ·

39. J. Dominguez, "The microbiology of vermicomposting," in Vermiculture Technology Earthworms, Organic Wastes, and Environmental Management, C. A. Edwards, N. Q. Arancon, and R. Sherman, Eds., 2011.

40. J. Dominguez, C. Edwards, and S. Subler, "Comparison of vermicomposting and composting," BioCycle, vol. 38, no. 4, pp. 57–59, 1997.

41. R. M. Atiyeh, J. Domínguez, S. Subler, and C. A. Edwards, "Changes in biochemical properties of cow manure during processing by earthworms (Eisenia andrei, Bouché) and the effects on seedling growth," Pedobiologia, vol. 44, no. 6, pp. 709–724, 2000. ·

42. V. K. Garg, Y. K. Yadav, A. Sheoran, S. Chand, and P. Kaushik, "Livestock excreta management through vermicomposting using an epigeic earthworm Eisenia foetida," Environmentalist, vol. 26, no. 4, pp. 269–276, 2006. · ·

43. J. Pathma and N. Sakthivel, "Microbial diversity of vermicompost bacteria that exhibit useful agricultural traits and waste management potential," SpringerPlus, vol. 1, article 26, 2012. ·

44. Albanell, J. Plaixats, and T. Cabrero, "Chemical changes during vermicomposting (Eisenia fetida) of sheep manure mixed with cotton industrial wastes," Biology

and Fertility of Soils, vol. 6, no. 3, pp. 266–269, 1988. · ·

45. R. P. Singh, A. Embrandiri, M. H. Ibrahim, and N. Esa, "Management of biomass residues generated from palm oil mill: vermicomposting a sustainable option," Resources, Conservation and Recycling, vol. 55, no. 4, pp. 423–434, 2011.
46. M. Vinceslas-Akpa and M. Loquet, "Organic matter transformations in lignocellulosic waste products composted or vermicomposted (Eisenia fetida andrei): chemical analysis and 13C CPMAS NMR spectroscopy," Soil Biology and Biochemistry, vol. 29, no. 3-4, pp. 751–758, 1997.
47. Chattopadhyay, "Use of vermicomposting biotechnology for recycling organic wastes in agriculture," International Journal of Recycling of Organic Waste in Agriculture, vol. 1, no. 1, pp. 1–6, 2012.

2
CHAPTER

VERMICOMPOSTING OF ANIMAL DUNG AND ITS LABORATORY EVALUATION

Hargopal Singh, Pritpal Singh and S.S. Hundal

Keywords: Animal dung, Earthworms, Nutrient content, Vermicomposting

INTRODUCTION

For centuries, the excreta of animals is being used as manure to supply nutrients to crops. Even today, the animal dung and leftover agro-wastes of organic origin, like fodder are collected and accumulated in manure heaps on a daily basis. This is left in the open for months to facilitate composting and during this period, the waste remains exposed to the vagaries of nature and experiences nutrient losses through volatilization and leaching. Though the State Government of Punjab had allotted a small piece of land to each family in a village for preparing quality manure, but the farmers continued to prepare manure by conventional means that leads to the production of seemingly poor quality manure because we do not take into consideration the environmental conditions (like pH, temperature) in which these decompositions take place. Appropriate disposal involves both maximum cost-effective recovery of recyclable constituents and transformation of non-recoverable material into forms, which do not present environmental hazards (Kale, 2004). In India, nearly 2000 million tons (MT) of animal waste, 300 MT of crop waste besides huge amount of agro industrial and domestic sewage waste is produced annually and therefore there is a tremendous scope for recycling of this waste using vermitechnology so that quality organic manure can be produced (Ramaswami, 1998; Mishra, 2001); it is becoming increasingly popular due to the emerging trend of organic

Citation: H.Singh et al., vermicomposting of animal dung and its laboratory evaluation, Vol. 5 No. 7 (July 2012) ISSN: 0974- 6846.

farming. Vermicomposting is an accelerated biotechnological process of composting of organic wastes that involves interaction between earthworms and micro-organisms. Utilization of earthworms for recycling of organic wastes is an important development in biological sciences and studies have documented vermicomposting as a low-cost technology for the processing or treatment of organic wastes to convert them into value added nutrient rich compost (Hand et al., 1988; Garg et al., 2006; Suthar, 2006, 2008; Singh et al., 2008). Vermicomposting is a mesophilic process utilizing worms and micro-organisms that are active at temperature range of 10°-32°C. Micro-organisms are responsible for the biochemical degradation of organic matter; the earthworms after passing the organic matter through their gizzard and gut increased the surface area for the microbial activity on the fragmented organic residues (Ndegwa & Thompson, 2001). As a result of this microbial conversion, both gut associated as well as the cast associated processes, the nutrients forms are much more soluble and available to plants than those in parent compounds.

Vermicompost is considered as an excellent product, since it is homogenous, has desired aesthetics, has reduced levels of contaminants and tends to hold more nutrients over a longer period without impacting the environment (Benitez et al., 2002; Dhiman, 2003; Suthar & Singh, 2008; Singh et al., 2007). Also, vermicompost is considered an inseparable component of sustainable farming because it promises to promote and sustain crop yields (Patil, 1995; Reddy & Ohkura, 2004; Sinha & Herat, 2009) because of increased nutrients content in the earthworm castings (Garg et al., 2006). Dissemination of information on organic farming in the country has enlightened the farmer to engage in and adopt vermitechnology as a part of organic farming practices (Kalra et al., 2008). Research on vermiculture and vermicomposting is leading towards appreciation of some of the finer and physical aspects of soil entity as well as adding to the other benefits derived from agriculture. The closely related Eisenia fetida and Eisenia andrei species are most commonly used for the management of organic wastes by vermicomposting and their biology on animal dung or sewage sludge, have been investigated (Hartenstein et al., 1979; Kalpan et al., 1980; Edwards, 1988; Reinecke & Vilijoen, 1990a; Butt, 1993; Elvira et al., 1996; Dominguez & Edwards, 1997; Dominguez et al., 1997; Mitchell, 1997; Dominguez et al., 2000; Bettiol, 2004).

Agricultural substrates and wastes like farm yard manure (FYM), animal dung, rice straw and plant litter are potential sources of organic nutrients which if converted into vermicompost improve soil and water conservation, sustain crop productivity and enhance crop yields (Hundal & Zinia, 2009).The present study was therefore, planned to evaluate the changes in the nutrient composition of vermicompost from animal dung and compare them with those of FYM.

MATERIALS AND METHODS

Experiment was carried out on the nutrient evaluation of vermicompost from FYM and compared it with farmyard manure (FYM) prepared without adding earthworms into it. Two consecutive vermicomposting cycles viz. September-December (C1) and January-April (C2) were carried out. Fresh dung was obtained from the dairy farm of Guru Angad Dev Veterinary and Animal Science University (GADVASU), Ludhiana,

India. The earthworms (Eisenia fetida) were procured from Amritsar Crown Caps Pvt. Ltd., Jandiala Guru, Amritsar, India. Microbial culture consisting of Trichoderma ressaie and Phanerochaeta used as starter material in a treatment was procured from the Department of Microbiology, PAU, Ludhiana. Fresh dung was added into the raised rectangular pits (3.35 X 1.20 X 0.25 m^3; length x width x height) constructed under galvanized iron sheet shed at a height of 11.5 feet. This provided 4.02 m^2 exposed top surface. The experiment was conducted in an ambient temperature of 25 ± 5°C. For C_1, dung was filled @ 500 kg pit^{-1} (fresh weight) (146 kg pit^{-1}, dry weight), after thermophilic stage of decomposition was over and temperature of the pit came down to 30°C. Earthworms were introduced @ 1.530 kg pit^{-1} (live weight) corresponding to approximately 4308 individuals on the next day of filling the pits, indicating earthworm stocking density of 1.522 kg-worms m^{-3} and feeding rate of 1.22 kg dung-$worm^{-1}$ day^{-1}, optimal levels for vermicomposting (Ndegwa et.al., 2000) in the respective treatments - this being considered as day 0 of the treatment. At the start of each cycle, the manure was manually mixed using a spade, and optimum moisture content (~60%) was maintained by sprinkling water in all the compost pits. The day-to-day activity and performance of earthworms was monitored. From each compost pit, six samples (each ~500g, fresh weight) were taken using stainless steel tube (0.32 m long, 1.5 inch i.d.), mixed thoroughly, screened through sieve (10 mm) and ~1 kg/ 1000g sub- sample was transferred to the laboratory after 30, 60 and 78 days of composting period. Thus, each sub-sample was a representative of six-grab samples taken from top to the bottom of each compost pit. Throughout the vermicomposting period, the material in the pits was turned at 7th, 21st, 36th , 51st and 81st day of composting to provide aeration and to ensure the homogeneity of the material inside the pit. Throughout the bioconversion period, pit moisture was monitored every 2 to 5 days, depending upon the atmospheric temperature and moisture loss, by adding enough water to obtain optimum moisture content. For the second cycle of vermicomposting (C_2) during January-April, fresh FYM was filled into same pits @ 500 kg pit^{-1} with same specifications and parameters as for C_1 cycle. The everyday activity and performance of termination of study period were analyzed for chemical parameters similar for C_1 cycle. The 6 treatments were as follows: (1) no earthworms, no moisture, no turning, no microbial culture ($E_0M_0T_0C_0$),which was considered as control (2) no earthworms, moisture, weekly turning, no microbial culture ($E_0M_1T_1C_0$), (3) Eisenia fetida earthworms, moisture, weekly turning, no microbial culture ($E_0M_1T_1C_0$), (4) Eisenia fetida earthworms, moisture, no turning, no microbial culture ($E_1M_1T_1C_0$), (5) no earthworms, moisture, weekly turning, microbial culture added ($E_0M_1T_1C_1$), (6) Eisenia fetida earthworms, moisture, weekly turning, microbial culture added ($E_1M_1T_1C_0$). Analytical methods The compost samples were analyzed for total organic carbon (TOC), total nitrogen (TN), total phosphorus (TP), total potassium (TK), total sulphur (TS), total zinc (T-Zn), total copper (T-Cu), total manganese (T-Mn) and total iron (T-Fe). The TOC was analyzed by dry combustion method at 500°C (Nelson & Sommers, 1982) and TN was determined by digestion with concentrated H_2SO_4 (Bremmer & Mulvaney, 1982). TP was analyzed after digesting the organics in 3:1 mixture of HNO_3: $HClO_4$ followed by calorimetric determination using molybdo- phosphoric acid method on Milton Roy spectrophotometer at 470 nm (Nelson & Sommers, 1982). The same extract was used for the TS and TK analysis of the organics. TS in the extract was determined by developing turbidity with the help of sodium-acetate acetic acid buffer and 30-60 mesh Barium Chloride crystals followed by suspension stabilization with 0.25% gum Acacia read at 420nm. Total micro-nutrients (Zn, Cu, Mn & Fe) were determined by means of Atomic Absorption Spectrophotometer (AAS) using specified lamp after digesting the samples HNO_3:$HClO_4$(2:1). Statistical analysis Statistical analysis of major and micronutrients in the vermicompost and FYM was carried out by ANOVA. Mean separation of different treatments was performed using the least significant difference (LSD) test at 0.05 level of probability.

RESULTS AND DISCUSSION

Vermicomposting is a method to treat faecal matter with water contents up to 85% but little is known about the environmental conditions in which worms treat the material effectively. The results have shown an interaction between different parameters.

Table 1. Nutrient composition of organic wastes used for vermicomposting.

Parameter	Organic wastes							
	FYM_1		FYM_2	RSC	RS	LL	FPP	BC
	C_1	C_2						
Ash (%)	48	52	68	44	93	121	70	40
TOC (%)	40.4	43.3	17.6	30.9	67.4	175	18.6	18.1
TN (%)	1.0	1.2	1.4	1.8	0.6	0.5	0.3	1.2
TP (%)	0.5	0.6	0.6	0.8	0.2	0.1	0.3	1.2
TK (%)	2.2	1.3	1.8	2.1	1.3	0.6	0.5	2.2
TS (%)	0.60	0.53	0.2	0.1	0.1	0.1	0.2	0.5
C:N	40.4	36.1	13.0	18.0	112.3	525	62	11.1
C:P	80.8	72.2	29.3	38.6	337.0	1050	62	15.1
C:S	67.3	81.7	80.0	309.0	518.5	1050	93	36.2
T-Zn ($\mu g\ g^{-1}$)	97	113	88	104	40	46	28	160
T-Cu ($\mu g\ g^{-1}$)	14	14	14	10	2	4	2	20
T-Mn ($\mu g\ g^{-1}$)	96	116	104	216	280	256	84	349
T-Fe ($\mu g\ g^{-1}$)	2723	3143	3336	5238	414	404	312	6438

Expressed on dry weight basis, FYM_1 and FYM_2 represents FYM used for first and second experiment, respectively

Nutrient content of animal dung

The TOC content of animal dung used for the experiment was 40.4% and 43.3 % in C1 and C2 composting cycles, respectively (Table 1). Total N was 1.0 and 1.2 %, TP was 0.5 and 0.6 % and TS was 0.60 and 0.53 % that leads to a C: N ratio of 40.4 and 36.1, C: P ratio of 80.8 and 72.2, and C: S ratio of 67.3 and 81.7 used for C1 and C2 cycle of composting, respectively. Total K in the animal dung used for C1 and C2 composting cycles was 2.2 and 1.3 %. Similarly the content of total Zn, Cu, Mn and Fe in the animal dung used for 1st cycle was 97, 14, 96 and 2732 $\mu g\ g^{-1}$ and in 2nd cycle was 113, 14, 116 and 3143 $\mu g\ g^{-1}$, respectively. Gupta (1998) and Hamoda et al. (1998) interpreted that various organic materials have been successfully composted with C/N ratio between 17-78, however, a much narrow range of 25 to 35 are desirable indicating optimum initiating parameters.

Nutrient composition of compost

Table 2. Average (3 sampling periods) Nutrient composition (dry weight basis, mean of three samplings) of animal dung manure (FYM1)-vermicompost.

Treatment	Major-nutrients					C:N	Micro-nutrients			
	TOC (%)	TN (%)	TP (%)	TK (%)	TS (%)		T-Zn ($\mu g\ g^{-1}$)	T-Cu ($\mu g\ g^{-1}$)	T-Mn ($\mu g\ g^{-1}$)	T-Fe ($\mu g\ g^{-1}$)
September-December Cycle (C_1)										
$E_0M_0T_0C_0$	24.4	1.1	2.4	0.6	0.5	22	126	23	190	6373
$E_0M_1T_1C_0$	25.4	1.2	2.5	0.6	0.4	21	114	18	158	6324
$E_1M_1T_1C_0$	23.4	1.0	2.3	0.6	0.5	24	106	20	170	7261
$E_1M_1T_0C_0$	22.6	0.9	2.2	0.5	0.5	23	107	21	165	7146
$E_0M_1T_1C_1$	22.8	1.1	2.1	0.5	0.6	22	103	19	159	7367
$E_1M_1T_1C_1$	24.2	1.1	2.2	0.5	0.5	22	102	19	155	6865
LSD(0.05)	NS	NS	NS	0.2	NS	NS	NS	NS	NS	NS
January-April Cycle (C_2)										
$E_0M_0T_0C_0$	19.2	1.5	0.6	1.4	0.6	13	90	15	114	3036
$E_0M_1T_1C_0$	18.0	2.0	0.7	1.5	0.7	9	93	15	119	3466
$E_1M_1T_1C_0$	21.5	1.5	0.7	1.4	0.7	15	94	15	127	3546
$E_1M_1T_0C_0$	20.9	1.6	0.6	1.5	0.6	13	91	13	118	3924
$E_0M_1T_1C_1$	17.1	1.9	0.6	1.5	0.7	9	106	15	107	2706
$E_1M_1T_1C_1$	17.5	1.8	0.7	1.4	0.6	10	86	14	116	3138
LSSD (0.05)	2.4	0.3	NS	NS	NS	3.0	NS	NS	NS	NS

The average nutrient composition of vermicompost and FYM have been shown in Table 2. The content of TOC varied non-significantly ($p \leq 0.05$) among different treatments during C1 after 78 days of composting. However, vermicomposting FYM with Eisenia fetida resulted in significant ($p \leq 0.05$) decrease in TOC from 21.5% in E1M1T1C0 to 17.1% in E0M1T1C1 treatment during C2. During C2, even the traditional composting (E0M0T0C0 and E0M1T1C0) appreciably reduced the TOC content (Table 2). During C1, TN and TP content in compost among different treatments differed non-significantly and TN and TP content remained near to 1.1 and 1.0% respectively. Interestingly, TN and TP content for the two composting cycles was though non-significant; ($p \leq 0.05$) in organics undergone traditional composting than from organics vermicomposted using worms. During C2, significantly ($p \leq 0.05$) higher TN content in E0M1T1C0 over E1M1T1C0 treatment was observed. Like TN and TP, other major and micro nutrients also differed non-significantly ($p \leq 0.05$) among different treatments, for two consecutive composting cycles (Table 2). Intense microbial activity can alter the degradation of soil organic matter, resulting in an accelerated decomposition of indigenous organic matter (Sikora & Yakovchenko, 1996). Results revealed that composting the organics by adopting the scientific methods is equally good as vermicompost as far as nutrient composition of the organics produced by the two different procedures is concerned. Our results found support from the work of other researchers who have reported T-Fe and T-Mn content of most of the composts ranged between 3428-9963 and 264-540 mg kg-1 , respectively (Forster et al., 1993; Kadalli et al., 2000; Chastain et al., 2006). Similarly, T-Zn and T-Cu content in most of the composts varied from 200- 400 and 60-1200 mg kg-1 in the experiments reported by Genevini et al., (1997), Gies (1997) and Vogtman et al., (1993), respectively. Plant available N in manures ranges from 40 to 70% of their TN content (Pratt et al., 1973) and when the manures are composted N availability decreases by as much as half as a result of immobilization in organic matter, including microbial biomass, transforming the compost into a slowly available

N source (Eghball & Power, 1999). Usually initial C: N ratio has been reported as a key factor for controlling N losses through ammonia within a composting process (Sikora & Sowers, 1985; Sikora, 1999).

According to Iglesias–Jimenez and Perez-Garcia (1992), a C: N ratio lower than 12 indicates a good degree of maturity. Benito et al. (2003) expressed a value around 30 as more adequate for pruning waste. Manna (2004) reported that TN content of the mature compost should be more than 1.0% and TK between 1.5 to 2.0%. Umrit and Friesen (1994), however, observed P mineralization from organics with C:P ratio of 227, slightly greater than the breakout point suggested by others. Composted organic solid waste contained between 2 to 16% of their TP as rapidly exchangeable inorganic-P, between 40 and 77% of their TP as slowly exchangeable or not exchangeable inorganic P, probably in the form of condensed calcium phosphates and some organic P (Frossard et. al., 2002). Evidences suggest that manure- P may be equally or even more available than fertilizer-P (Gale et al., 2000); therefore these organic wastes have the potential to meet the partial or complete P requirement of crops (Singh, 2004; Singh et al., 2010) on application to the soils. Compost output and growth rate of earthworms. The compost output varied from 53-76 % during Sep- Dec cycle and from 39-86 % during Jan-Apr cycle with minimum output (53% in Sep-Dec cycle and 39 % in Jan- Apr cycle) from control treatment. On the other hand the maximum output varies from 76-86% in pits where earthworms or earthworms along with microbial culture and moisture were added. The increase in the number and weight of earthworms was respectively 91% and 39 % in the earthworm treatment and 74% and 28 % in the treatment where earthworms along with microbial culture were added. Activity of earthworms started in the 1st week of March with the rise of temperature to optimize earthworms' growth. It was 33 and 45 times in case of earthworm treatment and 43 and 48 times in case of earthworm along with microbial culture treatment The results put forth the concept that no appreciable change in nutrient composition in the vermicompost over FYM that is produced by only thermophilic microorganisms without adding earthworms under similar conditions. It can be concluded that composting the animal dung by scientific means is as efficient as vermicomposting the animal dung.

REFERENCES

1. Benitez E, Nogales R, Masiciandro G and Ceccanti B (2002) Isolation of isoelectric focusing of humic- urease complexes from earthworm (Eisenia fetida) Processed sewage sludges. Biol. Fert. Soils. 31. 489- 493.

2. Benito MA, Masaguer A, Moliner Arrigo N and Palma (2003) Chemical and microbiological parameters for the characterization of the stability and maturity of pruning waste compost. Biol. Fertil. Soils. 37, 184-189. Bettiol W (2004) Effect of sewage sludge on the incidence of corn stalk rot caused by Fusarium' Summa Phytopathol. 30, 16-22.

3. Bremmer JM and Mulvaney CS (1982) Nitrogen-total: (In) Methods of Soil Analysis, Part-2. Chemical and Microbial Properties. Page A L (ed), Agronomy Monograph No.9 (2nd ed.) ASA, SSA, Madison, WI, USA. pp: 595-624.

4. Butt KR (1993) Utilization of solid paper mill sludge and spent brewery yeast as

a feed for soil-dwelling earthworms. Biores. Technol. 44, 105-107.

5. Chastain JP, Frase CC and Moore KP (2006) Composition of animal manure compost and utilization implications. ASABE paper No. 064054. St. Joseph, Mich.: ASABE. Available at: asabe.org.

6. Dhiman N (2003) Ecotaxonomy and biology of terrestrial earthworms of North Indian States. (Unpublished) Ph.D., Thesis., Punjab Agri. Univ., Ludhiana.

7. Dominguez J and Edwards CA and Webster M (2000) Vermicomposting of sewage sludge: Effect of bulking materials on the growth and reproduction the earthworm Eisenia andrei. Pedobiologia. 44, 1- 13.

8. Edwards CA (1988) Breakdown of animal, vegetable and industrial organic wastes by earthworms. In: Earthworms in Waste and in Environment. SPB Acad. Publi., The Hague, The Netherlands. pp: 21-31.

9. Eghball B and Power JF (1999) Phosphorus-and nitrogen-based manure and compost based applications: Crop production and soil phosphorus. Soil Sci. Soc. American J. 63, 895-901.

10. Elvira CA, Dominguez J and Neuhauser EF (1996) The potential use of Perionyx excavatus (Perr.) (Megascolecidae) in organic waste management. Biol. Fertil. Soils. 27, 155-161.

11. Forster JC, Wolfgang Z and Wurdinger E (1993) Comparison of chemical and microbiological methods for the characterization of the maturity of composts from contrasting sources. Bio. Fertil. Soils. 16, 93-99. Frossard E, Skrabal P, Sinaj S, Bangerter F and Traore O (2002) Form and exchangeability of inorganic phosphate in composted solid organic wastes. Nutrient Cycling in Agroecosystem. 62, 103-113.

12. Gale PM, Mullen MD, Cierlik C, Tyler DD, Dcuk BN, Kirchner M and Mc Clure (2000) Phosphorus distribution and availability in response to dairy manure application. Commun. Soil Sci. Plant Anal. 31, 553-565.

13. Garg P, Gupta A and Satya S (2006) Vermicomposting of different types of wastes using Eisenia fetida: A comparative study. Biores. Technol. 97, 391-395.

14. Genevini PL, Adani F, Borio D and Tambone F (1997) Heavy metal content in selected european commercial compost. Compost Sci. Uti. 5, 31-39.

15. Gies G (1997) Developing compost standards in Europe. Biocycle. 38, 82-83.

16. Hand P, Hayes WA, Frankland JC and Satchell JE (1988) The vermicomposting of cow slurry. Pedobiologia. 31, 199-209.

17. Hartenstein R, Neuhauser EF and Kaplan DL (1979) Reproductive potential of the earthworm Eisenia fetida. Oecologia. 43, 329-340.

18. Hundal SS and Dhingra Z (2009) Growth and reproduction of Eudrilus eugeniae in Agricultural wastes. GEOBIOS 36, 9-12.

19. Iglesias–Jimenez E and Perez-Garcia V (1992) Determination of maturity indices for city refuse compost. Agri. Ecosys. Environ. 38, 331-343.

20. Kadalli GG, Suseeladevi L, Siddaramappa R and Patil CR (2000) Quality and

efficiency of value added coir dust based compost. J. Ind. Soc. Soil Sci. 48, 141-144.

21. Kale RD (2004) The use of earthworms: Nature's gift for utilization of organic wastes in Asia. In: C. A. Edwards (ed.). Earthworm Ecol. (2nd ed.) CRC Press LLC USA.
22. Kalra RK, Zinia and Hundal SS (2008) Adoption of Vermiculture as a segment of Organic farming: A Case Study. Ind. J. Ecol. 35(2), 143-145.
23. Kaplan DL, Hartenstein R, Neuhauser EF and Malechi MR (1980) Physiochemical requirements in the environment of the earthworm Eisenia fetida. Soil Biol. Biochem. 12, 347-352.
24. Manna MC (2004) Evaluation of compost maturity, stability and quality. Short course on assessment of compost quality for agriculture crop production. Division of Soil Biology. Indian Instit. Soil Sci., Bhopal (India) pp: 20-25.
25. Mitchell A (1997) Production of Eisenia fetida and vermicompost from feed-lot cattle manure. Soil Bio. Biochem. 29, 763-766.
26. Mishra B (2001) Organic waste recycling in aqua farming. Yojana, Nov. pp: 419-425. 29.Ndegwa PM and Thompson SA (2001) Integrating composting and vermicomposting in treatment and bioconversion of biosolids. Biores. Technol. 76, 107- 112.
27. Ndegwa PM, Thompson SA and Das KC (2000) Effect of stocking density and feeding rate on vermicomposting of biosolids. Biores. Technol. 71, 5- 12.
28. Nelson DN and Sommers LE (1982) Total C, organic C and organic matter. In: Page AL, Muller RH and Keeney DR (eds) Methods of soil analysis. American Soc. Agronomy, Madison, Wisconsin, USA. pp: 539- 579.
29. Patil MP (1995) Integrated nutrient management in commercial vegetable. M.Sc. Agri. Thesis, Univ. Agric. Sci. Dharwad (India).
30. Pratt PS, Broadbent SP and Martin JP (1973) Using organic waste as N-fertilizer. Calif. Agric. 27, 10-13.
31. Ramaswami PP (1998) Recycling of wastes as manure – potentials and prospects. In: Summer School on Integrated Soil Productivity Managt. Sustainable Agricul. ICAR, New Delhi. pp: 83-91.
32. Reddy MV and Ohkura K (2004) Vermicomposting of rice straw and its effects on sorghum yield. Trop. Ecol. 45, 27-31.
33. Reinecke AJ and Vijioen SA (1990) The influence of worm density on growth and cocoon production of the compost worm Eisenia fetida (Oligochaeta). Revue d' Ecologie et Biologie du Sol. 27, 221- 230.
34. Singh H, Hundal SS and Singh P (2008) Vermicomposting of rice straw compost and farm yard manure with two epigeic earthworm species, J. Res. (PAU). 43(182), 51-55 PAU.
35. Singh H, Hundal SS and Singh P (2010) Vermicomposting potential of Eisenia fetida for different agro-organic wastes. Indian. J. Ecol. 37(1), 60-64.

36. Singh P (2004) Phosphorus availability in recent floodplain soils as influenced by their physical and chemical parameters and application of pressmud. M.Sc. Thesis (Unpublished) Punjab Agricultural University, Ludhiana, Punjab.

37. Singh R, Sharma RR and Tyagi SK (2007) Pre- harvest foliar application of calcium and boron influences physiology disorders, fruit yield and quality of strawberry (Fragaria x ananassa Duch.). Scientia Horticult. 112, 215-220.

38. Sinha RK and Herat S (2009) Vermiculture and sustainable agriculture. Americ. Eurasian J. Agric. Environ. Sci. 5(S), 01-55.

39. Sikora LJ (1999) MSW compost reduces nitrogen volatilization during dairy manure composting. Compost Sci. Uti. 7, 34-41.

40. Suthar S (2006, Nutrient changes and biodynamics of epigeic earthworm, Perionyx excavatus (Perrier) during recycling of some agricultural wastes. Bioresource Tech doi. 10 1016/j.biotech).

41. Suthar S (2008) Bioconversion of post harvest crop residues and cattle shed manure into value added products using earthworm Eudrilus eugeniae (Kinberg). Ecol. Eng. 32, 206-314.

42. Suthar S and Singh S (2008) Feasibility of vermicomposting in bio-stabilization sludge from a distillery industry. Sci. Total Environ. 393, 327- 343.

43. Umrit G and Friesen DK (1994) The effect of C:P ratio of plant residues added to soils of contrasting phosphate sorption capacities on P uptake by Panicum maximum (Jacq.) Plant Soil. 158, 275-285.

44. Vogtman H, Fricke K and Turk T (1993) Quality, physical characteristics, nutrient content, heavy metals and organic chemicals in Biogenic waste compost. Compost Sci. Uti. 1, 69-87.

3
CHAPTER

MANAGING AGRICULTURAL PHOSPHORUS TO MINIMIZE WATER QUALITY IMPACTS

Andrew Sharpley

Keyword: Fertilizer managemen, conservation practice,; runof,; erosio,; eutrophication

INTRODUCTION

The accelerated eutrophication of freshwaters and to a lesser extent some coastal waters is primarily driven by phosphorus (P) inputs. While efforts to identify and limit point source inputs of P to surface waters have seen some success, nonpoint sources have remained elusive and more difficult to identify, quantify, target, and remediate. Thus, attention has focused more on non-point source reduction, particularly the role of agriculture. This attention has increased over the last 10 years due to the fact that impaired water quality has not seen as great an improvement as expected after the adoption of conservation practices.

The increased use of fertilizer and manures has contributed to a steady increase in con (*Zea ays*) and soyben (*Glycinemax*) yields in Brazil ([USDA, 2015a] and [2015b]) (Figure 1A). In many cases, record corn and soybean yields have been achieved since 2012 and are expected to surpass U.S. soybean production for the first time in 2015 ([USDA, 2015c]).

Along with increased crop production, there have been tremendous changes in agricultural production systems in most developed countries, with crop and animal operations evolving into spatially separate production systems. In Brazil, there has been a 4.5-fold increase in fertilizer P use on agricultural lands since 1975 (Figure 1A). Up to now there has been a 128 % increase in the amount of P produced in manure by beef, 13 % increase for swine, and 88 % increase for poultry (Figure 1B; [IBGE, 2014]).

Citation: Sharpley, Andrew. (2016). Managing agricultural phosphorus to minimize water quality impacts. Scientia Agricola, 73(1), 1-8.https://dx.doi.org/10.1590/0103-9016-2015-0107.

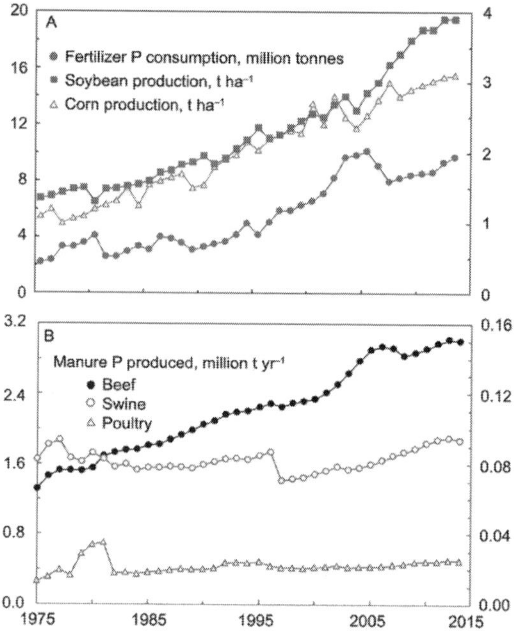

Figur 1: Fertilizer P consumption and soybean and corn production between 1975 and 2014 (upper graph; A) and manure P produced by livestock over the same period P (lower graph; B) in Brazil. Data from the Brazilian Institute of Geography and Statistics, 2014 and the U.S. Department of Economic Research Service, 20.

The fact that Brazil has almost twice the available water resources than the next ranked country, Russia, and over 2.5 times more than the U.S., highlights the tremendous potential for the continued growth of Brazilian agriculture and its productivity (Figure 2; Food and Agriculture Organization, 2015). Thus, the wise use of water and nutrients now will minimize the risk of water quality degradation issues that the U.S. and China, for instance, now face. This sustainable or conservationist approach to nutrient and agricultural management will also, importantly, limit external pressures that would be levelled at agricutlural stakeholders to mitigate degradation. This has certainly become an issue that is hotly and fractiously debated worldwide, where P impairment of surface waters persists, such as in Asia (Dai et al., 2011; Li et al., 2015; Novotny et al., 2010; Sun et al., 2012), Europe (Hilton et al., 2006; Withers and Jarvie, 2008), and U.S. (Dubrovsky et al., 2010; National Research Council, 2008).

While grazing-based beef production is an extensive system, where manure is deposited over grazed pastures, confined or housed swine and poultry are intensive systems, producing large amounts of manure in localized areas. The large increase in size and number of animal production systems creates localized surpluses of P-rich manure due to the major input of nutrients in animal feed from grain-producing areas. This accumulation leads to an overall increase in the risk of P runoff. Management of fertilizer to maximize grain and livestock production, while minimizing environmental impacts, has evolved into nutrient and conservation management strategies (Figure 3). Strategically targeting these management strategies to critical areas of the landscape and to regions of the country, where there is the potential for accumulation of nutrients, is key to minimizing the impact of agricultural P-related water quality issues.

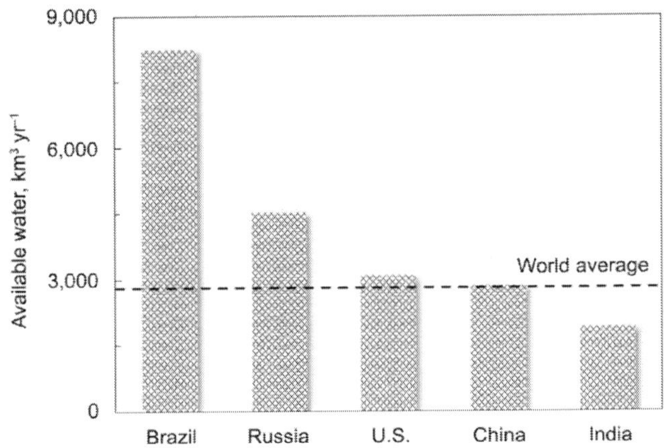

Figue 2: Water resource availability for several countries, showing that Brazil has twice the volume of water resources than any other couny.

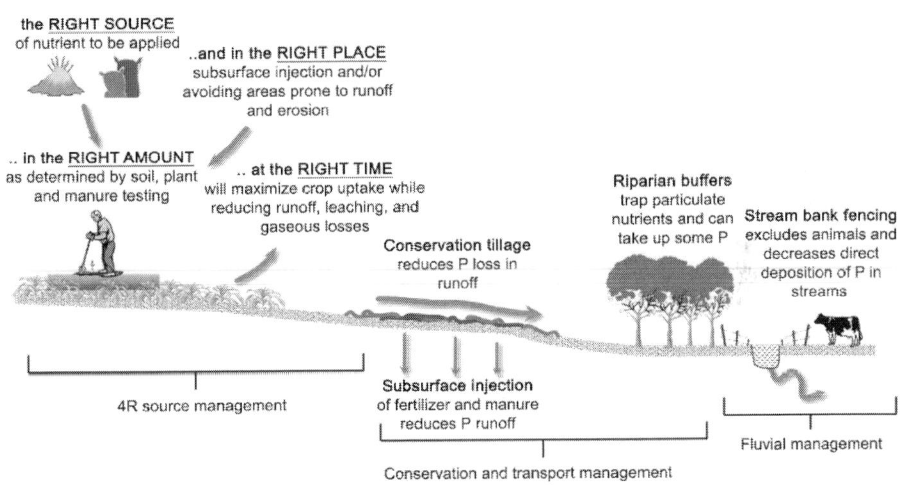

Figre 3: "4R" nutrient source management and conservation management of P in agricultural production systems.

This paper reviews the management of P applications in mineral and manure forms to agricultural land in Brazil, in the context of examples and lessons learnt from other countries. Brazilian agriculture has tremendous potential for growth, along with an opportunity to continue this growth in an environmentally sustainable manner. As several land management terms used in this review might differ slightly between the Brazil and the U.S., they are defined herein for clarification. No-till (also called zero tillage or direct drilling) is a way of growing crops from year to year without disturbing the soil through tillage (i.e., plowing). Minimum tillage is the least amount possible of cultivation or soil disturbance incurred to prepare a suitable seedbed. This means that physically, the soil is not inverted, which is usually desirable, as intensive tillage tends to break down soil structure. Finally, conventional tillage is a tillage system using cultivation as the major means of seedbed preparation and weed control.

NURIENT MMANAGEMENT

Careful nutrient management planning on a field-by-field and farm basis is a major component of any remedial action plan to minimize the risk of nutrient loss from agricultural land. This basically follows the "4R" nutrient management approach advocated by the IFA (2009) and IPNI (2014), which is adding P t the R ight rate to match crop needs, in the Right source,at the Right time, andin the Right place (Figure 3).

The Right Rate

Fertilizer P rates are usually established by crop need and modified by the amount already in the soil, as determined by established soil P test methods. In the case of commercial P fertilizer, applications can be tailored to match crop needs and minimize excessive soil P accumulation, because of the economic incentive of avoiding applying too much costly fertilizers.

The Right Source

Fertilizer nutrients can be formulated to match crop needs; however, manures have more P than N compared to crop needs. For instance, the ratio of N:P in manure (2 to 4:1) is three to four times lower than that taken up by major grain and hay crops (8:1). As a result, applications of manure to meet crop N needs, apply three to four times more P than annual crop needs. Repeatedly applying manure at rates to provide sufficient N, will increase soil P levels and the risk of P runoff. On the other hand, application of manure to meet crop P needs will apply insufficient N, putting an economic burden on farmers to purchase costly mineral N.

To address this, there needs to be a greater exchange and coordination of information among researchers, farm advisors, and fertilizer consultants, to increase the awareness and availability of different fertilizer formulations. Awareness of the range of nutrient formulations commercially available to farmers can help farmers tailor N, P, and K additions to crop needs and to what is already in the soil and available for plant uptake.

The Right Time

Many studies show the loss of P in runoff relates directly to the timing of applied P (Dampney et al., 2000; Sharpley et al., 2007; Sims and Kleinman, 2005). Avoiding applications within a few days of expected rainfall can minimize runoff losses due to the sorption of added P with soil. Several studies show, for example, reductions in N and P losses with an increase in the length of time between manure application and surface runoff (Djodjic et al., 2000; Edwards and Daniel, 1993; Sharpley, 1997; Westerman et al., 1983). These reductions can be attributed to the reaction generated when P is added to the soil and dilution of applied P by in-filtrating water from rainfall that did not cause surface runoff. However, timing the application of P to avoid times of the year when the risk of rain-induced runoff is greatest, must be tempered with providing the nutrients at the time of most active plant uptake.

The Right Place

Because of the relative immobility of P in soil, placement of P fertilizer is generally more critical for plant availability than is the case for N fertilizer. The incorporation of manure into the soil profile, either by tillage or subsurface placement, reduces the potential for P runoff. For example, Mueller et al. (1984) showed incorporation of dairy manure by chisel plowing reduced total P loss in runoff from corn 20-fold, compared to no-till areas receiving surface applications.

CONSERVATION AND TRANSPORT MANAGEMENT

Transport management refers to efforts to control the movement of P from soils to sensitive locations such as bodies of fresh water. Phosphorus loss via surface runoff and erosion may be reduced by conservation tillage and crop residue management, buffer strips, riparian zones, terracing, contour tillage, cover crops, and impoundments or ponds. These practices tend to reduce rainfall impact on the soil surface, reduce runoff volume and velocity, and increase soil resistance to erosion. However, none of these measures should be relied on as the sole or primary practice to reduce P runoff.

Adoption of conservation measures has been enocuraged in many areas. In the U.S. for example, Gulf of Mexico hypoxia has led to a large scale initiative to implement conservation practices in the contributing Mississippi River Basin in the U.S. (Dale et al., 2010). The Mississippi River Basin drains 41 % of the U.S. (or 1,245,000 miles^2). Thus, spatially expansive sources of P and the relative contributions of municipalities, industries, and agriculture, create system complexities that should not limit action by any one entity. Recent model estimates (USGS-SPARROW; Alexander et al., 2008) suggest that up to 85 % of the P and N entering the Gulf of Mexico originate from agriculture. Government agencies charged with implementing and checking compliance with conservation programs used these model estimates to prioritize allocation of conservation cost-share funding under the 2009 Healthy Mississippi River Basin Initiative ($320 million over 5 years) to 41 of the top contributing sub-watersheds in 13 states along the Mississippi River corridor (USDA, 2009). One unique aspect of this program has been the provision of financial incentives to farmers to conduct edge-of-field monitoring to quantify the effectiveness of conservation practices to reduce nutrient runoff. These initiatives have been very successful in getting conservation systems implemented on a large acreage of agricultural lands to help farmers address water quality concerns. However, a transparent framework is needed to document and verify on-the-ground conservation practices to ensure farmers are credited with accurate reductions.

LEGACIES OF PAST MANAGEMENT

Past land use or land management activities can lead to a long-term legacy of P in watersheds (Figure 4). This legacy of P refers to P stored in surface soils, ditches, riparian zones, wetlands, stream, and lake sediments from prior land and nutrient management (Meals et al., 2010; Sharpley et al., 2013). The stored P can be subsequently re-released as the P storage capacity gradually becomes saturated, or after a change in land use, or land management. Lag times associated with release of P from the P legacy stores may help explain difficulties in detecting water quality improvements at a watershed scale. For example, where soil P test levels have risen to more than 10 times crop sufficiency levels, it can take a decade or more to 'draw down' soil P reserves to levels where dissolved P in runoff is substantially reduced (Cox et al., 1981; Shulte et al., 2010). However, by developing watershed-scale monitoring that identifies local-scale improvements and associated time lags in water quality as they occur, watershed planners can start to better understand and plan mitigation strategies.

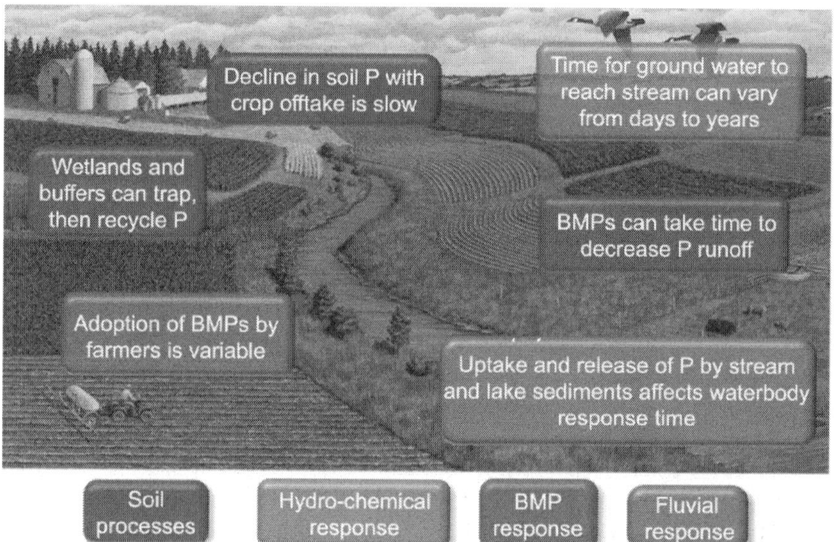

Figure 4: Conceptual representation of natural, managed, soil, and Best Management Practices processes influencing the lag time for system response.

The richly documented history of water quality in the Lake Erie Watershed, outbreaks of harmful algal blooms, and land management in the Watershed over the last 50 years, provides an excellent example of how well-meaning conservation strategies can result in intended and unintended consequences on the fate of P and transport. Steady declines in P inputs from predominantly agricultural watersheds were measured between 1980 and 1995 with the adoption of conservationist practices, such as nutrient management planning that reduced fertilizer and manure applications to corn and soybeans and a transition to no-till cropping.

However, simply implementing conservation practices may not always provide the desired outcomes. For example, despite widespread adoption of conservation in the Lake Erie Watershed, increased losses of P in agricultural runoff over the last decade, have resulted from complex, dynamic, but predictable factors. These include the accumulation of P at the soil surface, a fall in the application of fertilizer, continued surface broadcasting of P, a focus on implementing conservation practices that reduce erosion and P particulate loss, a rapid rise in tile drainage fuelled by higher grain prices, and release and remobilization of fluvial P. For instance, there are now more fields with tile drainage that connect to ditches and streams, contributing source areas of the legacy of P to Lake Erie. The combination of these factors created a "perfect P loss situation," which, along with more intense summer rains, increased P inputs to Lake Erie to record levels in 2010, culminating in the 2014 toxic bloom and water crisis in Toledo, OH. In the year that followed, there were at least 25 theories proposed as to what may have contributed to the harmful and nuisance algal blooms (Smith et al., 2015). However, scientifically valid remedial strategies were not likely to be readily adopted by farmers due to several logistical, practical, and cost limitations (Daloğlu et al., 2012; Michalak et al., 2013; Sharpley et al., 2012). Clearly, the research community needs to work closely with the farming community to generate innovative support, stewardship, and reward programs that will empower change.

FUTURE CHALLENGES FOR FERTILIZER MANAGEMENT AND WATER QUALITY

Fertilizer-P use efficiency

Several reviews and status perspectives of global cycling of P, have acknowledged that the biogeochemical P cycle has failed on a global, national, regional, and farm scale (Sharpley and Jarvie, 2012; Jarvie et al., 2013; Haygarth et al., 2014). The failed P cycle in mining, fertilizer production, and land application of P in livestock manure, food waste, and human waste is the underlying problem since it limits sustainable P use. Contributing to a solution is the need to Realign the inputs of P, Reuse P from manures and residuals (Bonvin et al., 2015; Stutter, 2015), Recover P from waste, Redefine systems, and Reduce P losses. This was defined as the 5R's of P management stewardship by Schoumans et al., (2015), which is an update of the "4R" nutrient management stewardship, specifically formulated to address the failed P cycle, which is the root cause of P related use-impairments of water from accelerated eutrophication. Thus, increasing the efficiency of P use within agricultural systems, which include intensively concentrated and spatially separated livestock and arable operations is critical to resolving the current unsustainable P cycle.

The development and use of slow release fertilizers, mixing mineral and organic P fertilizers, fertigation, and the advancement of biofertilizers (i.e., inoculation of seed with effective P solubilizing microorganisms), are important research opportunities that may hold promise. The development of fertigation and biofertilizers have been reviewed elsewhere (Bauder et al., 2014; Cruz et al., 2012; Stewart and Gordon, 2008; Yara, 2014). The main points relevant to the present discussion of environmental stewardship are that as the costs of these products decrease, they will provide farmers new and real opportunities to closely tailor fertilizer nutrient applications to crop needs. The use of a reactive phosphate rock, such as North Carolina Rock Phosphate, has been shown to be another way to sustain crop productivity on certain soil types, while at the same time, minimizing P runoff (Shigaki et al., 2006, 2011).

Defining P transport pathways

When P is land applied in the form of either mineral or organic fertilizer, surface and subsurface hydrological pathways transport P from the land surface, where it is a valuable resource used to achieve and maintain optimum plant yields, to receiving waters where it can be a major contributor of water-use impairment. Hydro-logical pathways along which P is transported are often complex and tortuous, involving multiple contributions from surface and subsurface transfers (Gburek and Sharpley, 1998; Heathwaite and Dils, 2000). The complexity of subsurface drainage and difficulties in deconvoluting flow pathways were identified as key gaps in our understanding of P transport and fate, particularly in groundwater-dominated watersheds.

It is clear that hydrological pathways along which P is transmitted, do not simply transport P conservatively, but act as a series of reactive conduits, mediating P flux transformations through retention and recycling of P, on a range of timescales from years to centuries (Jarvie et al., 2014; Sharpley et al., 2013). While inherent hydro-logical pathways are to a large extent a function of site topography and geology, they must be quantified to reliably estimate the timing and extent of P loss reductions that can be achieved with land management changes.

Modeling nonpoint P transport on a watershed scale

Nonpoint source models represent mathematical descriptions, ranging from simple (risk assessment indices) to more complex scientific understanding about chemical, physical, and biological processes, which determine P movements within a watershed. Use of non-point source models provides a method of simulating the risk of P loss or estimating the relative effects of change in climate, land use, and land management practices on sediment and nutrient loss. Models have strong appeal to policymakers and managers, because models provide answers much more quickly than can be obtained from field measurements and monitoring. However, this appeal can bring with it false confidence and misconceptions about the reliability of the predictions (Boesch et al., 2001).

Thus, it is important that model developers clearly define what the model is useful for and what it is not designed to do. Likewise, model users must decide what they want to accomplish with a model. For example, one must consider the scale (field, watershed, or basin), time (flow event, annual, or multi-year), and level of accuracy (0.1 or 10 kg ha^{-1} yr^{-1}) that needs to be simulated, as well as the amount and quality of data available. The predictive uncertainty of models can result from imperfect mathematial representations of the real world, numeric approximations in equations, inaccurate model input data, and uncertainties in input data (Harmel et al., 2006). Nevertheless, because models play an important role in scientific studies and policy decision-making, every effort must be made to obtain reasonable estimates of model uncertainties and make the estimates clear to all those using the model output. In other words, we should not be using models to predict the effect of changing land management on the magnitude of P loss to an accuracy of say 0.1 kg P $ha^{-1}yr^{-1}$, when model uncertainty can be as much as ± 50 %.

Implementing appropriate conservatoin practices

While there are many Best Management (or conservation) Practices (BMPs) that can be implemented over a wide range of scales to minimize the loss of P from agriculture to surface and ground waters (Figure 4), cost-benefit estimates are needed for more widespread adoption. If a farmer does not receive financial support to implement a conservation practice, then it is not likely to be adopted voluntarily unless it either makes or saves the farmer money.

For real and lasting changes to occur in agricultural systems, balancing production and environmental stewardship constraints, there needs to be greater consideration of socioeconomic drivers of what, how, and why some conservation practices are adopted and others are not (Kleinman et al., 2015). There should be greater emphasis on education, outreach, and consumer-driven programs rather than assuming that farmers will absorb the total costs associated with implementing remedial practices. Bearing this in mind, except for decreasing off-farm import of P and increasing on-farm P-use efficiency, conservation practices are thus only a temporary measure for minimizing P loss and receiving water impacts.

There needs to be a discussion of how regional, as well as national agricultural infrastructures, can control P inputs to farming systems and assess large-scale nutrient balances. However, it must be kept in mind that some P loss cannot be avoided, such as during large storm events. Thus, there needs to be an emphasis on quantifying background or baseline losses of P, to better define what reductions or targeted P loss goals might be achievable. For example, while pristine water conditions might be most desirable, watersheds are naturally leaky systems and some loss of P will have to be accepted where anthropogenic inputs of P to a watershd have occurred (Haygarth et al., 2014).

CONCLUSIONS

The failed cycling of P in mining, fertilizer production, and land application as an essential plant nutrient to the recovery and reuse of P in human waste, livestock manure, and food waste is the underlying problem. This limits sustainable P use and is the root cause of P related use-impairments of water from accelerated eutrophication. Thus, increasing the efficiency of P use within agricultural systems that include intensively concentrated and spatially separated livestock and arable operations, is a critical component in resolving the failed P cycle situation (Haygarth et al., 2014).

Pressures placed on farmers to maximize yields in response to increasing demand for cheap food, feed, and fuel due to an increasing global population that is becoming more affluent are on the increase. To a large extent this has been made possible by the development and use of fertilizer. At the same time, increased pressures are being placed on farmers to be environmental stewards. However, despite a long history of soil and water P research, management questions still exist and water-use impairment continues as a result of P enrichment of soil-water systems. Thus, we need to accept that current research knowledge has to be translated and transferred better to farmers, while at the same time realizing that farm management decisions are largely driven by competitive economics.

The recent involvement of farmers in on-farm environmental stewardship programs and more site monitoring of edge-of-field P losses, will provide documentation on how successful they have been at conserving on-farm P. While many conservation programs have led to more widespread adoption of conservation practices aimed at reducing P loss, on-farm research and monitoring will help farmers document and receive credit for their efforts. This is particularly important to the voluntary adoption of conservation practices, which tend to have less tracking by and accountability to Federal agencies, but are, nevertheless, as effective as regulated practices.

REFERENCES

1. Alexander, R.B.; Smith, R.A.; Schwarz, G.E.; Boyer, E.W.; Nolan, J.V.; Brakebill, J.W. 2008. Differences in phosphorus and nitrogen delivery to the Gulf of Mexico from the Mississippi River Basin. Environmental Science & Technology 42: 822-830. [Links]
2. Bauder, T.A.; Waskom, R.M.; Andales, A. 2014. Nitrogen and Irrigation Management. Colorado State University Extension, Ft. Collins, CO, USA. (Fact Sheet, 0.514). Available at: http:// www.ext.colostate.edu/pubs/crops/00514.pdf [Accessed Sept. 22, 2015] [Links]
3. Boesch, D.F.; Brinsfield, R.B.; Magnien, R.E. 2001. Chesapeake Bay eutrophication: scientific understanding, ecosystem restoration, and challenges for agriculture. Journal of Environmental Quality 30: 303-320. [Links]
4. Bonvin, C.; Etter, B.; Udert, K.M.; Frossard, E.; Nanzer, S.; Tamburini, F.; Oberson, A. 2015. Plant uptake of phosphorus and nitrogen recycled from synthetic source-separated urine. Ambio 44: 217-227. [Links]
5. Cox, F.R.; Kamprath, E.J.; McCollum, R.E. 1981. A descriptive model of soil test nutrient levels following fertilization. Soil Science Society of America 45: 529-532. [Links]
6. Cruz, P.C.; Banayo, N.P.; Marundan, S.R.; Magnaye, A.M. 2012. Bio-inoculant and foliar fertilizer in combination with soil-applied fertilizer on the yield of lowland rice. Philippine Journal of Crop Science 37: 85-94. [Links]
7. Dai, Z.; Du, J.; Zhang, X.; Su, N.; Li, J. 2011. Variation of riverine material loads and environmental consequences on the Changjiang (Yangtze) estuary in recent decades (1955-2008). Environmental Science & Technology 45: 223-227. [Links]
8. Dale, V.H.; Wright, D.; Kling, C.L.; Boynton, W.; Meyer, J.L.; Mankin, K.; Sanders, J.; Opaluch, J.; Conley, D.J.; Stallworth, H.; Crumpton, W.; Paerl, H.; Armitage, T.; Reckhow, K.; David, M.; Gilbert, D.; Sharpley, A.N.; Wangsness, D.; Howarth, R.W.; Simpson, T.W.; Bianchi, T.; Blumberg, A.; Lowrance, R.; Snyder, C.S. 2010. Hypoxia in the northern Gulf of Mexico. Springer Science, New York, N Y, USA. (Springer Series on Environmental Management). [Links]
9. Daloğlu, I.; Cho, K.H.; Scavia, D. 2012. Evaluating causes of trends in long-term dissolved reactive phosphorus loads to Lake Erie. Environmental Science &

Technology 46: 10660-10666. [Links]

10. Dampney, P.M.R.; Lord, E.I.; Chambers, B.J. 2000. Development of advice for farmers and advisors. Soil Use and Management 16: 162-166. [Links]
11. Djodjic, F.; Ulen, B.; Bergstrom, L. 2000. Temporal and spatial variations of phosphorus losses and drainage in a structured clay soil. Water Research 34: 1687-1695. [Links]
12. Dubrovsky, N.M.; Burow, K.R.; Clark, G.M.; Gronberg, J.M.; Hamilton, P.A.; Hitt, K.J.; Mueller, D.K.; Munn, M.D.; Nolan, B.T.; Puckett, L.J.; Rupert, M.G.; Short, T.M.; Spahr, N.E.; Sprague, L.A.; Wilber, W.G. 2010. The Quality of our Nation›s Waters-Nutrients in the Nation›s Streams and Groundwater, 1992-2004. U.S. Geological Survey, Washington, DC, USA. (Circular, 1350). Available at: http:// water.usgs.gov/nawqa/nutrients/pubs/circ1350 [Accessed Sept. 22, 2015] [Links]
13. Edwards, D.R.; Daniel, T.C. 1993. Drying interval effects on runoff from fescue plots receiving swine manure. Transactions of the American Society of Agricultural and Biological Engineers. 36: 1673-1678. [Links]
14. Gburek, W.J.; Sharpley, A.N. 1998. Hydrologic controls on phosphorus loss from upland agricultural watersheds. Journal of Environmental Quality 27: 267-277. [Links]
15. Harmel, R.D.; Cooper, R.J.; Slade, R.M.; Haney, R.L.; Arnold, J.G. 2006. Cumulative uncertainty in measured streamflow and water quality data for small watersheds. Transactions of the American Society of Agricultural and Biological Engineers 49: 689-701. [Links]
16. Haygarth, P.M.; Jarvie, H.P.; Powers, S.M.; Sharpley, A.N.; Elser, J.J.; Shen, J.; Peterson, H.M. Chan, N.I.; Howden, N.J.K.; Burt,T.; Worrall, F.; Zhang, F.; Liu, X. 2014. Sustainable phosphorus management and the need for a long-term perspective: The legacy hypothesis. Environmental Science & Technology 48: 8417-8419. [Links]
17. Heathwaite, A.L.; Dils, R.M. 2000. Characterising phosphorus loss in surface and subsurface hydrological pathways. Science of The Total Environment 251: 523-538. [Links]
18. Hilton, J.; O›Hare, M.; Bowes, M.J.; Jones, J.I. 2006. How green is my river? A new paradigm of eutrophication in rivers. Science of The Total Environment 365: 66-83. [Links]
19. Food and Agriculture Organization [FAO]. 2015. World water resources by country. Aquastat Programme of FAO›s Land and Water Development Division. Water Reports 23. Available at:ftp://ftp.fao.org/docrep/fao/005/y4473E/y4473E00.pdf [Accessed Sept. 22, 2015] [Links]
20. Instituto Brasileiro de Geografia e Estatística [IBGE]. 2014. Available at: http:// www.sidra.ibge.gov.br/bda/tabela/ listabl.asp?z=t&c=73 [Accessed Sept. 22, 2015] (in Portuguese). [Links]
21. International Fertilizer Association [IFA]. 2009. The Global "4R" nutrient

stewardship framework: developing fertilizer best management practices for delivering economic, social and environmental benefits. IFA, Paris, France. Available at: https://www.ipni.net/ipniweb/portal/4r.nsf/0/BAB4157B4888 71A385257DF100739D94/$FILE/The%20Global%204R%20 Nutrient%20 Stewardship%20Framework.pdf [Accessed Sept. 22, 2015] [Links]

22. International Plant Nutrition Institute [IPNI]. 2014. 4R Nutrient stewardship portal. Available at: http://www.ipni.net/4R [Accessed Sept. 22, 2015] [Links]

23. Jarvie H.P.; Sharpley, A.N.; Withers, P.J.; Scott, J.T.; Haggard, B.E.; Neal, C. 2013. Phosphorus mitigation to control river eutrophication: murky waters, inconvenient truths and 'Post-Normal' science. Journal of Environmental Quality 42: 295-304. [Links]

24. Jarvie, H.P.; Sharpley, A.N.; Brahana, V.; Simmons, T.; Price, A.; Neal, C.; Lawlor, A.J.; Sleep, D.; Thacker, S.; Haggard, B.E. 2014. Phosphorus retention and remobilization along hydrological pathways in karst terrain. Environmental Science & Technology 48: 4860-4868. [Links]

25. Kleinman, P.J.A.; Sharpley, A.N.; Withers, P.J.A.; Bergström, L.; Johnson, L.; Doody, D.G. 2015. From sacred cows to sacrificial lambs: implementing agricultural phosphorus science and management to combat eutrophication. Ambio 44: 297-310. [Links]

26. Li, H.; Liu, J.; Li, G.; Shen, J.; Bergström, L.; Zhang, F. 2015. Past, present and future use of phosphorus in Chinese agriculture and its influence on phosphorus losses. Ambio 44: 274-285. [Links]

27. Meals, D.W.; Dressing, S.A.; Davenport, T.E. 2010. Lag time in water quality response to best management practices: Review. Journal of Environmental Quality 39: 85-96. [Links]

28. Michalak, A.M.; Anderson, E.J.; Beletsky, D.; Boland, S.; Bosch, N.S.; Bridgeman, T.B., Chaffin, J.D.; Cho, K.; Confesor, R.; Daloğlu, I.; DePinto, J.V.; Evans, M.A.; Fahnenstiel, G.L.; He, L.; Ho, J.C.; Jenkins, L.; Johengen, T.H.; Kuo, K.C.; LaPorte, E.; Liu, X.; McWilliams, M.R.; Moore, M.R.; Posselt, D.J.; Richards, R.P.; Scavia, D.; Steiner, A.L.; Verhamme, E.; Wright, D.M.; Zagorski, M.A. 2013. Record-setting algal bloom in Lake Erie caused by agricultural and meteorological trends consistent with expected future conditions. Proceedings of the National Academy of Sciences 110: 6448-6452. [Links]

29. Mueller, D.H.; Wendt, R.C.; Daniel, T.C. 1984. Phosphorus losses as affected by tillage and manure application. Soil Science Society of America 48: 901-905. [Links]

30. National Research Council. 2008. Nutrient control actions for improving water quality in the Mississippi River Basin and Northern Gulf of Mexico. Committee on the Mississippi River and the Clean Water Act: Scientific, Modeling and Technical Aspects of Nutrient Pollutant Load Allocation and Implementation. National Research Council, Washington, DC, USA. Available at: http://www.nap.edu/catalog/12544.html [Accessed Sept. 22, 2015] [Links]

31. Novotny, V.; Wang, X.; Englande, A.J.; Bedoya, D.; Promakasikorn, L.; Tirado,

R. 2010. Comparative assessment of pollution by the use of industrial agricultural fertilizers in four rapidly developing Asian countries. Environment, Development and Sustainability 12: 491-509. [Links]

32. Schoumans, O.F.; Bouraoui, F.; Kabbe, C.; Oenema, O.; van Dijk, K.C. 2015. Phosphorus management in Europe in a changing world. Ambio 44: 180-192. [Links]

33. Sharpley, A.N. 1997. Rainfall frequency and nitrogen and phosphorus in runoff from soil amended with poultry litter. Journal of Environmental Quality 26: 1127-1132. [Links]

34. Sharpley, A.N.; Herron, S.; Daniel, T. 2007. Overcoming the challenges of phosphorus-based management in poultry farming. Journal of Soil and Water Conservation 62: 375-389. [Links]

35. Sharpley, A.N.; Richards, R.P.; Herron, S.; Baker, D.B. 2012. Can production and environmental goals coexist in phosphorus-based farm management? Journal of Soil and Water Conservation 67: 149-193. [Links]

36. Sharpley, A.; Jarvie, H.P.; Buda, A.; May, L.; Spears, B.; Kleinman, P. 2013. Phosphorus legacy: Overcoming the effects of past management practices to mitigate future water quality impairment. Journal of Environmental Quality 42: 1308-1326. [Links]

37. Shigaki, F.; Sharpley, A.N. 2011. Phosphorus source and soil properties effects on phosphorus availability. Soil Science 176: 502-507. [Links]

38. Shigaki, F.; Sharpley, A.N.; Prochnow, L.I. 2006. Source-related transport of phosphorus in surface runoff. Journal of Environmental Quality 35: 2229-2235. [Links]

39. Shulte, R.P.O.; Melland, A.R.; Fenton, O.; Herlhy, M.; Richards, K.; Jordan, P. 2010. Modelling soil phosphorus decline: Expectations of Water Framework Directive policies. Environmental Science & Policy 13: 472-484. [Links]

40. Sims, J.T.; Kleinman, P.J.A. 2005. Managing agricultural phosphorus for environmental protection. p. 1021-1068. In: Sims, J.T.; Sharpley, A.N., eds. Phosphorus: agriculture and the environment. American Society of Agronomy, Madison, WI, USA. [Links]

41. Smith, D.R.; King, K.W.; Williams, M.R. 2015. What is causing the harmful algal blooms in Lake Erie? Journal of Soil and Water Conservation 70: 27A-29A. [Links]

42. Stewart, W.M.; Gordon, W.B. 2008. Fertilizing for Irrigated Corn: Guide to Best Management Practices. International Plant Nutrition Institute, Peacht ree Corners, GA, USA. Available at: http://www.ipni.net/ipniweb/portal.nsf/0/9bbc50427c6469ae8 52574f200162796/$FILE/ATT4QHV2/Irrigated_%20 Corn.pdf [Accessed Sept. 22, 2015] [Links]

43. Stutter, M.I. 2015. The composition, leaching and sorption behavior of some alternative sources of phosphorus for soils. Ambio 44: 207-216. [Links]

44. Sun, B.; Zhang, L.; Yang, L.; Zhang, F.; Norse, D.; Zhu, Z. 2012. Agricultural

non-point source pollution in China: causes and mitigation measures. Ambio 41: 370-379. [Links]

45. United States Department of Agriculture [USDA]. 2013. Fertilizer use and prices. Economic Research Service, Washington, DC, USA. Available at: http://www.ers.usda.gov/data-products/ fertilizer-use-and-price.aspx# [Accessed Sept. 22, 2015] [Links]
46. United States Department of Agriculture [USDA]. 2015a. Global Agricultural Information Network: Grain and Feed Update. Foreign Agricultural Service, Washington, DC, USA. (GAIN Report BR0958). [Links]
47. United States Department of Agriculture [USDA]. Global Agricultural Information Network: Oilseeds and Products Update. Foreign Agricultural Service, Washington, DC, USA. (GAIN Report BR0957). [Links]
48. United States Department of Agriculture [USDA]. 2015c. World Agricultural Production. Foreign Agricultural Service, Washington, DC, USA. Available at: http://apps.fas.usda.gov/ psdonline/circulars/production.pdf [Accessed Sept. 22, 2015] [Links]
49. United States Department of Agriculture [USDA]. 2009. Mississippi River Basin Initiative. Natural Resources Conservation Service, Washington, DC, USA. Available at: http://www.nrcs.usda.gov/wps/portal/nrcs/detailfull/national/home/?cid=stelprdb1048200 [Accessed Sept. 22, 2015] [Links]
50. Westerman, P.W.; Donnely, T.L.; Overcash, M.R. 1983. Erosion of soil and poultry manure: a laboratory study. Transactions of the ASAE 26: 1070-1078, 1084. [Links]
51. Withers, P.J.A.; Jarvie, H.P. 2008. Delivery and cycling of phosphorus in UK rivers: Implications for catchment management. Science of the Total Environment 400: 379-395. [Links]
52. Yara. 2014. Liquid fertilizer solutions. Available at: http://yara. co.uk/crop-nutrition/knowledge/yara-chafer-literature/liquid-fertilizer-brochure/ [Accessed Sept. 22, 2015] [Links]

4

CHAPTER

SPATIAL NITRIFICATIONS OF MICROBIAL PROCESSES DURING COMPOSTING OF SWINE, COW AND CHICKEN MANURE

Ke Wang, Weiguang Li, Xiangkun Li & Nanqi Ren

INTRODUCTION

Livestock waste is an important environmental pollution source as well as a valuable nutrient resource1. Aerobic composting is a traditional method of biologically transforming livestock feces to organic fertilizer in the presence of oxygen2. Forced aeration, oxygen concentration control and a large amount of bulking agents are used in the commercial composting processes to provide adequate oxygen and free airspace in this solid incubation matrix3·4·5·6. Although the microenvironment in the interior of the livestock manure particles is unclear, the composting is commonly regarded as an aerobic bioprocess, and is called aerobic composting7·8.

Improving the efficiency of organic stabilizing is an important goal in the composting operations9. The composting process could be divided into two stages: the active stage (or the thermophilic stage) and the curing stage. Generally, these two stages require a period of more than two months period before reaching biological stabilization of the product10 and this long treatment period essentially impacts the application of this technology. To reduce the composting period, a lot of research has focused on optimizing incubation conditions such as temperature, aeration rate, C/N ratio, moisture, porosity, pH, adding bulking agent and turning frequency to provide the optimal environment for microbial metabolism11·12·13·14·15. However, the composting period has not been reduced over the last three decades.

Severe nitrogen loss is another bottleneck for the composting treatment of the livestock manure16·17. A large amount of NH_3 released from composting pile during

Citation: Wang, K. et al. Spatial nitrifications of microbial processes during composting of swine, cow and chicken manure. Sci. Rep. 5, 14932; doi: 10.1038/srep14932 (2015).

the thermophilic period, which not only reduces the fertilizer effect but also causes acidification and eutrophication. The low C/N ratio of livestock manure is regarded as the main reason leading to the nitrogen loss in the literature18·19. Thus, a large amount of carbon-rich bulking agents, such as woodchips, straw and sawdust, is input into the feedstock before the livestock manure composting20·21·22·23·24. Although the consumption of the bulking agents greatly increases the operational cost, the issue of nitrogen loss has not been well resolved. It was reported that the ammonification and nitrification of microorganisms play important roles on the nitrogen loss in the composting matrix25·26·27·28. However, the manure particle is the basic unit of the composting matrix, and the spatial distribution characteristics of microbial nitrogen transformation in the interior of manure particle have not been understood.

Oxygen level not only affects the bio-stabilization rate of the organics but also the transformation process of the nitrogen in the composting process29·30·31·32. However, electron acceptors (oxygen) and donors (soluble organic matter and reduced nitrogen) are subject to mass transfer limitations. Oxygen transfer process in the composting matrix can be divided into the interparticle stage (the space between particles) and the interior particle stage (inside the particle)6. The bulking agent provides the structural support to create interparticle voids that act as the air-flow channels. However, the spatial stratifications of the microbial activity, dissolved oxygen (DO) and chemical changes (e.g., NH_4^+ and NO_3^- concentration) in the interior of the manure particle are not clear. This information is quite necessary for an understanding of the process of nitrogen transformation and organic biodegradation in composting system.

In this study, oxygen transfer characteristics and spatial distribution of microbial nitrogen transformation inside chicken, cow and swine manures were investigated during composting. The chemical gradients of DO, oxidation–reduction potential (ORP), and NH_4^+ and NO_3^-concentration were determined for the first time in the surface layer of three manures using microelectrodes during the composting period. The effects of oxygen permeation depth, ORP, C/N, pH, dissolved organic carbon, temperature and moisture on NH_3, CO_2 and CH_4emission rate of the three manures were evaluated. The relationships between oxygen transfer and the manure stabilization process and between microbial stratification and chemical gradients were elucidated.

RESULTS

NH_4^+ in the micro-profile of manure particles

The C/N ratios in the raw swine and chicken manure (Day-0) were closed (8.0 and 8.3), but their difference in NH_3 emission rate was noticeable. The cumulative amount of the NH_3 emission from the chicken manure (189.4 mmol) was substantially higher than that from the swine manure (59.2 mmol) and from the cow manure (12.3 mmol) during 30 days of composting (Fig. 1a). This result was consistent with the order of the NH_4^+/ NH_3 content in the three manures (Fig. 1b). The spatial difference in the NH_4^+ concentration in the interior of the manure particles was not significant, which was in-site detected by NH_4^+ microelectrode (Fig. 1c–e). This finding may be attributed to the dynamic balance of NH_3 volatilization and re-absorption in the liquid phase.

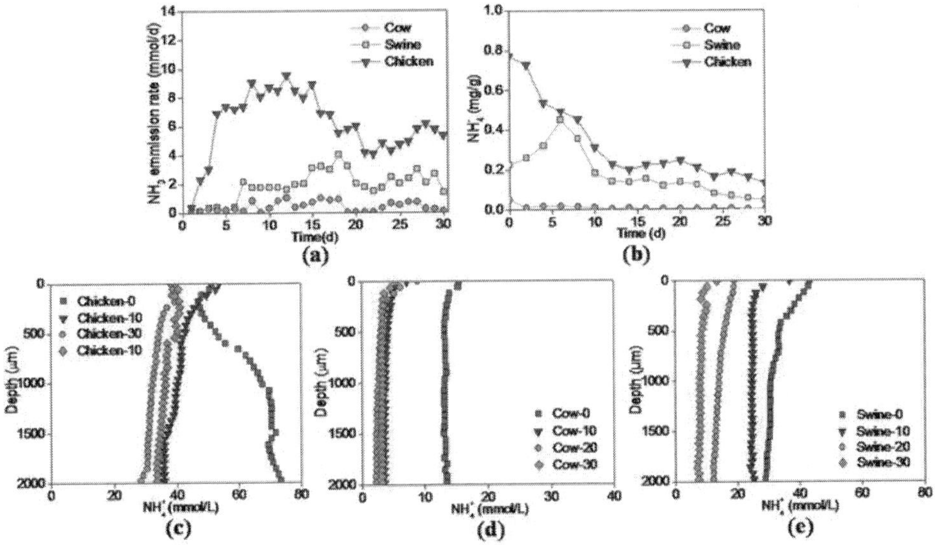

Figure 1: Variation of (**a**) NH_3 emission rate and (**b**) NH_4^+/NH_3 content of the swine, cow, chicken manures during composting; Variation of NH_4^+ concentration with depth in the microprofile of the manure particles of chicken (**c**), cow (**d**) and swine (**e**) on Day-0, Day-10, Day-20 and Day-30.

Although a mass of NH_4^+/NH3 was released into the liquid phase from biological ammonification25,33, the NH_4^+/NH3 content was synchronously reduced in the three manures as NH3 emission during the initial period. The ammonification also increased the pH value of the three manures. The pH value increased from 8.1 to 9.0 in the swine manure, from 8.0 to 8.6 in the cow manure, and from 8.4 to 9.4 in the chicken manure during the first 10 days, respectively. The increase in the pH value significantly promoted the transformation from NH_4^+ to NH3 in the NH_4^+/NH3 pool, and improved NH3 emission24,25,26,27,28. Furthermore, the moisture reduced by 59.1% in the swine manure, 51.7% in the cow manure and 64.9% in the chicken manure during the first 10 days. Meanwhile, the solubility of NH3 in water decreased from 520 to 200 g NH3 (kg water)$^{-1}$ as the incubation temperature rose from 20 to 55 °C. Due to the moisture loss and temperature decrease, the maximum NH3 dissolving capacity was reduced by 84.3% 81.4% and 86.5% in the swine, cow and chicken manures during the first 10 days, respectively. As a result, the NH_4^+/NH3 content dropped from 771.6 to 312.1 mg kg−1 in the chicken manure, from 450.8 to 184.5 mg kg−1 in the swine manure and from 10.1 to 6.4 mg kg−1 in the cow manure. Therefore, ammonification, moisture loss and temperature increase are important causes of the severe NH3 emissions.

CO_2 and CH_4

Dissolved organic matters are the direct carbon sources for the microbes in the composting matrix. The content of dissolved organic carbon in the chicken manure (47.6 to 30.1 mg g^{-1}) was consistently higher than that in the swine manure (29.7 to 16.3 mg g^{-1}) and the cow manure (31.7 to 10.6 mg g^{-1}) during the entire test (Fig. 2a). The significant decline in the dissolved organic carbon content was mainly ascribed to rapid biodegradation of dissolved organic matters in the three manures. The content of dissolved organic carbon has a negative correlation with the biological stability of the compost34. This means that the cow manure had the highest stability among the three manures on Day-30.

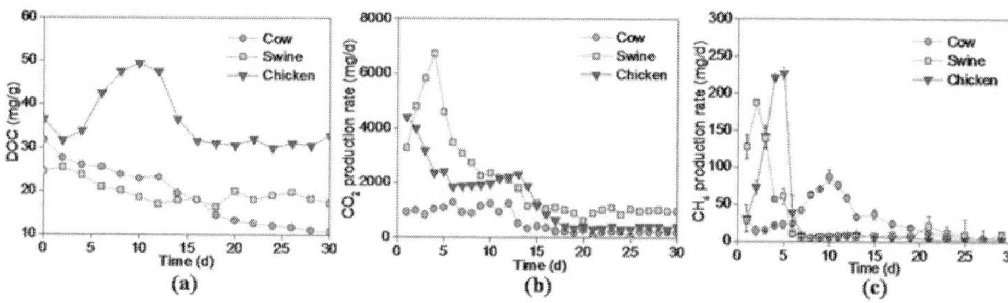

Figure 2: Variation of dissolved organic carbon content (**a**), CO_2 emission rate (**b**) and CH_4 emission rate (**c**) of the swine, cow, chicken manures during composting.

The evolution of CO_2 emission rate showed that the mineralization rate of the three livestock manures significantly decreased with composting time (Fig. 2b). The CO_2 production rate was reduced by 90.4% for the swine manure, 85.6% for the chicken manure and 81.6% for the chicken manure during the first 15 days. This result could be related to the decrease in available organic content and the moisture. The cumulative amount of CO_2 emissions was 62.0 g, 41.9 g and 16.9 g for the swine, chicken and cow manure within the entire test, respectively.

CH_4 emission from the three animals' manures was mainly occurred in the initial period of composting (Fig. 2c). This result was also supported by the previous literatures35·36. Although the swine manure had the highest CO_2 emission rate among the three animals' manures, the chicken manure had the highest CH_4 emission rate during the initial period. The peak values of CH_4 emission rate were presented on Day-2 (187.2 mg d^{-1}), Day-5 (226.9 mg d^{-1}) and Day-10 (87.5 mg d^{-1}) for the swine, chicken and cow manures, respectively. The cumulative amounts of CH_4 emissions were 819.0 mg, 711.1 mg and 676.7 mg for the swine, chicken and cow manures during the incubation, respectively. Because the methane is produced in a strictly anaerobic environment, the methane generated from the three manures indicated the existing of the anaerobic environment inside the three animals' manures.

DO in the micro-profile of manure particles

The depth profiles of the DO concentration in the surface layer of the three manures were measured by the DO microelectrode on Day-0, Day-10, Day-20 and Day-30 (Fig. 3). The DO concentrations began to slowly decrease at 0.4–0.3 mm above the surface of the manure particles. This was probably caused by resistance to mass transfer or the diluting effects of water vapor releasing from the manure. The DO concentration rapidly dropped with the depth in the surface layer of the three raw manures on Day-0. The DO was depleted in the top 0.30 mm of raw swine manure, 0.42 mm of raw chicken manure and 0.50 mm of raw cow manure, and the rest of the raw manure particles were anoxic. After 20 days forced-aeration composting, oxygen penetrated 0.78 mm in the chicken manure and 2.10 mm in the cow manure on Day-20, whereas the oxygen penetration depth in the swine manure still remained approximate 0.30 mm (Fig. 3b). These findings indicate that the aerobic zone was consistently limited to the surface layer of the three manure particles over the active stage. Oxygen diffusion in the manure particles appears to be the critical rate-limiting step for the oxygen transfer process in the composting matrix. The average diameter of the manure particles was measured to be larger than 8 mm on Day-20, indicating more than 75% of the swine manure particle, 50% of the chicken manure particle and 10% of the cow manure particle was the anoxic zone. Composting has generally been regarded as an aerobic bioprocess. However, this work confirmed that a significant volume of anoxic zone existed in the interior of the three manures during the active stage.

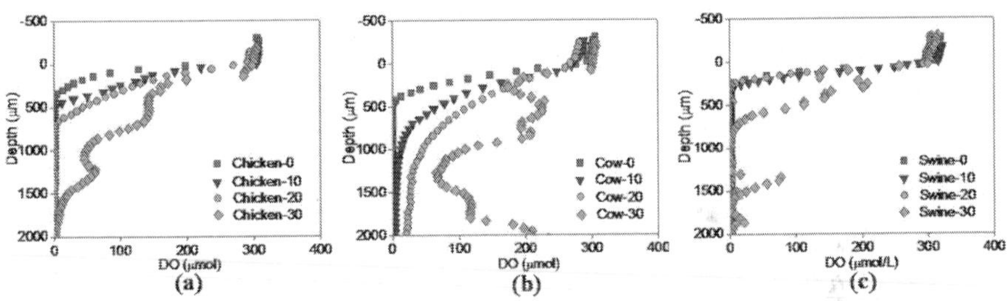

Figure 3: Variation of DO with depth in the microprofiles of the manure particles of chicken (**a**), cow (**b**) and swine (**c**) on Day-0, Day-10, Day-20 and Day-30.

The DO gradient inside the manure particles mainly depends on the balance between the oxygen diffusion rate and microbial oxygen utilization rate. The close depths of oxygen permeation in the three raw manures suggested their closed value of oxygen utilization rate at the beginning of composting. According to the Fick's first law37, the average oxygen flux in the aerobic layer was 0.556, 0.214 and 0.140 μmol $(cm^2 \cdot h)^{-1}$ in the swine, chicken and cow manure on Day-20, respectively. The oxygen utilization rate in the surface of the swine and chicken manure was significantly higher than that of the cow manure. This result was consistent with the order of their CO_2 production rate. The differences in oxygen utilization rate in the three manures were related to their substrate condition. The low available substrate in the cow manure reduced the microbial oxygen utilization rate on Day-20. Hence, the cow manure had the largest oxygen penetration depth. Moreover, the chicken manure had a relatively higher oxygen penetration depth than did the swine manure on Day-20, although the former had the highest content of dissolved organic matters and NH_4^+/NH_3 as the electron donor. The high NH_4^+/NH_3 concentration and pH value possibly impacted the microbial activity in the chicken manure during the active stage.

Oxygen penetration depth apparently increased in the three manures during the period from Day-20 to Day-30. The moisture content decreased to 55.3% in the swine manure, 66.7% in the cow manure and 58.1% in the chicken manure on Day-30 due to water evaporation. The decrease in moisture content not only increased the porosity of the manure but also essentially limited microbial activity. It has been reported that 50% is the lower limit of the moisture13. The moisture content in the surface layer of the manure particles was lower than that in the core zone. Thus, the metabolic activity of the microbes was firstly limited by decreased moisture in the surface layer. Furthermore, as the amount of available substrate decreased, microbial oxygen utilization rates were significantly reduced in the three animals' manures. Therefore, the increased porosity and decreased oxygen utilization rate greatly improved the oxygen penetration depth in the manures during the later period.

ORP in the micro-profile of manure particles

The ORP, as a non-biological factor, affects strongly on biological and chemical processes. The decrease in the ORP value with depth was mainly attributed to decreased DO in the three manures (Fig. 4). The cow manure had the highest ORP level among the three manures, which was in connection with it's the lowest available substrate and NH_4^+/NH_3content as well as the largest oxygen penetration depth. It is worth noting that the ORP value in the anoxic zone of the three manures intensively decreased first and then gradually increased with composting time. During the initial 10 days, the ORP value at the 2 mm depth decreased from −50 to −120 mV in the chicken manure, from 300 to −200 mV in swine manure and from 450 to 230 mV in the cow manure. The rapid drop in the ORP in the initial period could be related to three factors: the microbial consumption of oxidizing substances, the release of reducing substances in the biodegradation process and the concentration of reducing substances in the liquid phase of the manure caused by the moisture loss. However, with the increase in oxygen penetration depth, the ORP values in the three manures all became higher than 400 mV on Day-30.

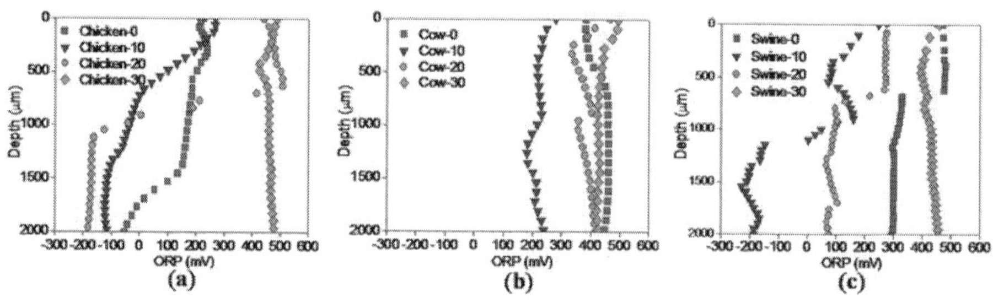

Figure 4: Variation of ORP with depth in the microprofiles of the manure particles of chicken (**a**), cow (**b**) and swine (**c**) on Day-0, Day-10, Day-20 and Day-30.

Metabolic mechanisms of organics would change with dramatic variation of the ORP level. The rapid decrease in ORP value inside the manures provided a redox potential condition for methanogenesis. That is why the CH_4 emission from the three animals' manures was only detected in the initial period of composting (Fig. 2b). Furthermore, the chicken manure had the lowest ORP level and the highest CH_4 emission rate among the three manures, and the cow manure had the highest ORP level and the lowest CH_4 emission rate. It indicated a significant negative correlation between ORP level and CH_4 production. As the low-ORP microenvironment gradually disappeared, the methane producing process was stopped in the later stage.

NO_3^- in the micro-profile of manure particles

The NO_3^- concentration in the core zone of the three manures first decreased and then gradually increased with composting time (Fig. 5a–c). During the first 20 days, the NO_3^- concentration in the 2 mm depth region was reduced from 15.8 to 6.6 mmol L^{-1} for the chicken manure, from 11.9 to 4.1 mmol L^{-1} for the swine manure and from 3.1 to 1.5 mmol L^{-1} for the cow manure, indicating that the NO_3^- could be used as the electron acceptor by the microbes in the anoxic zone, and that the denitrification occurred in the active composting.

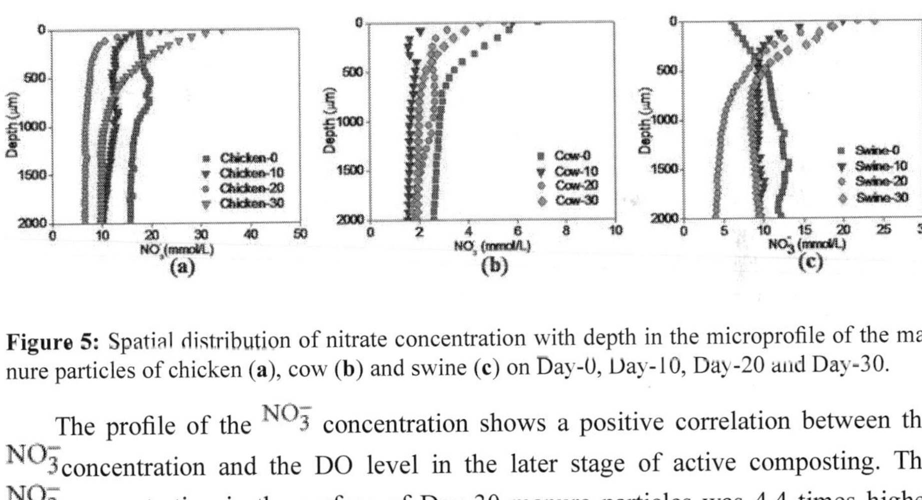

Figure 5: Spatial distribution of nitrate concentration with depth in the microprofile of the manure particles of chicken (**a**), cow (**b**) and swine (**c**) on Day-0, Day-10, Day-20 and Day-30.

The profile of the NO_3^- concentration shows a positive correlation between the NO_3^- concentration and the DO level in the later stage of active composting. The NO_3^- concentration in the surface of Day-30 manure particles was 4.4 times higher for swine manure, 3.5 times higher for chicken manure and 2.4 times higher for cow manure than that at 2 mm below the suface of the manure particles, respectively. This finding indicates that the nitrification process firstly occurred in the surface aerobic layer of the manure. No apparent improvement in the NO_3^- concentration was observed in the core zone of the three manures because biological nitrification would be inhibited under the anoxic and low ORP conditions. With the increase in the aerobic zone ratio in the later period, the increment of DO concentration provided the electron acceptors in the nitrification. That is why the increase in NO_3^- concentration was only reported in the curing stage.

DISCUSSION

It is commonly known that organic mineralization rate under the aerobic condition is essentially faster than that under anoxic condition. The aerobic metabolism in the surface layer released more respiration heat, which increased the temperature of the composting system and substantially improved organic hydrolysis rate and anoxic metabolism efficiency in the internal zone. A portion of soluble substrate hydrolyzed in the anaerobic core possible diffused and was oxidized in the surface aerobic layer38. As a result, oxygen penetration was limited to the surface layer of the manure particles due to microbial oxygen utilization during the active stage. Oxygen diffusion efficiency in the manures cannot be directly improved though increasing aeration rate. Hence, the aerobic-anoxic dual structure has a long duration inside the manure particles during the active stage even under a high aeration rate (Fig. 6). The low DO level limits the organic mineralization rate in the anoxic zone. Therefore, the high ratio of anoxic zone in the manure particle could be one of the major reasons leading to the long treatment period required for active composting.

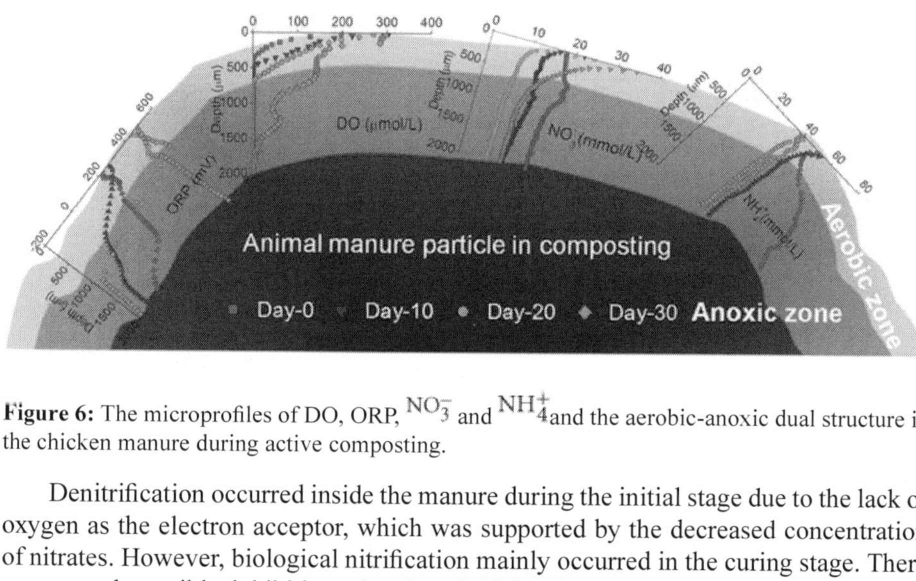

Figure 6: The microprofiles of DO, ORP, NO_3^- and NH_4^+ and the aerobic-anoxic dual structure in the chicken manure during active composting.

Denitrification occurred inside the manure during the initial stage due to the lack of oxygen as the electron acceptor, which was supported by the decreased concentration of nitrates. However, biological nitrification mainly occurred in the curing stage. There are several possible inhibitions for the nitrifying bacteria in the active composting stage. These are high temperature, low DO and ORP level in the manure particle, high /NH3 and available substrate content and competition among different microbial groups24,25,26,27,28. These factors could substantially retard the growth of nitrifying bacteria. Although the forced aeration maintained a high oxygen level in the air space of the matrix, the nitrate concentration was still quite low inside the manure during this period. With the decrease in temperature, /NH3 and dissolved organic matter concentration in the later stage, biological nitrification began in the surface layer of the manure particles due to the higher DO and ORP level. Afterwards, the concentration significantly increased in the inner zone of the manure particles with the improvement of aerobic conditions during the curing stage.

Although low C/N ratio and high pH values are regarded as the main reasons for the nitrogen loss in the previous literature, our work suggests that ammonification, moisture loss, anoxic and high temperature conditions are important reasons leading to severe NH3 emission during the thermophilic stage. First, most of the proteins were hydrolyzed in the initial period and released a large amount of /NH3, which increased the pH value in the liquid phase of the manure. Hence, the increased pH was the result of /NH3 production rather than a cause. Second, the anoxic and high-temperature environment inhibited the biological nitrification process. The carbon sources in the organic bulking agent could not be utilized by the microbes inside the manure particles due to the space separation. Third, the high proportion of water emissions substantially concentrated the /NH3 in the liquid phase. Meanwhile, the high-temperature condition reduced the /NH3 solubility in the water. Consequently, a large proportion of NH3 was released from the manure particles with water evaporation.

METHODS

Experimental methodology

Cow, swine and chicken feces were collected from Harbin Central Red Farm in northeast China. Moisture and volatile solids contents were 75.4% and 73.9% for the swine feces, 86.7% and 86.6% for the cow feces, 79.3% and 81.2% for the chicken feces. The pH and C/N ratio were 8.1 and 8.0 for the swine feces, 8.0 and 17.7 for the cow feces, and 8.4 and 8.4 for the chicken feces, respectively. Pumice, a porous volcanic rock, was used as a bulking agent in the composting matrix to avoid the influence of organic bulking agents on the analysis of NH_3 emission and CO_2 production from the animal manure. The chemical and physical properties of the pumice were reported in our pervious literature39. Briefly, the high porosity of pumice was 71.8–81%. The initial moisture and water-absorbing capacity of pumice were 8% and 51–72%, respectively. The volume weight and compressible strength of the pumice were 0.34–0.49 g cm^{-3} and 14–38 kg cm^{-2}. The raw cow, swine and chicken feces (300 g each) were composted separately with 250 g of pumice for 30 days in three cylindrical glass reactors (5L volume, 20 cm diameter × 27 cm height). These three composting reactors were placed in a constant-temperature water bath at 55 °C for the first 20 days and at 45 °C for the next 10 days. Fresh air was blown into the composted materials from the bottom of the three reactors. To avoid the influence of the low oxygen concentration in the interparticle voids on oxygen diffusion efficiency in the manure particle, a high aeration rate of 0.2 L min^{-1} is adopted to improve air exchange and oxygen level in each composting reactor.

Chemical analysis

The samples (2 g × 3) was dried at 105 °C and mass loss was calculated every 2 h. The moisture content of the manure samples at the different stages was determined until mass loss got inferior to 0.5% of mass loss of the previous. The volatile solids content of the dry manure samples was determined by the weight loss at 550 °C for 2 h. The pH was measured using a pH meter by dissolving 1 g manure sample in 10 mL of DI water. Manure samples were dried, ground and screened through a 200 mesh screen, and 2 g of the residues was extracted with 50 mL of DI water at 25 °C for 24 h. The aqueous extract of the manure samples was filtered through a 0.45 μm membrane. The dissolved organic carbon concentration of the aqueous extract was determined by a TOC analyzer (TOC-V_{CPN}, Shimadzu). The NH_4^+/NH_3 content in the wet manure samples was analyzed by distillation and titration using $H_2SO_4$24. The NH_3 and CO_2 releasing from the three composting reactors were absorbed by the absorption bottles filled KOH (4 mol L^{-1}) and H_3BO_3 (1 mol L^{-1}) in sequence. The NH_3 emission rate and CO_2production rate of the three manures were determined by chemical titration of the solutions in the six absorption bottles every day. The CH_4 concentration in the exhaust gas was determined three times a day by using the GC system with a flame ionization detector (6890 Series, Agilent Technologies).

Microelectrode measurements

Four microelectrodes (DO, ORP, NO_3^- and NH_4^+) were used to investigate the chemical stratification inside the swine, cow and chicken manure particles during the composting. DO concentration profiles were obtained by using a Clark-type microelectrode with a tip diameter of 20–30 μm (OX-25, Unisense). A two-point calibration of the DO microelectrode was made with O_2-saturated bulk water and N_2-bubbled water. The microprofile of ORP in the particles was measure by ORP microelectrode (RD-25, Unisense) and the reference electrode (REF-25, Unisense) with a tip diameter of 20–30 μm. The ORP microelectrode was calibrated with 100 ml of pH=4 and pH=7 buffer solution mixed with 1 g of quinhydrone, respectively. Liquid ion-exchanging membrane microsensors for NO_3^- and NH_4^+ were prepared in our lab. The ionophores of NO_3^- and NH_4^+ in the liquid-membrane of the microsensors were from Fluka (Sigma-aldrich, USA). The detailed manufacturing procedures were described as before40·41·42. The microelectrodes of $NO_3^-$and NH_4^+ were calibrated in a dilution series (10^{-2} to 10^{-6} mol L^{-1}) of $NH_4^+$and NO_3^- before measurement43·44.

The livestock manure samples were taken from the three composting reactors at the four different periods of Day-0, Day-10, Day-20 and Day-30. The concentrations of DO, ORP, NO_3^- and NH_4^+ in the 2 mm depth surface layer of the manure particles were *in-situ* measured by the four microelectrodes. The four microelectrodes were combined with a computerized depth control and data acquisition (MicroProfiling System, Unisense). The microelectrodes were separately lowered into the manure particles by a computer controlled micromanipulator. All measurements were taken at spatial intervals of 20 μm through the manure particles. The microelectrodes were calibrated before each test. Spatial resolution of the microelectrodes is approximately twice the tip diameter42. The net oxygen fluxes in the manure particles were calculated based on the Fick's first law of diffusion $J=-D_s dC/dx$ (J: net flux; Ds: molecular diffusion coefficient of compound C in water; dC/dx: concentration gradient in the boundary layer at the surface of particle). The oxygen diffusion coefficient (2.24×10^5) was used here from previous study37·44.

ACKNOWLEDGEMENTS

This work was financially supported by National Natural Science Foundation of China (51408151, 51278146), Major Science and Technology Program for Water Pollution Control and Treatment (2013ZX07201007), China Postdoctoral Science Foundation (2012M520755), China Postdoctoral Science special Foundation (2014T70357), Postdoctoral Science Special Foundation of Hei Longjiang (LBH-TZ0510) and Specialized Research Fund for the Doctoral Program of Higher Education (20132302120075).

CONTRIBUTIONS

K.W. wrote the main manuscript text, conceived and planned the experiments. X.L. prepared figures 1–6. N.R. and W.L. provided significant feedback to the design and interpretation of the data. All authors reviewed the manuscript.

COMPETING INTERESTS

The authors declare no competing financial interests.

CORRESPONDING AUTHORS

Correspondence to Ke Wang or Xiangkun Li.

REFERENCES

1. Adhikari, B. K., Barrington, S., Martinez, J. & King, S. Effectiveness of three bulking agents for food waste composting. Waste Manage. 29, 197–203 (2009).
2. Agnew, J. M. & Leonard, J. J. The physical properties of compost. Compost Sci. Util. 11, 238–264 (2003).
3. Beck-Friis, B., Smars, S., Jonsson, H., Eklind, Y. & Kirchmann, H. Composting of source-separated household organics at different oxygen levels: Gaining an understanding of the emission dynamics. Compost Sci. Util. 11, 41–50 (2003).
4. Bernal, M. P., Alburquerque, J. A. & Moral, R. Composting of animal manures and chemical criteria for compost maturity assessment. A review. Bioresour. Technol. 100, 5444–5453 (2009).
5. Blazy, V. et al. Process conditions influence on pig slaughter house compost quality under forced aeration. Waste Biomass Valor. 5, 451–468 (2014).
6. Debeer, D. & Sweerts, J. P. R. Measurement of nitrate gradients with an ion-selective microelectrode. Anal. Chim. Acta 219, 351–356 (1989).
7. Debeer, D. & Vandenheuvel, J. C. Response of ammonium selective microelectrodes based on the neutral carrier nonactin. Talanta 35, 728–730 (1988).
8. Eftoda, G. & McCartney, D. Determining the critical bulking agent requirement for municipal biosolids composting. Compost Sci. Util. 12, 208–218 (2004).
9. Eftoda, G. & McCartney, D. Determining the critical bulking agent requirement

for municipal biosolids composting. Compost Sci. Util. 12, 208–218 (2004).

10. Ekinci, K., Keener, H. M. & Elwell, D. L. Composting short paper fiber with broiler litter and additives part I: Effects of initial pH and carbon/nitrogen ratio on ammonia emission. Compost Sci. Util. 8, 160–172 (2000).
11. Fukumoto, Y., Osada, T., Hanajima, D. & Haga, K. Patterns and quantities of NH3, N2O and CH4 emissions during swine manure composting without forced aeration--effect of compost pile scale. Bioresour. Technol. 89, 109–114 (2003).
12. Ge, J. Y., Huang, G. Q., Yang, Z. L., Huang, J. & Han, L. J. Characterization of the dynamic thickness of the aerobic layer during pig manure aerobic composting by fourier transform infrared microspectroscopy, Environ. Sci. Technol. 48, 5043−5050 (2014).
13. Gieseke, A. & de Deer, D. in Molecular Microbial Ecology Manual 2nd edn (eds Kowalchuk, G. A. et al.) Section 8, 3483–3514 (Springer, 2004).
14. Guardia, A. D. et al. Comparison of five organic wastes regarding their behavior during composting: Part 2, nitrogen dynamic. Waste Manage. 30, 415–425 (2010).
15. Guardia, A. D., Petiot, C., Rogeau, D. & Druilhe, C. Influence of aeration rate on nitrogen dynamics during composting. Waste Manage. 28, 575–587 (2008).
16. Haug, R. T. in The Practical Handbook of Compost Engineering Ch.7, 261–286 (CRC Press, Florida, 1993).
17. Iqbal, M. K., Shafiq, T. & Ahmed, K. Characterization of bulking agents and its effects on physical properties of compost. Bioresour. Technol. 101, 1913–1919 (2010).
18. Jessica, K. L. & David C. V. Improving Waste management strategies for small livestock farms. Environ. Sci. Technol. 47, 11940–11941 (2013).
19. Kulcu, R. & Yaldiz, O. Determination of aeration rate and kinetics of composting some agricultural wastes. Bioresour. Technol. 93, 49–57 (2004).
20. Li, B. K. & Bishop, P. L. Micro-profiles of activated sludge floc determined using microelectrodes. Water Res. 38, 1248–1258 (2004).
21. Liang, C., Das, K. C. & McClendon, R. W. The influence of temperature and moisture contents regimes on the aerobic microbial activity of a biosolids composting blend. Bioresour. Technol. 86, 131–137 (2003).
22. Liu, D. et al. Changes in biochemical and microbiological parameters during the period of rapid composting of dairy manure with rice chaff. Bioresour. Technol. 102, 9040–9049 (2011).
23. Lorenzen, J. Biosensor determination of the microscale distribution of nitrate, nitrate assimilation, nitrification, and denitrification in a diatom-inhabited freshwater sediment. Appl. Environ. Microbiol. 64, 3264–3269 (1998).
24. Nakasaki, K., Ohtaki, A., Takemoto, M. & Fujiwara, S. Production of well-matured compost from night-soil sludge by an extremely short period of thermophilic composting. Waste Manage. 31, 495–501 (2011).
25. Ogunwande, G. A., Osunade, J. A. & Adekalu, K. O. Nitrogen loss in chicken

litter compost as affected by carbon to nitrogen ratio and turning frequency. Bioresour. Technol. 99, 7495–7503 (2008).

26. Revsbech, N. P., Christensen, P. B., Nielson L. P. & Sorensen, J. Denitrification in a trickling filter biofilm studies by a microsensor for oxygen and nitrous oxide. Water Res. 23, 867–71 (1989).
27. Said-Pullicino, D., Erriquens, F. G. & Gigliotti, G. Changes in the chemical characteristics of water-extractable organic matter during composting and their influence on compost stability and maturity. Bioresour. Technol. 98, 1822–1831 (2007).
28. Sanchez-Monedero, M. A., Roig, A. & Paredes, C. Nitrogen transformation during organic waste composting by the Rutgers system and its effects on pH, EC and maturity of the composting mixtures. Bioresour. Technol. 78, 301–308 (2001).
29. Smet, E., Van Langenhove, H. & De Bo, I. The emission of volatile compounds during the aerobic and the combined anaerobic/aerobic composting of biowaste. Atmos. Environ. 33, 1295–1303 (1999).
30. Steger, K., Eklind, Y., Olsson, J. & Sundh, I. Microbial community growth and utilization of carbon constituents during thermophilic composting at different oxygen levels. Microbial Ecol. 50, 163–171 (2005).
31. Szanto, G. L., Hamelers, H. V., Rulkens, W. H. & Veeken A. H. M. NH3, N2O and CH4 emissions during passively aerated composting of straw-rich pig manure. Bioresour. Technol. 98, 2659–2670 (2007).
32. Tiquia, S. M. & Tam, N. F. Y. Co-composting of spent pig litter and sludge with forced-aeration. Bioresour. Technol. 72, 1–7 (2000).
33. Tiquia, S. M. & Tam, N. F. Y. Fate of nitrogen during composting of chicken litter. Environ. Pollut. 110, 535–541 (2000).
34. Wang, K. et al. Spatial distribution of dynamics characteristic in the intermittent aeration static composting of sewage sludge. Bioresour. Technol. 102, 5528–5532 (2011).
35. Wang, W. D., Wang, X. F. & Liu, J. B. Effect of oxygen concentration on the composting process and maturity. Compost Sci. Util. 15, 184–190 (2007).
36. Xiao, Y. et al. Continuous thermophilic composting (CTC) for rapid biodegradation and maturation of organic municipal solid waste. Bioresour. Technol. 100, 4807–4813 (2009).
37. Yamada, Y. & Kawase, Y. Aerobic composting of waste activated sludge: Kinetic analysis for microbiological reaction and oxygen consumption. Waste Manage. 26, 49–61 (2006).
38. Yanez, R., Alonso, J. L. & Diaz, M. J. Influence of bulking agent on sewage sludge composting process. Bioresour. Technol. 100, 5827–5833 (2009).
39. Zeng, Y. et al. Impact of the addition of a nitrifying activated sludge on ammonia oxidation during composting of residual household wastes. J. Appl. Microbiol. 117, 1674–1688 (2014).

40. Zeng, Y., Guardia, A. D., Daumoin, M. & Benoist, J. C. Characterizing the transformation and transfer of nitrogen during the aerobic treatment of organic wastes and digestates. Waste Manage. 32, 2239–2247 (2012).

41. Zeng, Y., Guardia, A. D., Ziebal, C., De Macedo, F. J. & Dabert, P. Impact of biodegradation of organic matters on ammonia oxidation in compost. Bioresour. Technol. 136, 49–57 (2013).

42. Zeng, Y., Guardia, A. D., Ziebal, C., De Macedo, F. J. & Dabert, P. Nitrification and microbiological evolution during aerobic treatment of municipal solid wastes. Bioresour. Technol. 110, 144–152 (2012).

43. Zeng, Y., Guardia, A. D., Ziebal, C., De Macedo, F. J. & De Guardia, A. Nitrogen dynamic and microbiological evolution during aerobic treatment of digested sludge. Waste Biomass Valor. 5, 441–450 (2014).

44. Zhu, N. W. Effect of low initial C/N ratio on aerobic composting of swine manure with rice straw. Bioresour. Technol. 98, 9–13 (2007).

5

CHAPTER

MAPPING OF BIOMASS FLUXES: A METHOD FOR OPTIMIZING BIOGAS-REFINERY OF LIVESTOCK EFFLUENTS

Francesca Nardin and Fabrizio Mazzetto

Keywords: Biogas, bio-waste manure, farm setting, feeding solutions, housing, N-balance

INTRODUCTION

Following the concept of sustainable development as defined by the Brundtland Commission, energy systems should be ecologically, socially and economically sustainable, so that the present generation can meet its energy needs without compromising the ability of future generations to meet their energy and other needs [1]. Thereby, energy production via biomass could be regarded as a promising approach, which helps to preserve non-renewable resources, improves energy security, mitigates the greenhouse effect and promotes regional development [2,3,4]. According to AEBIOM (European Biomass Association) (2013) [5], bioenergy represented 68% of the total gross inland consumption of renewables in Europe in 2011. Use of renewable biomass (including energy crops and organic wastes) as an energy resource is not only "greener" with respect to most pollutants, but its use represents a closed balanced carbon cycle with respect to atmospheric carbon dioxide. A third concern is the recognized need for efficient methods for treatment and disposal of large quantities of municipal, industrial and agricultural organic wastes [6].

Citation: Francesca Nardin and Fabrizio Mazzetto. Mapping of Biomass Fluxes: A Method for Optimizing Biogas-Refinery of Livestock Effluents. doi:10.3390/su6095920.

Abbasi and Abbasi [7,8] give a rather isolated standing about biomass used for renewable energy. In their review, the authors state that biomass appears to be a very attractive source of renewable energy, only if seen from the limited perspective of standing crop, theoretical replenishing ability and 'carbon neutral' character as a fuel. Biomass energy is indeed a sustainable option, and it has proven to be so for thousands of years, but only as long as it is used to a very limited extent. Other forms of bioenergy raise perplexity. A review of studies comparing bio-ethanol systems to conventional fuels on a life cycle basis affirm that the balance of environmental impacts of current liquid fuels from biomass is ambiguous. Apart from the definite advantage of biofuels, their impacts on acidification, human and ecological toxicity and eutrophication have been evaluated more often unfavorably than favorably [9]. Even their economic viability is still a critical issue to large-scale commercialization [10]. According to Awudu and Zhang [11], due to the nature of the biofuel supply chain, many uncertainties exist. The uncertainties include, but are not limited, to raw material supply, finished goods demand, price, pre-treatment production, yield and transportation uncertainties. In order to achieve optimal performance, the decisions of the biofuel supply-chain's management should incorporate these uncertainties. The net energy yield per hectare as the difference between the total energy gain from biofuel and biogas minus the fossil fuel input into the production process is 178 GJ (1 gigajoule = 278 kWh) in the case of biogas and, thus, more favorable in comparison to liquid fuels. Plant oil (35 GJ), sugar beet (88 GJ), grain (30 GJ) and lignocellulose (18 GJ) are less efficient with regard to the area under cultivation. Even the biomass-to-liquid (BtL) technology generating the so-called second-generation biofuels, which is not yet applicable to series production, yields just 135 GJ net energy per hectare. The BtL technique makes use of the whole plant, while first-generation biofuels only use the starchy and oleic parts of the feedstock [12,13]. These findings indicate in certain respects the environmental and economic advantageousness of biogas [14].

Biogas from Anaerobic Digestion (AD)

Over the last few decades, interest has increased in the potential for biofuel as a means of reducing dependency on fossil fuels and developing environmentally-friendly and renewable energy. Biomass includes all plant and plant-derived materials, including animal manure, not just starch, sugar and oil crops already used for food and energy [10]. The biomass resource base is composed of a wide variety of forestry and agricultural resources, industrial-process residues and municipal-solid and urban-wood residues. Biofuels produced from biomass are classified according to three generations of processing technology, while Awudu and Zhang [11] classify biofuels in four categories. According to Chynoweth *et al.* [6], biomass may be converted to a variety of energy forms, including heat (via burning), steam, electricity, hydrogen, ethanol, methanol and methane. The selection of a product for conversion is dependent upon a number of factors, including the need for direct heat or steam, conversion efficiencies, energy transport, transformation and use hardware, economies of scale and the environmental impact of conversion-process waste streams and product use.

Under most circumstances, methane is an ideal fuel. Biogas is produced by anaerobic digestion or intensive fermentation of organic matter (OM). Feedstocks of this process are energy crops, but also manure, sewage sludge, municipal solid and biodegradable waste [12,15,16,17].

Anaerobic digestion is applied in Europe to the treatment of agricultural waste mainly in Germany, Austria, Denmark, Italy, the Netherlands and Spain [18]. Anaerobic digestion (AD) of dairy manure is an attractive treatment that provides benefits, such as pollution control, odors and pathogen level reduction, nutrient recovery and energy production [19,20,21,22,23]. Biogas is a combustible gas consisting of methane (CH_4, 50%–80%), carbon dioxide (CO_2, 20%–50%) and small amounts of other gases and trace elements. The digestate is the decomposed substrate, rich in macro- and micro-nutrients and therefore valued as plant fertilizer. Every biomass that is suitable to be fermented is named the "substrate". When used in a fully-engineered system, AD technology not only prevents pollution, but also allows for sustainable energy production, as well as for compost and nutrient recovery [24]. Nowadays, decentralized farm-based manure facilities represent probably the most common AD technology in low-income agricultural countries (e.g., China and India) to provide biogas for cooking and lighting [25]. Like natural gas, biogas has a wide variety of uses. Braun [20] summarized the range of biogas use options for energy production. The simplest energy application of biogas is heat production through natural gas boilers modified for raw biogas. Cleaned biogas can be inserted into municipal gas pipeline grids for cooking use [26]. Currently, biogas is mainly used for power generation through combined heat and power units with gaseous fuel engines or for feeding biogas grids [27].

If sufficiently cleaned of water, hydrogen sulfide and carbon dioxide, biogas acquires the same characteristics as natural gas and can be fed into the local gas distribution grid [12,13,14,15,16,17,18,19,20,21,22,23,24,25,26,27,28]. Power generation from biogas with even higher efficiency may offer fuel cell technology in the future [27]. Biogas emerges in the current scenario as the "renewable of the future", though requiring deep research and efforts to be improved and possibly integrated with other renewables. There are several issues that need to be studied and assessed more in depth [8]. Some of them are the improvement of the LCA of technologies, the development of models on energy and environmental balance, simultaneous monitoring of the critical parameters of the anaerobic digestion process, optimization of mixed substrates co-digestion and integration and innovation of pre- or post-treatment technology. The full benefits of AD have not been realized, even because of the perceived costs involved and because the system is not well known to the agricultural, industrial and engineering communities.

MATERIALS AND METHOD

Nowadays, a whole farm perspective is necessary to deal with the environmental concerns due to livestock activities. The approach has to consider the strong links among feeding, housing, treatment processes, storage conditions and field application practices along the manure management chain, because it affects soil, air and water quality, the crop growth and, consequently, farm income. The selection of livestock manure management options is becoming a strategically important task that farmers and public policy makers have to face [26]. In Europe and developed countries in general, one of the common tendencies of animal production activities is to intensify the animal production in relatively small areas, in order to reduce the production costs [29]. In those areas, the production of manure is consequently huge, and land application, the traditional dairy manure management strategy, is posing serious environmental pollution problems [30,31].

This paper presents the topic of the management of livestock effluents and, therefore, nutrients (particularly N) in the framework of biogas supply chain. The bio-refinery will be analyzed as a unique system, from the farm to the biomass produced and sent to anaerobic digestion, focusing on the fate/change of the flow of material and nutrient content through the system. Modifications might be due to feed (V1), housing type (V1), effluents amounts and storage (V2) and the presence of co-substrates (V2); those are the variables described in this paper (Vn). Farm/medium-scale biogas plants usually work in the following schematic way, as proposed by [32]. Livestock effluents (V2) eventually mixed with other agricultural residues, produced by animals bred under specific conditions (V1), are collected and sent to the digester tank (V4). Pre-treatments (V3) and post-treatments (V5) can respectively precede and follow this step, according to the composition, nutrients content, volume and fate/use of the effluents. In the reactor, in the absence of oxygen and under carefully controlled conditions, anaerobic microorganisms start the transformation and the decomposition of substrates, and anaerobic digestion takes place. The resulting biogas (V4) is collected, stored and can be transformed into heat or electric power, or it can be used directly as a gaseous fuel. Digestate, the solid rejected part coming out from the digester, is usually stored and used as a fertilizer or soil conditioner (V6). Inputs and outputs are different for each phase of the biogas supply-chain. We gave a description of inputs and outputs for V1 and V2 by arbitrarily considering four cattle farms (Table 1). For each category, we considered the most feasible solutions, in order to highlight possible/best practices for improving the system. We always gave description of inputs and outputs within each single step according to the final scope of the bio-energy refinery: to minimize environmental impacts, to maximize beneficial externalities and to optimize material-flow in order to avoid surplus/deficit of nutrients.

RESULTS AND DISCUSSION

Farm Step

"Typical farm" groups have been traditionally categorized. A shrewd evaluation of the flow of material, which passes through the biogas refinery, is strongly affected by the characteristics of the livestock bred. In general, it is influenced by the setting of the stall (intensive, integrated or extensive) and, in particular, by the feed supply, housing types and effluent management and storage on the farm.

According to [33], the indirect estimation of nitrogen content (N_c) can be made according to nitrogen balance (Equation1):

$$N_c = N(effluent) = N(excreted) - N(volatilized)$$

$$N(excreted) = N(feeding) - N(animal\ products,\ milk,\ eggs,\ meal)$$
$$N(volatilized) = (0.1\ N_{feces}) \times F_s + N_{urine} \times F_l$$

where:

-

N_{feces}: (the amount of feed provided by the ration (kg) × crude protein (g kg $^{-1}$ TS (total solids))) – (the amount of feed provided by the ration (kg) × digestible protein (g kg^{-1} TS))

-

N_{urine}: $N_{sup} - N_{feces} - N_m - N_g$

-

F_s: factor of solid nitrogen losses = $F_s = 1 - (1 - L_d/100) \times (1 - L_s/100) \times (1 - L_a/100 \times R)$

-

L_d: direct losses during first days after excretion = 4% in stables with full floor drained; 9% in stables with full floor; 10% in stables with slatted floor; 10%–15% in stables with permanent litter material.

-

L_s: losses from storage/processing = 1%–6%, depending on straw content; if composted, 35%.

-

L_a: losses from application on crops = 80%–100% manure.

-

R: factor of reduction for application types = 0.05 for injection in the soil; 0.1 if immediately plowed; 0.5–0.1 if later interred, depending on the environmental conditions and the type of soil (higher on dry, sandy soils).

-

F_l: factor of liquid nitrogen losses = $F_l = 1 - (1 - L_d/100) \times (1 - L_s/100) \times (1 - L_a/100 \times R)$

-

L_d: direct losses during first days after excretion = 4% in stables with full floor drained; 9% in stables with full floor; 10% in stables with slatted floor; 10%–15% in stables with permanent litter material.

-

L_s: losses from storage/processing = slurry, 3%–7% in open tank; 1% in closed tank; dense slurry (feces + urine), 2%–5% in open tank; 0.7% in closed tank.

-

L_a: losses from application on crops = 50%–100% slurry.

-

R: factor of reduction for application types = 0.05 for injection in the soil; 0.1 if immediately plowed; 0.5–0.1 if later interred depending on the environmental conditions and the type of soil (higher on dry, sandy soils).

-

N_{sup}: (amount of nitrogen in feed ration): amount of feed provided by the ration (kg) × N_{cp}.

-

N_{cp}: crude protein (g kg^{-1} TS)/6.25.

-

N_m: kg of milk produced.

-

N_g: kg of meal produced

A huge amount of data is necessary in order to precisely estimate the actual nitrogen content of the effluents produced at the farm. Three main aspects related to the "farm step" can be identified: (1) farm setting; (2) housing type; and (3) feeding type and breeding management. Effluents storage conditions are also essential for determining nitrogen losses after excretion. Animal manure collected in housing systems has to be stored inside or outside the stall until its application in the field. The size and nature of storage depend largely on the value of the nutrients in the manure and on the regulatory climate [34]. Often, the storage capacity is designed to allow timely spreading of the manure in the field, *i.e.*, during the growing season when the crop can utilize the plant nutrients. Loss of NH_4^+ via volatilization from animal houses, manure stores and applied manure will reduce the fertilizer value of animal fertilizers applied in the field. In addition, the variability of NH_3 emission will cause variability and uncertainty in the fertilizer efficiency of the manure, reducing farmers' confidence in manures as a source of N for crops. This fact may lead them to oversupply the crops with N, in this way leading to reduced crop quality and increased losses of nitrogen to the environment by leaching of nitrate (NO_3^-) and emission of nitrous oxide (N_2O) [34,35]. A reliable estimate of these emissions relies on a precise description of the processes involved in the transfer of NH_3 from the manure to the atmosphere. Because of its complexity, we did not address this topic. Average data are given for completeness. According to Fixen [36], almost 30% of the N excreted is lost from livestock buildings or during storage. Approximately 19% is lost via emissions of ammonia (NH_3), 7% via emissions of other N gases and 4% via leaching and run-off. A further 19% is lost via NH_3 emissions after application of the manure to land. The results show that only *ca*. 52% or less of the N excreted is potentially recycled as a plant nutrient [36]. These emissions constitute an important loss (up to 83%) of valuable N fertilizer.

Farm Setting

Three main divisions of farm-system exist: intensive, integrated and extensive, the latter including organic management [37]. This separation summarizes the three main approaches to nutrient management in farming. Intensive farm systems seek to maximize yield through best management practices, which involve the efficient use of all inputs, including fertilizers and plant protection chemicals, crops and crop rotations, livestock breeding programs and, often, precision agricultural techniques. Integrated farming aims at using fewer inputs, especially less fertilizer and pesticide, accepting slightly smaller yields and gross profit, while maintaining net profitability through the reduced costs of inputs. Extensive and organic systems rely on low rates of inputs to attempt to balance offtake in products. Recycling of animal manures and composted wastes is central to organic management systems. An example of "typical farm" groups, which will be used in the following paragraph, as well, is given below. The estimation of effluent amounts for each farm is the first step in the evaluation of material and nutrient management and allocation (Table 1).

Table 1: Typical farm" groups' characteristics, in terms of animals bred, livestock size and effluents produced

Parameters	Animal	U.M.	Farm setting			
Type			Extensive	Extensive-integrated	Integrated-Intensive	Intensive
Size of Herd	Dairy cattle [b]		Min	Small	Medium	Max
		No.	≤ 20	90	320	≥ 550
Daily effluents [a]		kg	≤ 900	4050	14,400	$\geq 24,750$
		m^3	≤ 1	4.5	16	≥ 27.5

[a] Free housing (bunk), bedding (rump to rump), straw, based on [33]; [b] no differences in terms of the cattle category are considered here.

Herd Composition

According to Goulding *et al.* [37], the number of replacement cows needed to ensure adequate substitution of reformed cows depends on several factors, among which the mean reproductive parameters of the herd are particularly important. This quantity is calculated by dividing the average number of parts per cow per year by the average number of parts per cow per life. The analysis assumes that calves and young heifers represent 20% of the herd, heifers 11.6% of the herd pregnant heifers 12.7% of the herd and productive cows 55% of the herd (of which, only 85% is dairy cows). The composition of herds is crucial to evaluate the biogas productive potential. The productive cows, in particular dairy cows, are those that need the greatest feeding ration and, consequently, have the most significant amount of effluents excreted (Table 2). Similarly, the amount of straw or another litter material used in the different phases is critical for the definition of biogas productive potential.

Table 2: Typical characteristics of cattle categories in terms of life weight and livestock effluents produced. Data updated and averaged on [38,39,40,41]

Animal category	Life weight	Livestock effluents [a]	
	kg	kg day^{-1}	m^3 day^{-1} [b]
Calf (0–3 months)	70	4.4	0.005
Calf (3–6 months)	140	8.6	0.009
Heifer (6–12 months)	230	20	0.022
Heifer (12–24 months)	390	34	0.037
Veal (12 months)	400	23	0.025
Dairy cow (>12 months)	500–600	50–60	0.055–0.066

[a] Livestock effluents = slurry + manure, with free housing (bunk), bedding (rump to rump), straw, based on [33]; [b] a density of 900 kg/m^3 is assumed.

A first rough estimation of the nitrogen content of the effluents produced in the farm is based on tabled average value (Table 3). Average nitrogen content is about 3.5‰ for dairy cows livestock effluents, as the average value from literature reviewed data (Table 4).

Table 3: Typical characteristics of effluents, in terms of total solids, volatile solids, total nitrogen and ammonia content. Data averaged based on [38,40,41,42].

Effluents	Total solids (TS)	Volatile solids (VS)	Total nitrogen (NTK)	Ammonia nitrogen ($N\text{-}NH_4$)
	%	%TS	g/kg	%NTK
Dairy cows	4–11	65–85	2.5–4.5	40–65
Pigs	2–6	40–60	2–5	70–85
Agricultural residues	5–10	65–80	3.5–7	30–65

Table 4: Typical herd composition of "typical farm" groups considered in this review, amount of livestock effluents produced according to the animal category-specific effluent production and nitrogen excretion, by using the data of Table 1 and Table 2.

Farm setting	No. of animals	Calf, young heifers [a]	Heifers, veal [b]	Pregnant heifers [c]	Productive cows [d]	Livestock effluents [e]		Nitrogen content [f]
		No.	No.	No.	No.	kg day^{-1}	m^3 day^{-1}	kg day^{-1}
Extensive	20	4	2	3	11	787	0.87	2.75
Extensive-integrated	90	18	10	12	50	3257.2	3.62	11.40
Integrated-Intensive	320	64	37	41	177	11,524.6	12.81	40.34
Intensive	550	110	65	71	304	19,853	22.06	69.49

[a] The average value of 6.5 kg day^{-1} of effluent produced is considered (Table 2); [b] the average value of 27 kg day^{-1} of effluent produced is considered (Table 2); [c] a value of 34 kg day^{-1} of effluent produced is considered (Table 2); [d] the average value of 55 kg day^{-1} of effluent produced is considered (Table 2); [e] livestock effluents = slurry + manure, averaged on free housing (bunk), bedding (rump to rump), straw, based on [33]; [f] the value of 3.5 g N kg^{-1} of effluent produced is assumed (Table 3).

Housing

The correct evaluation of livestock effluents quantities, divided into manure and slurry, is crucial. On the one hand, it is central for the proper determination of the maximum BMP (biomethanization potential) and the consequent sizing of anaerobic digesters. BMP is the methane yield, reflecting the destruction of organic materials, according to the methane potential of each organic component in the volatile solids (VS) [43]. On the other hand, it is necessary for the precise determination of the actual amount of N_c in the effluents and, consequently, for the adequate determination of quantities spreadable over available acres. Housing types for dairy cows are classified into two major groups. Fixed housing is especially used in small herds (up to 35–40 cows). In this case, the productive cows are tied in a box, while calves and young heifers are free. Free housing is the most recently introduced type of housing, and it can be considered today a practically forced choice. This type of housing can be conducted with different configurations, which primarily differ in the presence of bunks for animal rest, in the use of straw or other litter materials and in the presence of areas with more or fewer quantities of straw. Webb *et al.* [44] found that housing systems with deep litter emit more NH_3 than tied stalls. This is due to the smaller emitting surface area in a tied stall. Different housing types affect mainly the distribution between the slurry and manure of the effluents produced. The main housing techniques and the effects that these may have on the management of effluents and on nutrient content are summarized in Table 5.

Table 5: Main housing techniques related to the amount of effluents produced (manure and slurry) and the nitrogen excreted, per main cattle category. Data are updated and averaged based on [36,38,41]

Housing					Nitrogen to field (net of losses)	
Animal	**Housing**	**Presence/Type of bedding**	**Litter material** [a]	**Life weight**	**N Slurry**	**N Manure**
				kg	kg day^{-1}	kg day^{-1}
Dairy cow	Fixed housing	Bedding	Straw	600	0.064	0.163
		No bedding	-		0.227	-
	Free housing (bunk)	Bedding (rump to rump)	Straw		0.139	0.088
		Bedding (head to head)			0.088	0.139
		No bedding	-		0.227	-
	Free housing (Litter)	Bedding removed (3 months)	Straw		0.083	0.144
		Bedding removed (30–60 days)			0.102	0.125
		Inclined bedding continuously discharged			0.064	0.163
Veal and heifer	Free housing (bunk)	Slatted floor	Straw	300	0.098	-
	Free housing (Litter)	Only in rest area (removal at end of the cycle)	Straw		0.050	0.048
		Also in feeding area (frequent removal)			0.014	0.084
		Inclined bedding continuously discharged			0.014	0.084
Calf	Free housing (Litter)	Bedding	Straw	125	0.007	0.034
		Slatted floor			0.041	-
		Single box (water cleaning)			0.024	0.017

[a] Straw is assumed as the litter material, but also sawdust can be used.

Feeding and Breeding Management

The best ways to reduce the amount of nutrients excreted by animals include matching the amount consumed to that needed, in order to meet the animals' requirements and to increase the efficiency of the utilization of the nutrients consumed [45]. Lazzerini *et al.* [33] underlines the poor efficiency of N conversion into product (25%–45%) and the large amounts of N in animal diets, which are converted into excreta. Methods currently used to obtain a reduction/recovery/control of the nitrogen load, related to this first step of the chain, are usually named "rational management of effluents" [46]. The N abatement techniques are based on the reduction of protein nitrogen in the diet and/or the efficiency's increase of nitrogen use. In fact, with the optimization of the protein ration, a reduction of 10% of the nitrogen excreted for each percentage point of protein reduction in the ration can be achieved. The formulation of the ration with a "safety margin" of nutrients (from 30 up to 50%) results in excretion of the excess N and P. The goal of efficient and productive feeding of animals, within economic and environmental constraints, is to provide essential available nutrients for maintenance and production while minimizing excess amounts. Formulating the protein content of cattle rations, to meet the animals' requirements for essential amino acids, can reduce N excretion by 15%, compared with more traditional feeding. It is possible to reduce N excretion up to 40%–50% by reducing the crude protein of the ration and adding supplemental amino acids. By reducing the administered nitrogen, the amount of nitrogen to the field can be reduced from 18% to 30% if compared with a standard diet [47]. A "low protein diet" can reduce the volume of slurry from 9.5% to 11.3%. Reduced dietary N input resulted in reduced CH_4 and N_2O emissions. However, in order to transfer a number of scientific findings to the operational level, proper application of precision farming technologies must be applied, particularly with regard to the strict record of the amount of feed assumed by animals [47].

Several techniques can be used in feed preparation, handling and delivery that can affect animal performance and, consequently, nutrient excretion. For instance, ensiling forages and cracking grains increases the digestibility of the ration for beef and dairy cattle. Similarly, lactating dairy cows can be fed with specific rations at separate periods in their milk production cycle. Consequently, fewer nutrients are wasted and excreted (from 5 to 10% less). This evidence requires grouping or penning animals together of similar ages, gender, weights and (or) production or management groups. Regulation and maintenance of feeders, bunks and waterers minimize the spillage of feed and water into the manure storage. Several new technologies are being developed and tested to enhance the nutrient content or utilization of feed ingredients or to alter the availability of nutrients in current commercial feeds. Within those are enzymes, genetically-modified feed ingredients and feed processing technologies to enhance the availability of nutrients to meet the needs of specific animals and reduce the excretion of nutrients. These new technologies will provide nutrients in proper balance that will allow "precision-feeding" of animals.

Table 6 summarizes the effects of different feeding types on the effluents amount and the nitrogen excreted, for each "typical farm" group. This overview should give an idea of the importance of managing the feeding step well, at different farm size levels.

Farm setting and herd composition are even taken into account according to Table 1 and Table 2.

Table 6: The main feeding solutions, compared with a regular diet, related to the amount of effluents produced (manure and slurry) and nitrogen excreted, per "typical farm" group. Data elaborated based on [46,47,48]

		Regular diet [b]			Hypo-protein diet [d]			Protein content on animals' needs [e]	Reduced Crude protein content [f]	Animals categories grouped [g]
Farm setting	No. of animals	Livestock effluents [a]		Nitrogen content	Livestock effluents [c]		Nitrogen content			
		kg day^{-1}	m^3 day^{-1}	kg day^{-1}	kg day^{-1}	m^3 day^{-1}	kg day^{-1}			
Extensive	20	621	0.69	3.15	557	0.62	2.40	2.68	1.73	2.96
Extensive -integrated	90	2808	3.12	14.24	2518.77	2.80	10.83	12.11	7.83	13.17
Integrated-Intensive	320	9945	11.05	50.45	8,920.66	9.91	38.34	42.88	27.75	46.66
Intensive	550	17,118	19.02	86.85	15,354.85	17.06	66	73.82	47.76	80.33

[a] Considering free housing (bunk), bedding (rump to rump), straw. The values of effluents produced are slightly lower and of nitrogen content slightly higher than the values in Table 4, since here, differences between the manure and slurry produced, according to Table 5, are considered; [b] considering a value of 3.5 g N kg^{-1} of effluent produced for a regular diet; [c] an average reduction of 11.3% of slurry excreted is assumed; [d] an average reduction of 24% of N excreted is assumed; [e] an average reduction of 15% of N excreted is assumed; [f] an average reduction of 45% of N excreted is assumed; [g] an average reduction of 7.5% of N excreted is assumed.

Biomass Step

Choice of Organic Matrices

Incineration or thermal gasification of biomass is the most efficient method to produce energy only if the dry matter (TS) content of the biomass is high, *i.e.*, above 40% [34]. Untreated animal manure typically has TS content much below 40%, as shown in Table 3. Therefore, the alternative anaerobic digestion process is a very good option, as a wet biomass is perfectly suited for the anaerobic biogas process [34]. Bio-waste contains a high amount of only partially-oxidized organic components. By fully oxidizing these components, the stored energy might be released and used [34].

ccording to Raju [49], the substrate used to produce biogas in a digester is an important aspect that determines the organic loading rate (OLR) (kg VS m^{-3} day^{-1}), the amount of biogas (m^3 ton^{-1} fresh substrate) and the final BMP (m^3 CH_4kg^{-1} VS).

Biogas yields (m^3 ton^{-1} VS) for categories of substrates suitable for on-farm AD vary from 200 to 550 for livestock effluents, from 550 to 750 for biomass from dedicated crops, from 350 to 500 for agricultural residues and from 400 to 800 for agro-industrial and food production residues. If slaughterhouse bio-waste is added, from 600 to 950 m^3 ton^{-1} VS of biogas can be obtained. Sewage sludge and the organic fraction of urban waste provide from 250 to 350 and from 400 to 600 m^3 ton^{-1} VS of biogas [38].

The advantage of AD is that almost any biomass can be used as a substrate for biogas production [50]. This variety of substrates can be used as a mono-substrate or as mixed-substrates in co-digestion processes. Much research has gone into identifying species and cultivars of energy crops that can produce more biogas. However, energy crop production has been criticized for competing with food production for arable land [49]. Agricultural and livestock residues offer an excellent alternative to energy crops. The addition of agro-industrial or readily degradable waste from food processing is needed to give a sustainable energy production. Animal slurry has low organic matter content and about 30% of it is slowly degradable [34]. Agricultural residues, such as straw, rice husk or wood chips, often contain high concentrations of ligno-cellulose, which is difficult to degrade.

Animal manure contains more readily degradable organic materials, such as proteins and lipids, than other agricultural by-products, but it also has a high content of lignocellulose bio-fibers [43] (40%–50% of the total solids [51,52]), a considerable part of which are recalcitrant to anaerobic digestion. The use of such biomass for AD may require pre-treatment to improve their degradability. Moreover, animal manure has very variable characteristics, not only between animals (*i.e.*, pigs, cattle and poultry), but also within the same animal categories between countries and farms [53,54]. Those differences depend mainly on the farm production, on the animal feed composition and on the water consumption. Seasonal fluctuations can also be observed within the same farm.

Meadow grasses are a promising source of biomass [55] and a good option for AD, due to various reasons, such as availability, the possibility for nutrient transfer and low energy and chemical input requirements. Co-digestion of manure and biomass increases the methane yield when compared to digesting solely manure, but the results are sensitive to many operating parameters, not only related to the reactor type, but also the type of manure and biomass and the ripeness of the biomass [56].

Bringing biofuel production to the farm-scale provides an opportunity for the agricultural sector to reduce its reliance on imported fossil fuels, while improving the soil, water and air quality [57]. In effect, livestock/agricultural waste-to-bioenergy treatments have the potential to convert the treatment of livestock waste from a liability or cost component into a profit center that can: (1) generate annual revenues; (2) moderate the impacts of commodity prices; and (3) diversify farm income [58]. However, biogas production needs to be cost-effective, and excessive costs may provide an argument to (temporarily) loosen goals for a certain areas [59]. The analysis of costs and benefits associated with residual waste treatment options is a subject that has been discussed in a range of studies. Of course, different treatments will fare slightly better or worse depending on the composition of the material being treated [60]. The choice of organic matrices to be used in the process of AD is conducted following different management logics (Table 7).

(1) The choice of giving priority to farm effluents or waste: In recent times, this aspect has been associated with the logic of efficiency improvement of AD and, thus, to the agro-energy profitability increase of the farm.

(2) The ability to find dedicated biomass: It is usually easy to find dedicated biomass grown on the land of the same farm running the biogas plant. However, agreements for the provision of dedicated biomass from nearby farms are also common.

(3) The ability to find suitable products at low cost: Many experiences can be cited for use of by-products and waste present in large quantities near the biogas plant, often with little or no cost at all, with the exception of transportation costs. Therefore, the possibility to reduce the cost of biogas production makes this solution appealing, even if it must be accompanied by a careful combination of organic matrices.

(4) The cost of the organic matrix used: any choice of matrices for the process of AD must be assessed in relation to its value (€ ton^{-1} of substrate). For instance, many effluents, while presenting zero cost, have very low biogas-yields and/or induce difficulties in the process or the supply-chain.

(5) he productivity in terms of biogas ($m^3\ kg^{-1}$ VS) of the matrices available: the biogas-yield obviously depends on the type of matrix. To estimate the unit cost of biogas producible (€ m^{-3} of biogas), it is necessary to consider the matrices as a function of this aspect.

Table 7: The matrix summarizes biomass options and logics for choosing the substrate for the AD process

	Variable input	Description of input
Substrate	Livestock effluent Biomass from dedicated crops Sewage sludge Organic fraction of urban waste Agricultural residues Agro-industrial and food production residues	Choice of organic matrices to be used in the process of AD (both as mono-substrate or in co-digestion)
Depending on:	Choice of giving priority to farm effluents or waste Ability to find dedicated biomass Ability to find suitable products at low cost Cost of the organic matrix used Productivity in terms of biogas	Factors affecting choices, based on specific management logics

According to [58], manure processing by anaerobic digestion and converting the biogas into electricity with a combined heat and power (CHP) engine show negative costs, *i.e.*, economic benefits. Investment is recovered by electricity sales, savings on heating and "green power" and CHP support. However, costs are only negative if support schemes continue to exist and if digestate can be used as fertilizer for croplands without further treatments. A precise cost-benefit analysis (CBA) should combine mass flow data with unit cost factors in order to calculate the environmental and financial costs for each scenario. Capital investment, operational costs for process feedstock, greenhouse gases (GHGs) and air quality externalities, power supply, logistics and fertilizer costs are important factors in performing CBA. The definition of the investment costs' reference standard is complicated. In general, for most of the plants, a range between 250 and 700 € per cubic meter of anaerobic digester can be defined, or 2500–7500 € per kW of electricity installed in CHP. The threshold for making a profitable biogas plant is usually about 50–100 kWe of installed power.

In selecting options for biogas use, some important aspects must be taken into account. The main form of energy consumed by the farm: it is important to consider the type of user, consumption trends (mechanical energy for stationary or mobile applications, cooling energy, low-temperature thermal energy, steam) and the energy sources used. This assessment may lead to solutions of biogas use other than electricity generation. Consumption trend during the year: biogas production, if well planned, is constant throughout the year, while energy consumption frequently has an irregular trend or is characterized by periodicity. Efficient use of biogas should be oriented to serve users continually as much as possible, in order to minimize expensive fuel storage. Exploitation of biomethane as fuel: the European Directive 2003/55 authorizes the placing of other types of gas in natural gas networks. For bio-methane production, water, sulfur compounds, halogenated molecules, carbon dioxide, oxygen and metals must be eliminated.

Considering that 23 MJ m^{-3} biogas (65% methane), 11.5 MJth (thermal energy) and 8.74 MJe (electricity) from a CHP engine can be obtained or with 1 N m^3 biogas, it is possible to produce, on average, 1.5–2 kWh of electricity and 2–3 kWh of thermal energy [38]. According to these values and considering that average biogas yields for cattle effluents range from 200 to 400 m^3 ton^{-1} VS, extensive farms can convert 42 m^3 day^{-1} of biogas in about 4 kWe, while intensive farms will obtain 1155 m^3 day^{-1} of biogas and up to 93.5 kWe. In integrated farms, according to more extensive or intensive settings, values from 320 to 672 m^3 day^{-1} of biogas can be produced, obtaining average values from 15.3 to 54.4 kWe.

Characteristics of Livestock Effluents Compared to Other Substrates

Above the other factors, the productivity of biogas in terms of BMP (m^3 CH_4 kg^{-1} VS) of different substrates is crucial. BMP is the most important parameter to assess the quality of feedstock among other biological and physiochemical parameters in AD. It is used to design real scale biogas digesters. It can be used for kinetic modelling to predict biogas yield, and it has furthermore been used as the most relevant indicator for assessing digestibility [34]. Table 8 presents BMP and TBMP (theoretical BMP) values for different substrates, in order to compare the biogas production from various biomasses used in the AD process. TBMP is a stoichiometric approach, which assumes that all VS can be converted to methane. Anaerobic digestibility (BMP × $TBMP^{-1}$ × 100) depends on the physical resistance of each biomass's component (*i.e.*, lignin, lipids, protein) of VS against hydrolysis. TBMP is a useful tool for assessing the energy potentials of biomass having a larger fraction of readily degradable VS that can typically be found in non-waste biomass and energy crops not containing lignocelluloses. For these biomasses, the measured BMP is close to TBMP. However, most of the waste-biomass, such as animal slurry, present critical digestibility below 50% of TBMP, and many plant biomasses have a low digestibility due to lignocellulose [43,44,45,46,47,48,49,50,51,52,53].

Table 9 summarizes the substrates usable/available for each "typical farm" group, based on management logics, according to the considerations given in this paragraph. In particular, the pros and cons of the use of livestock effluents in the AD process are summarized in Table 10.

Table 8: Biomethanization potential (BMP) (CH_4NL kg^{-1} VS), theoretical BMP (TBMP) (CH_4NL kg^{-1} VS) and the digestibility of different biomass [43,44,45,46,47,48,49,50,51,52,53]

Biomass	BMP	TBMP	BMP/TBMP × 100
	CH_4NL kg^{-1} VS	CH_4NL kg^{-1} VS	-
Piglet manure	417	449	92.9
Sow manure	213	537	39.7
Pig fattener manure	345	527	65.5
Deep litter manure	237	442	53.6
Cattle manure	223	523	42.6
Maize (corn)	478	501	95.3
Maize (leaves)	402	445	90.4
Lawns + clover	328	484	67.6
Lawns	339	505	67.1
Wheat straw	227	477	47.6
Tufted hair-grass	235	458	51.3
Ivy	231	486	47.5
Willow	164	513	32
Birch tree (branch + leaves)	240	546	44
Common reed	188	486	38.6

Table 9: Biomasses available/feasible for AD for each "typical farm" group

Parameters	Animal	U.M.	Farm setting			
Type			Extensive	Extensive-integrated	Integrated-Intensive	Intensive
Size of			Min	Small	Medium	Max
Herd	Dairy Cattle [a]	No.	≤20	90	320	≥550
Daily effluents [b]		Kg	≤621	2808	9945	≥17,118
		m^3	≤0.69	3.12	11.05	≥19.02
Other biomasses for AD	Biomass from dedicated crops	feasible/available	No	No	Yes	Yes
	Sewage sludge		No	No	Only if locally produced	Yes
	Organic fraction of urban waste		Only on farm produced	Only if locally produced	Yes	Yes
	Agricultural residues		Yes	Yes	Yes	Yes
	Agro-industrial and food production residues		No	Only if locally produced	Yes	Yes

[a] The cattle categories are considered in the general definition "dairy cattle", as described in Table 3; [b] the amount of effluents produced per day is based on the assumptions made in Table 6.

Table 10: Pros and cons of livestock effluents used in the AD process

Livestock effluents: pros and cons related to their use in the AD process	
PROS	CONS
Very low or null matrix costs	Low methane yields
Availability in neighboring areas	Reduction of carbonaceous components contribution to soil
Valorization of waste and potential integration of income for livestock farm	Possible formation of floating crusting waste with high TS% manure types
Presence of pre-existing storage facilities	
Creation of potential easier outlets for the digestate, when spreading is expected in non-farm soils	
Stabilization of effluents with odorous emissions reduction	
Possibility to use different matrixes in the processes of co-digestion	

The main characteristics of manure that have to be taken into account are the volume and chemical characteristics. The determination of volume is required for the purposes of re-sizing the storage capacity, to correctly set the fertilization plans, to check if the volumes of effluents produced differ from those calculated indirectly and for the size of the effluent treatment plants. The determination of chemical characteristics is necessary for the purposes of their correct use for crop fertilization and the identification and design of the treatment techniques and size.

Angelidaki *et al.* [50] reviewed useful values for cattle manure characteristics. Typically, 60%–85% of the total N present in the effluent is in organic form. Most of this organic N is from the fecal material (N_{feces}). The urine contributes some organic compounds; 60%–90% of the total N in urine (N_{urine}) is urea, which is rapidly hydrolysed to ammonium-N. Boe *et al.* [52] found a mean nutrient content in cattle slurry from dairy cows of 1120 mg (NH_4-N) L^{-1} and 1380 mg (P_2O_5) L^{-1}, while Lübken *et al.* [61] found 3090 mg (NH_4-N) L^{-1} and 74,500 mg (COD) L^{-1} (chemical oxygen demand).

The composition of the material affects the rate of degradation (degradability), having decreasing values respectively for proteins, lipids, cellulose and, lignin. Cattle effluents have a greater content of cellulosic material: their degradation rate will be lower, for instance, than the one of pig slurry, which is richer in lipids. The presence of toxic elements for microbial metabolism is also an important factor. Within these are often micronutrients, such as sodium (Na), potassium (K), calcium (Ca), magnesium (Mg), ammonium (NH_4^+) and sulfur (S), which, if present in excess, can induce the inhibition of the process. Heavy metals, such as copper (Cu), chromium (Cr), nickel (Ni), zinc (Zn) and lead (Pb), if present in concentrations greater than 1 mg L^{-1}, can lead to severe damage, both to human metabolism and crops.

Anaerobic digestion of animal manure can be inhibited by ammonia (NH_3) and hydrogen sulfide (H_2S). The former is generated by the fermentation of organic nitrogen (urea and proteins) and the latter by sulfate reduction. Because of high nitrogen concentration in the manure and quite high pH values, free ammonia (FA) is considered as the main inhibitor, even if the inhibition thresholds are very dependent on the inoculum and its adaptation to the inhibitor [62]. However, FA concentrations above 1500 mg N L^{-1} are not desirable. This fact highlights the importance of balancing the N content of manure while maintaining a proper C/N ratio. It is necessary to maintain the adequate composition of the feedstock for efficient plant operation, so that the C/N ratio in the feed remains within the desired range. During anaerobic digestion, microorganisms utilize carbon 25–30 times faster than nitrogen. Thus, to meet this requirement, microbes need a 20–30:1 ratio of C to N with the largest percentage of the carbon being readily degradable [63].

Cattle manure presents 10–30 as the C/N ratio [34], but in the case of co-substrate added to AD, the real ratio has to be defined by summing the C/N(Xn) multiplied by the weight of each material Xn and dividing the sum by the total weight of feedstock.

Some recommendations can be extrapolated from the excursus on farm and livestock effluent biomass characteristics. Livestock activities should better integrate other agricultural and agri-food activities, in order to re-cycle both inputs and outputs. Animal production increase should no longer be stressed by genetic improvement of animals nor a strong increase of the "average daily gain". The human and economic pressure on cereals will compete more and more with livestock production. Cereals used for animal feeding should be replaced by co- or by-products from agri-food activities, allowing a reduction of the cost of animal feeding and the development of recycling systems of so far unused products. In particular, specific hypo-protein diets and housing modalities that allow a lower nitrogen content excretion and more valuable livestock-biomass should be integrated with the farm, particularly for larger ones.

CONCLUSIONS

Effluent management methods that are more feasible and available at different levels have to be integrated into the animal husbandry sector, in order to protect the environment and to allow one to switch back to a recycling view of manure handling. The ideal situation would be to work at the same time on both inputs and outputs of livestock production and its integration in its "regional" or geographical aspects. However, to reach such a goal, all treatment aspects have to be considered, not only the constraints, whatever they are (environmental, sanitary), but also the overall consequences, integrating economical parameters, like the cost of livestock buildings, evolution and depletion of fuel energy, phosphorus and, maybe, cheap cereals. Possible stronger policies on environmental protection have to be integrated, such as the necessity to include new "emerging" pollutant-like antibiotics, endocrine disrupters and antibiotic-resistant pathogens. The development of such new systems will require the development of new measuring devices and global methods to assess the viability of the production chain and food supply. Evidently, this raises the awareness that for each farm, solutions have to be evaluated, actively considering the features and the inherent vulnerability of the territory on which the effluents are disposed. Intensive farms are expected to be able to apply and integrate in the livestock activities and facilities the best available techniques, through which the nutrient amount contained in the effluents can be significantly reduced up to 70%. In fact, the cost of the most sophisticated and efficient treatments cannot be supported by small farms. However, extensive and/or organic farms usually better integrate livestock activities and cultivation activities. Integrated farms may present characteristics more close to extensive farms or to intensive ones. They represent a "typical farm" group, which is more easily expected to calibrate and adapt different treatment and management solutions, balancing farm-production profit, energy self-sufficiency, environmental protection and animal husbandry well-being.

AUTHOR CONTRIBUTIONS

All authors contributed extensively to the work presented in this article. Francesca Nardin wrote the paper with input from Fabrizio Mazzetto, discussing the results, implications and commenting on the manuscript at all stages.

REFERENCES AND NOTES

1. World Commission on Environment and Development. Our Common Future; Oxford University Press: Oxford, UK, 1987.
2. McCormick, K.; Kåberger, T. Key barriers for bioenergy in Europe: Economic conditions, know-how and institutional capacity, and supply chain co-ordination. Biomass Bioenerg. 2007, 31, 443–452.
3. Ecosense, Forum Nachhaltige Entwicklung der Deutschen Wirtschaft (Forum for Sustainable Development of German Business). Available online: http://www.econsense.de/sites/all/files/Schriftenreihe_1_Klimafaktor_Biokraftstoff.pdf (accessed on 23 March 2014).
4. Torretta, V.; Conti, F.; Leonardi, M.; Ruggieri, G. Energy recovery from sludge and sustainable development: a tanzanian case study. Sustainability 2012, 4, 2661–2672.
5. European Biomass Association (AEBIOM). European Bioenergy Outlook 2013; AEBIOM: Brussels, Belgium, 2013.
6. Chynoweth, C.P.; Owens, J.M.; Legrand, R. Renewable methane from anaerobic digestion of biomass. Renew. Energ. 2001, 22, 1–8.
7. Abbasi, S.A.; Abbasi, N. The likely adverse environmental impacts of renewable

energy resources. Appl. Energ. 2000, 65, 121–144.

8. Abbasi, S.A.; Abbasi, N. Biomass energy and the environmental impacts associated with its production and utilization. Renew. Sust. Energ. Rev. 2010, 14, 919–937.

9. Von Blottnitz, H.; Curran, M.A. A review of assessments conducted on bio-ethanol as a transportation fuel from a net energy, greenhouse gas, and environmental life cycle perspective. J Cleaner Prod. 2007, 15, 607–619.

10. An, H.; Wilhelm, E.W.; Searcy, S.W. Biofuel and petroleum-based fuel supply chain research: A literature review. Biomass Bioenerg. 2011, 35, 3763–3774.

11. Awudu, I.; Zhang, J. Uncertainties and sustainability concepts in biofuel supply chain management: A review. Renew Sust. Energ. Rev. 2012, 16, 1359–1368.

12. Elz, D. Bioenergy systems. Q. J. Intern. Agricul. 2007, 46, 325–332.

13. Torres, E.A.; Cerqueira, G.S.; Ferrer, T.M.; Quintella, C.M.; Raboni, M.; Torretta, V.; Urbini, G. Recovery of different waste vegetable oils for biodiesel production: A pilot experience in Bahia State, Brazil. Waste Manage. 2013, 33, 2670–2674.

14. Gold, S.; Seuring, S. Effective Governance for Securing Feedstock Supply of Biogas Plants. In Proceedings of Knowledge Collaboration & Learning for Sustainable Innovation ERSCP-EMSU conference, Delft, The Netherlands, 25–29 October 2010.

15. Martinez, S.; Torretta, V.; Minguela, J.; Siñeriz, F.; Raboni, M.; Copelli, S.; Rada, E.C.; Ragazzi, M. Treatment of slaughterhouse wastewaters using anaerobic filters. Environ. Technol. 2014, 35, 322–332.

16. Rada, E.C.; Ragazzi, M.; Torretta, V. Laboratory-scale anaerobic sequencing batch reactor for treatment of stillage from fruit distillation. Water Sci. Technol. 2013, 67, 1068–1074.

17. Andreottola, G.; Ragazzi, M.; Foladori, P.; Villa, R.; Langone, M.; Rada, E.C. The unit integrated approach for OFMSW treatment. UPB Sci. Bull. 2012, 74, 19–26.

18. Observ'ER. The State of Renewable energies in Europe. Available online: http://www.energies-renouvelables.org/observ-er/stat_baro/barobilan/barobilan13-gb.pdf (accessed on 23 March 2014).

19. Amon, T.; Amon, B.; Kryvoruchko, V.; Zollitsch, W.; Mayer, K.; Gruber, L. Biogas production from maize and dairy cattle manure—Influence of biomass composition on the methane yield. Agr. Ecosyst. Environ. 2007, 118, 173–182.

20. Braun, R. Anaerobic digestion: A multi-faceted process for energy, environmental management and rural development. In Improvement of Crop Plants for Industrial End Uses; Ranalli, P., Ed.; Springer: Dordrecht, The Netherlands, 2007.

21. Hartmann, H.; Ahring, B.K. The Future of Biogas Production. In Proceedings of Risø International Energy Conference, Risø, Denmark, 23–25 May 2005.

22. Weiland, P. Biomass digestion in agriculture: A successful pathway for the energy production and waste treatment in Germany. Eng. Life Sci. 2006, 6, 302–309.

23. Clemens, J.; Trimborn, M.; Weiland, P.; Amon, B. Mitigation of greenhouse gas emissions by anaerobic digestion of cattle slurry. Agr. Eco. Environ. 2006, 112, 171–177.

24. Ragazzi, M.; Rada, E.C.; Schiavon, M.; Torretta, V. Unconventional parameter for a correct biostabilization plant design in agriculture areas. Mitt. Klosterneuburg 2014, 64, 2–14.

25. IEA. Energy and Poverty. 2002. Available online: http://www.worldenergyoutlook.org/media/weowebsite/energydevelopment/WEO2002Chapter13.pdf (accessed on 23 March 2014).

26. Savola, H. Biogas Systems in Finland and Sweden—Impacts of Government Policies on the Diffusion of Anaerobic Digestion Technology; The International Institute for Industrial Environmental Economics, Lund University: Lund, Sweden, 2006.

27. Hacker, V.; Rabenstein, G.; Baumgartner, W. Biogase für Brennstoffzellen (Biogas for fuel cell). Chemie Ingenieur Technik 2007, 79. Article 1333 (In German).

28. Schulz, W.; Hille, M. Untersuchung zur Aufbereitung von Biogas zur Erweiterung der Nutzungsmöglichkeiten (Enquiry concerning the refinement of biogas to enlarge its applicability); Bremer Energie-Institut (BEI): Bremen, Germany, 2003. (In German)

29. Tregaro, Y.; Lossouarn, J. Dynamique de la production porcine mondiale et européenne: enjeux de localisation et de relations au territoire (Global and European dynamics of the pig production: Issues of location and relationship to the territory). In Proceedings of Séminaire Porcherie Verte, La Rochelle, France, 7–8 December 2004. (In French).

30. Flotats, X.; Bonmatí, A.; Fernández, B.; Magrí, A. Manure treatment technologies: On-farm versus centralized strategies. NE. Spain case study. Bioresource Technol. 2009, 100, 5519–5526.

31. Poulsen, H.D.; Børsting, C.F.; Rom, H.B.; Sommer, S.G. Kvælstof, fosfor og kalium i husdyrgødning. (Nitrogen, phosphorus and potassium in animal manure: Norm Values 2000). Available online: http://anis.au.dk/fileadmin/DJF/Anis/normtal2000.pdf (accessed on 23 March 2014).

32. Nardin, F.; Mazzetto, F. The Manuresource concept through an optimized biogas refinery. UPB. Sci Bull. 2014. submitted.

33. Lazzerini, G.; Cecchi, S.; Zanchi, C.; Vazzana, C. Metodologia di gestione dei reflui zootecnici (Methodology of manure management). In Riduzione dell'inquinamento delle acque dai nitrati provenienti dall'agricoltura; Felici Editore: San Giuliano Terme, Italy, 2010; pp. 391–417. (In Italian)

34. Sommer, S.G.; Christensen, K.V.; Jensen, L.S. Environmental Technology for Treatment and Management of the Bio-Waste Manure; University of Southern Denmark, Faculty of Engineering, Institute of Chemical Engineering, Biotechnology and Environmental Engineering: Odense, Denmark, 2012.

35. Amon, B.; Kryvoruchko, V.; Amon, T.; Zechmeister-Boltenstern, S. Methane,

6

CHAPTER

MANUFACTURING AND UTILIZATION OF PELLETIZED LIVESTOCK MANURE COMPOST

Kwang-Hwa Jeong, Modabber Ahmed Khan, Ho Kang, Jung-Kon Kim, Jong-Hoon Kwag

Keywords: Compost, Livestock, Manure, Pelletizing system

INTRODUCTION

Livestock manure is an important resource for agriculture as it contains high level of nutrients and organic matter. The number of livestock industry in Korea has sharply increased over the past two decades, which results in greater awareness and concern for the proper storage, treatment, and utilization of livestock manure. The total amount of animal waste estimated in 2011 is about 46.5 million tons throughout the nation where 20.5 million tons (44% of total produced waste) originated from cattle. About 87% of the total produced animal waste is recycled to farmlands as compost and liquid organic fertilizer (Abedin and Kianmehr, 2012). The expansion of concentrated animal feeding operations has made it hard to reuse their manure because of the limited land area near the livestock farms, resulting in occasionally water, soil, and air pollution. Livestock manure is a valuable resource in improving soil structure and increasing vegetative cover, thereby reducing surface runoff and erosion potential. Most of the livestock excretions were used as compost in Korea and the government encourages the farmers to use livestock excretions as a fertilizer. But the total amount of arable land for applying livestock manure compost is decreasing every year. For instance, the total arable land was 2,167,000 hectares in 1983, which has decreased gradually and reached

Citation: HWA JEONG, Kwang et al. Manufacturing and Utilization of Pelletized Livestock Manure Compost. Global Journal of Animal Scientific Research, [S.l.], v. 3, n. 1, p. 11-19, sep. 2014. ISSN 2345-4385. Available at: <http://www.gjasr.com/index.php/GJASR/article/view/108/321>. Date accessed: 24 Aug. 2016.

to 1,729,982 hectares in 2012 (Ahn *et al.,* 2011). Another problem in utilizing the livestock manure compost is that rural population who are using compost made from livestock manure are becoming old. They are tending to use less compost because it is not so easy for the aged farmers to handle wet and heavy compost. Even livestock farms in some regions where the nitrogen content of the soil is low cannot find enough cropland on which to apply their compost. Groundwater has become polluted in these regions by nitrate due to heavy applications of composted livestock manure onto a limited area of cropland. Furthermore, the direct use of animal wastes acts health and environmental risks which should be treated accordingly. Stabilization deals with the decomposition of a waste substance to some extent where the hazards are limited and is normally reflected by decreasing microbial activity (AOAC, 1990). To reduce the environmental problems associated with the management of manure, stabilization is safer and suitable for application to soil (APHA, 2005). There are two factors that limit the application of ordinary composted livestock manure. The first problem is that composted livestock manure usually has high moisture content and a high volume per unit of weight, which makes it difficult and costly to transport (Tesfay, 2014). However, Korean government forces livestock manure composting companies to keep the moisture content below 55% in their commercial compost. The second problem is that the quality of the compost and its nutrient content are not constant that limits the efficient use of compost. The pelletizing technology used for composted livestock manure could be an effective solution to both problems (Hara, 1998). A pelletizing machine could be the way to make composted livestock manure into pellets with advantages in transportation, handling and storage (Benito *et al.,* 2003; Bhattacharya *et al.,* 1989; Carr, 1995; Hansen *et al.,* 1995; Hara, 1998). If this is to be done without adding any other materials, it is important to control the moisture content of the compost. As a result, the objective of this study was to develop a new type of pelletizing system for high quality pelletized compost.

MATERIALS AND METHODS

Cow dung mixed with sawdust was used and cured under the aerobic condition. At the initial stage of curing, the moisture content of cow dung was adjusted around 65% by mixing with saw dust. The air was supplied from the bottom of the experimental compost pile through pipeline connected with mechanical blower (Agnew, 2003). The experimental compost pile was periodically mixed homogeneously by a skid loader. The variations of temperature and ammonia emission were investigated daily to get the information of curing state of the compost pile. When the temperature of the compost pile was near to ambient temperature and not varied any more, the compost was manufactured into pelletized compost by pelletizing instrument. Pig manure and poultry manure were also treated with the same manner (Farhadi, 2014). To pelletize the livestock manure compost, a new type pelletizing instrument was designed and manufactured in this study. The pelletizing instrument shown in Figure 1 was equipped with a screw type compressing part, input control unit and extruding part, etc.

Figure 1: Image of experimental pelletizing instrument developed for this study.

Physico-chemical analysis of pelletized compost was conducted in accordance with standard methods (Jeong *et al.,* 2012; Peisker, 1994). To decide an optimum pelletizing instrument type, three types of pelletizing models were prepared. Screw extruding type pelletizing instrument, horizontal flat mill type pelletizing instrument and vertical ring mill type pelletizing instrument were selected as an early model of experimental instrument to decide ultimate experimental pelletizing instrument. Figure 2 represents three types of early models used in this study for investigation. Pelletized livestock manure compost was dried by horizontal tunnel type drier (Figure 3).

External appearance of experimental drier

Inner space for drying pelletized compost

Figure 2: Images of drier for pelletizing compost.

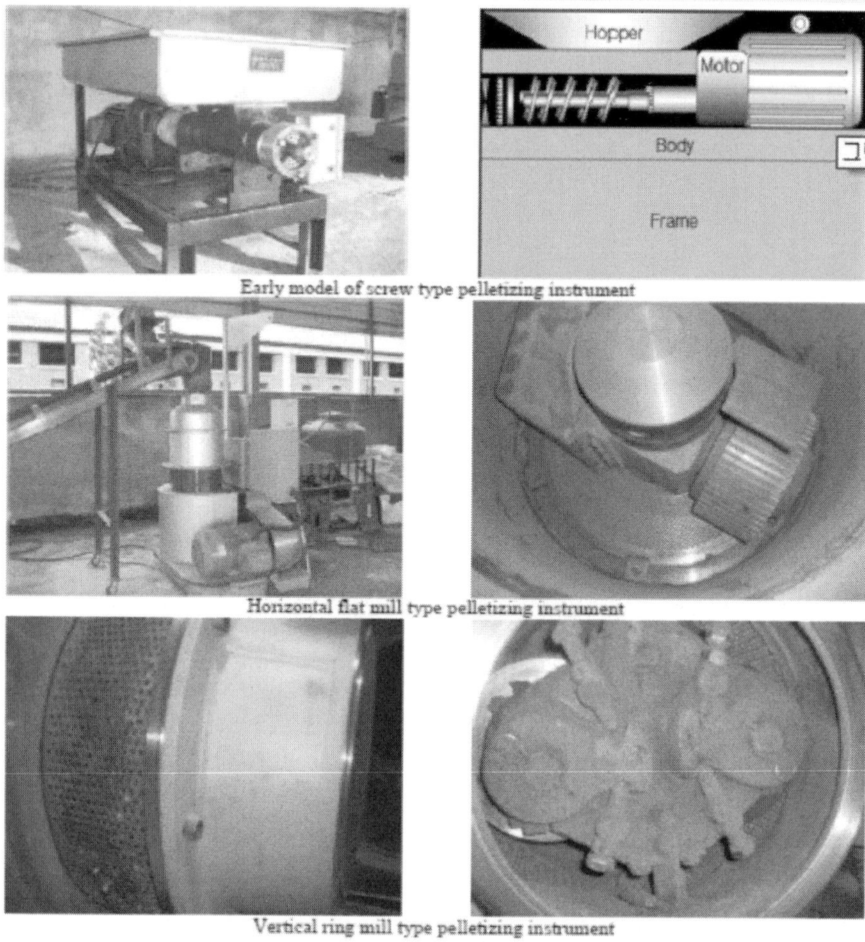

Figure 3: images of three types of early models used in this study for investigation.

RESULTS AND DISCUSSIONS

In Korea, livestock manure usually mixed with bulking agent such as sawdust, rice hulls and tiny woodchip; and then the mixed manure used to cure for dozens of days. During that period, the manure was generally turned over by the stirrer. Although after curing stage, the moisture content of cured livestock manure compost reached to 45-55%. This high moisture content of compost makes it difficult to produce pelletized compost. The moisture content ranges of livestock manure compost for this study is shown in Table 1.

Table 1: Moisture content ranges of livestock manure compost

parameters	Moisture content (%)
Cow manure compost	33-52
Pig manure compost	35-54
Poultry manure compost	38-51

Livestock manure composts was manufactured in this study. To test the effect of pelletizing process, adding water or drying process were used to adjust the moisture content of composts as stuff for pelletizing when needed. The moisture content of compost for this study was adjusted at 33-54%. The distribution of particle size of cow manure compost for pelletizing was from above 2mm to less than 0.1mm. However, most particle size of the compost was between 0.85-0.425mm (Table 2).

Table 2: Distribution of particle size of cow manure compost

Dimension	Percentage	Dimension	Percentage	Dimension	Percentage
> 2.0mm	14.6	1.0-0.85mm	5.5	0.3-0212mm	7.8
2.0-1.7mm	5.4	0.85-0.6mm	14.6	0.212-0.18mm	2.7
1.7-1.4mm	4.0	0.6-0.425mm	14.1	0.18-0.1mm	4.0
1.4-1.0	13.9	0.425-0.3mm	9.3	< 0.1mm	4.1

It is evident that the compost has various contents of N, C, S, H and C/N according to the particle size of compost, which is shown in Table 3. According to Table 4, the optimum moisture content was different in types of pelletizing instrument. The optimum moisture content in screw type pelletizing instrument was 45-50% whereas it was 20-25% in mill type pelletizing instrument.

Table 3: Contents of organic matter according to particle size of compost

Dimension	N (%)	C (%)	S (%)	H (%)	C/N
> 2mm	1.329	20.95	0.447	9.294	24.08
2.0 - 1.7mm	0.636	18.11	0.396	9.742	29.46
1.7 - 1.4mm	0.582	18.22	0.401	9.688	31.30
1.4 - 1.0mm	0.589	18.55	0.402	9.548	31.41
1.0 - 0.85mm	0.569	19.15	0.388	9.349	34.22
0.6 - 0.425mm	0.645	19.56	0.394	9.312	30.35
0.425 - 0.3mm	0.942	23.03	0.431	5.320	24.45
< 0.1mm	1.538	27.90	0.700	6.323	18.14

The resistance of compression stress of pelletized compost was higher in mill type pelletizing instrument than screw type pelletizing instrument. On the other hand, solubility of pelletized compost was higher in screw type pelletizing instrument.

Table 4: Processing condition of compost according to pelletizing instrument type.

Classification	Screw type pelletizing instrument	Mill type (flat, ring) pelletizing instrument
Optimum moisture content	about 45-50%	about 20-25%
Resistance of compression stress of pelletized compost	20 kg/cm	30 kg/cm
Solubility of pelletized compost	100% in 10h	90% in 10h

The moisture content of commercial compost in Korea is about 45-55% so the screw type pelletizing instrument was suitable for pelletizing of commercial compost. As a result, screw type of pelletizing instrument was selected as an ultimate experimental instrument for this study. Raw compost stuff and pelletized compost is shown in Figure 4.

Original compost

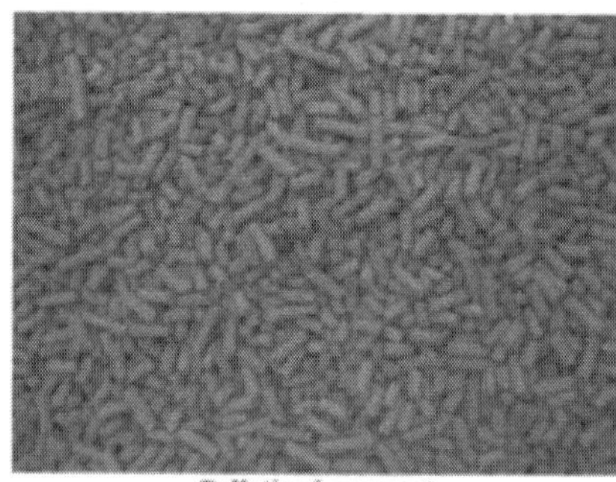

Pelletized compost

Figure 4: Changes of compost shape by pelletizing process.

Table 5 reveals the characteristics of cow manure mixed with sawdust, cow manure compost and pelletized cow manure compost.

Table 5: Characteristics of cow manure mixed with sawdust, cow manure compost and pelletized cow manure compost used in this study

Classification	NaCl	N	P	K	Cr	Cu	Ni	Cd	Pb	Zn
	mg/kg									
Cow manure mixed with sawdust	2.200	8148	3554	4957	1.17	28.82	4.82	0.53	3.68	89.43
Cow manure compost	2.200	11.304	4.636	5.134	1.99	39.77	5.29	0.62	8.48	130.94
Pelletized cow manure compost	2.900	10.996	5.948	5.563	2.44	42.68	7.50	0.76	6.81	147.84

The content of NaCl, N, P, K, Cr, Cu, Ni, Cd, Pb and Zn was higher in pelletized cow manure compost than cow manure mixed with sawdust and cow manure compost. The temperature of screw and ammonia emission was increased relatively to the revolution speed of screw of pelletizing instrument (Table 6).

Table 6: Emission of ammonia by revolution speed of screw of pelletizing instrument

Classification	10 RPM	15 RPM	20 RPM	30 RPM	50 RPM
Temperature of screw (°C	25	27	32	27	45
Ammonia emission (mg/L)	10	14	20	22	27

It is shown in Table 7 that pellet productivity was highest at 50 RPM but the processing effect was lowest due to the revolution speed of the screw in pelletizing process. However, the nutrients contents of compost were not changed by pelletizing process (Table 8).

Table 7: Effect of pelletizing process by revolution speed of screw of pelletizing instrument

Classification	10 RPM	15 RPM	20 RPM	30 RPM	50 RPM
Amount of pellet produced (kg/min)	0.5	0.7	1.1	1.6	2.5
Processing effect (%)	100	100	95	90	81

Table 8: Changes of nutrients of compost by pelletizing

Classification	OM (%)	N (%)	P_2O_5(%)
Swine manure compost	59.1	0.49	0.35
Pelletized swine manure compost	60.1	0.53	0.34

To investigate the pelletizing effect according to the thickness of compressing part, various thickness of compressing part were manufactured. Due to the thickness of compressing part of pelletizing instrument, the productivity and durability were different. The productivity of pellet was higher in thin compressing part than the thick compressing part. But the strengths of pellet were opposite. The compression part has two types of hole i.e. vertical and inclined. The edge of hole was cut and uncut as shown in Figure 5.

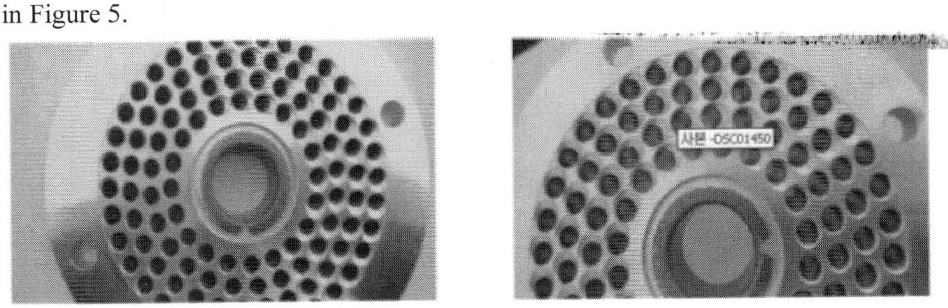

Figure 5: Images of compressing part of pelletizing instrument

As a result, pellet productivity was higher in inclined and cut type of compression part. Figure 6 shows the change of moisture content of livestock manure and compost due to the elapse of drying time. The moisture content of livestock manure and compost was decreased when the drying time increased. According to drying of pelletized compost in dryer, temperature of compost pellet and ammonia emissions is shown in Figure 7. The temperature of compost pellet and ammonia emission was decreased when the time increased.

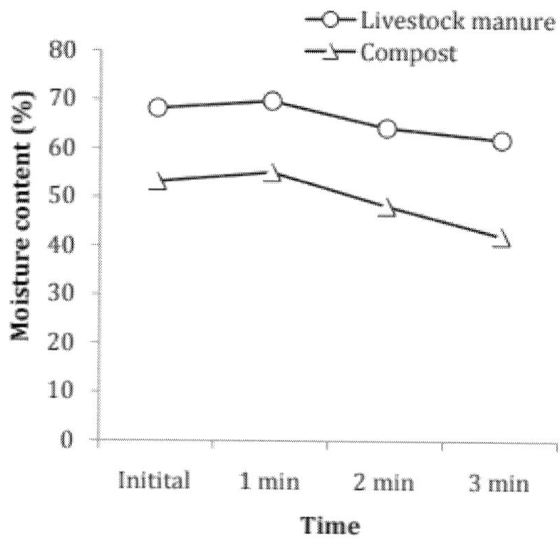

Figure 6: Changes in moisture content by drying.

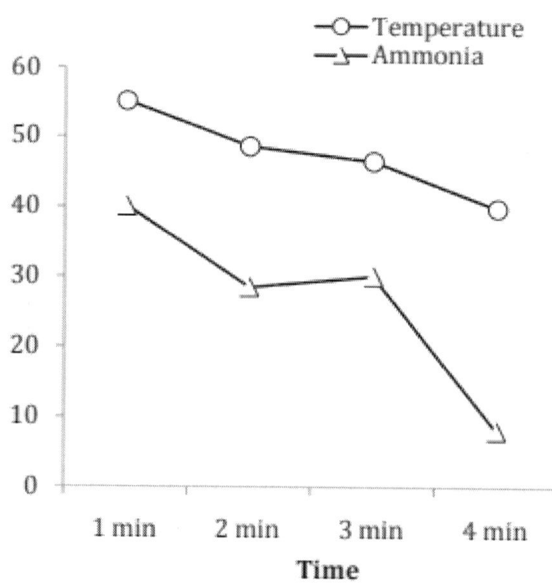

Figure 7: Comparison of temperature of compost and ammonia emissions by drying.

The comparisons of the emission of sulfur compounds by drying are shown in Table 9. Hydrogen sulfide (H2S) and methyl mercaptan (MM) were not detected in compost. Furthermore, dimethyl sulfide (DMS) was not detected after 1 min of drying time. Table 10 shows the changes of high heating value of compost pelletized and raw compost. According to the elapse of drying time the high heating value was highest after 3 min. Changes of temperature and ammonia concentration of compost during pelletizing process is shown in Table 11. During pelletizing of pig and cow manure compost, temperatures and ammonia emission concentrations were increased due to the frictions and compressions of pelletizing process. Ammonia emission concentrations were increased because of the increased temperature originated from the mechanical performance of the pelletizing process.

Table 9: Comparison of emission of sulfur compounds by drying

Classification	Sulfur compounds (ppb)		
	H_2S	MM	DMS
Compost	ND	ND	0.3582
Compost dried for 30 sec	ND	ND	1.9873
Compost dried for 1 min	ND	ND	ND
Compost dried for 2 min	ND	ND	ND

ND: not detected

Table 10: Changes of high heating value according to drying time

Classification	Initial	Compost dried for 1 min	Compost dried for 2 min	Compost dried for 3 min
	Kcal/kg			
Compost	3,113	3,002	3,022	3,256
Pelletized compost	2,866	3,185	3,044	3,032

Table 11: Changes of temperature and ammonia concentration of compost during pelletizing process

parameter	Pig manure compost		Cow manure compost	
	Raw compost	Pelletized compost	Raw compost	Pelletized compost
Temperature (℃)	27	47	25	51
NH_3 concentration (ppm)	25	45	12	39

Changes of volume of different pelletized composts with time have been depicted in Figure 8. The volume of the initial phase tended to decrease faster in case of oven-dried than air-dried pellet of pig and cow manure. After 30 days of drying, both oven and air-dried pellet reach to similar volume. In case of changes of weight, almost similar effects could be observed (Figure 9).

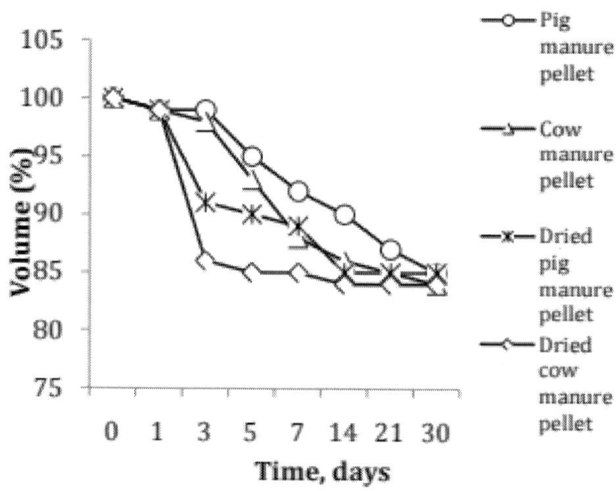

Figure8: Changes of volume of pelletized composts with time

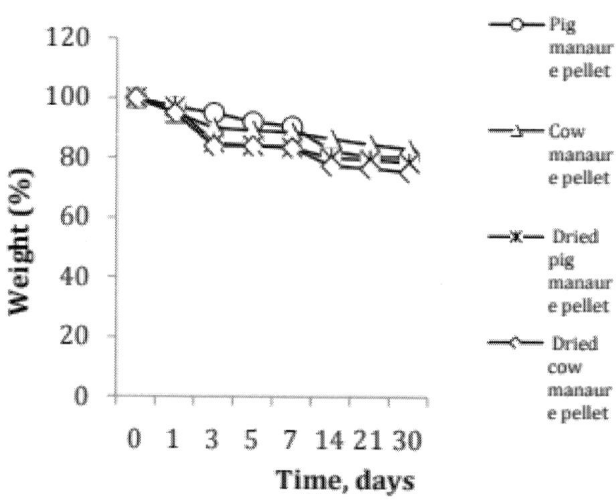

Figure 9: Changes of weight of pelletized composts with time

CONCLUSIONS

Pelletizing instrument in pilot scale was developed in this study. By developing this new system, it can directly manufacture wet compost into pelletized compost without pre-drying process of raw material. The pelletizing efficacy was affected by revolutions per minute of screw and the type of extruder. When the screw revolves at low speed, the yield of pellet was decreased. However, the solidity of pellet was improved under the low speed condition. The weight and volume of pellet were decreased through drying process, which may take it easy to transport and spread to land as fertilizer. Therefore, the quality of pelletized livestock manure compost could be rich organic compost.

REFERENCE

1. Abedin, Z. and M.H. Kianmehr. 2012. Livestock manure management and pelleting. *Agricultural Engineering International: CIGR Journal*. 14: 78-84.
2. Agnew, J.M., J.J. Leonard, J. Feddes, and Y. Feng. 2003. A modified air pycnometer for compost and air volume and density determination. *Canadian Biosystems Engineering*. 45: 627-635.
3. Ahn, H.K., J.H. Kim, J.H. Kwag, K.H. Jeong, D.Y. Choi, and Y.H. Yoo. 2011. Sustainable animal waste management and utilization in Korean livestock farms. Proceedings of International Seminar on Sustainable Resource Management of Livestock Wastes for Asian Small-scale Farmers, July 25-29, Ho Chi Minh City, Vietnam.
4. AOAC. 1990. Official Methods of Analysis, 15th ed., Association of Official Analytical Chemists, Washinton D.C. USA.
5. APHA. 2005. Standard Methods for the Examination of Water and Wastewater, 21st ed., American Public Health Association, New York, USA.
6. Benito, M., A. Masaguer, A. Moliner, N. Arrigo, and R.M. Palma. 2003. Chemical and microbiological parameters for the characterization of the stability and maturity of pruning waste compost. *Biology and Fertility of Soils*. 37: 184-189.
7. Bhattacharya, S.C., S. Sett, and R.M. Shrestha. 1989. State of the art of biomass densification, energy sources, division of energy technology. *Energy Sources*.11: 161-186.
8. Carr, L., R. Grover, B. Smith, T. Richard, and T. Halbach. 1995. *Commercial and on-farm production and marketing of animal waste compost products. In: Animal Waste and the Land Water Interface*, Steele, K. ed., Boca Raton: Lewis Publishers, pp. 485-492.
9. Farhadi, D. 2014. Evaluation of the physical and chemical properties of some agricultural wastes as poultry litter material. Global Journal of Animal Scientific Research. 2 (3): 270-276.
10. Hansen, R.C., K.M., Manel, H.M. Keener, H.A.J. Hoitink. 1995. The composting process-A natural way to recycle waste-OSUE, Bulletin 792. The Ohio State University, Columbus, OH.
11. Hara, M. 1998. Development of moulding technology of animal waste compost aim to distribution over distance. *Field and Soil*. 30: 40-46.
12. Hara, M. 1998. Trend of methods to produce pelletized compost and fertilization. *Agrotechnology*. 53 (10): 36-40. Jeong, K.H., M.A. Khan, H.K. Ahn, J. H. Kim, and Y.H. Yoo. 2012. Manufacturing and utilization of pelletized livestock manure compost in Korea. Proceedings of the ISWA World Solid Waste Congress, September 15-19, Florence, Italy.

13. Peisker, M. 1994. Influence of expansion on crude feed components. *Feed Mix.* 2: 26-31.
14. Tesfay, G. 2014. Dairy cattle production system in central zone of Tigray: in the case of Aksum and Adwa.

7

CHAPTER

ORGANIC FARMING IN THE NORDIC COUNTRIES – ANIMAL HEALTH AND PRODUCTION

S. M. Thamsborg

Keyword: Organic farming, animal health, animal welfare, food safety, diseases

INTRODUCTION

Organic farming (ecological agriculture or biological farming) is of growing importance in the agricultural sector of the Nordic countries. The reason for this being increasing production and market share of organic farm products. Organic farming is also to a large extent setting the scene in the public debate in areas like use of antibiotics in farm animals, alternative therapy, food safety and animal welfare and ethics. Another important aspect is the influence of organic standards on national legislation. As an example, several Danish organic standards preceded similar governmental legislation for the whole agricultural sector e.g. the ban on ruminant products in ruminant feed, animal housing and management according to physiological and ethological needs, and the ban on preventive use of antibiotics and anthelmintics.

As organic farming rely on product premiums in most countries, the market is very sensitive to consumers' perception of the production method and product quality. In a recent Danish survey of 1500 consumers' motives for buying organic products (phrased as an open question), own health, environmental considerations and animal welfare rated first, second and third (Suzanne Beckmann, upubl. data, 2000) thus confirming several earlier studies that organic products are perceived as healthier than conventional.

Citation: S. M. Thamsborg, Organic Farming in the Nordic Countries – Animal Health and Production, DOI: 10.1186/1751-0147-43-S1-S7.

The question of quality of animal food products and the major focus on animal welfare (and thus animal health) in organic farming make it relevant today to consider the state of the art and to discuss future veterinary challenges in this production system. The purpose of the present paper is to provide a framework for understanding of aims and basic concepts in organic farming, to give a brief overview of the present state, including selected problem areas relevant to veterinarians and to briefly speculate on future trends. Although organic farming has to be viewed as a whole (soil, plants and animals), the focus will be on livestock in organic farming.

DEFINITIONS

Aims in organic farming

Organic farming can be seen as an approach to agriculture where the aim is to create integrated, humane, environmentally and economically sustainable production systems [16]. It is a characteristic feature of organic farming that aims are formulated. These aims form background for the detailed standards or rules of practice. In 1989, organic movements in the Nordic countries agreed on the following rather broad description of organic farming:

Organic agriculture represents a self-supportive, well-balanced and sustainable agro-ecosystem. The system is based on the cyclical use of local and renewable resources as much as possible. Organic agriculture is based on a holistic view and includes the ecological, economical and social aspects of agricultural production in a local and global perspective. In organic agriculture, nature and the environment have their own value. Thus, humans have a moral responsibility to farm in a way that makes the cultural landscape an integrated and positive part of nature (translated from [13]).

It is interesting that animal welfare was not mentioned specifically in the definition although it became a hallmark in relation to livestock production. The principle aims are expressed more concisely in the standards document of International Federation of Organic Agriculture Movements (IFOAM), e.g.:

- to produce food of high quality in sufficient quantity
- to interact in a constructive and life-enhancing way with natural systems and cycles
- to encourage and enhance biological cycles within the farming system, involving microorganisms, soil flora and fauna, plants and animals
- to maintain and increase long-term fertility of soils
- to use, as far as possible, renewable resources in locally organised production systems
- to maintain the genetic diversity of the production system and its surroundings
- to create a harmonious balance between crop production and animal husbandry
- to give all livestock conditions of life with due consideration for the basic aspects of their innate behaviour [1]

Basic concepts

Three concepts fundamental to production in organic farming systems deserve further consideration, particularly in relation to livestock. Livestock plays an important role in *nutrient recycling* on the farm and contributes significantly towards the aim of operating as far as possible within a closed system. The closed cycle principle needs not necessarily apply at the individual farm but can apply at a regional level thus facilitating stockless farming. In organic farming, nitrogen self-sufficiency is secured through the use of legumes and biological nitrogen fixation as well as effective recycling of organic materials, including crop residues and livestock manure [15]. Maintaining and improving soil fertility through crop rotation, appropriate soil tillage, application of manure etc. are crucial to achieve this goal. Nitrogen is easily lost by leaching or volatilisation, and without careful management an organic system can contribute considerable amounts of nitrate to drainage water. A Danish study using a simple mathematical model for N-leaching and nutrient budgets concluded that organic farms on sandy soils with either plant production or dairy cattle had markedly less risk of nitrate leaching compared to similar conventional farms [6]. Conclusions regarding pig producing farms (sandy or clay soils) were less clear. Increased conversion to organic farming was advocated by the Bichel-commission in Denmark as one of the three best approaches to reduce the use of pesticides thereby improving environment [2]. A 100% conversion (non-voluntary) of Danish agricultural production was considered economically and politically unrealistic.

The concept of *prevention rather than treatment* is well known to most veterinarians. According to the present EU livestock regulation No. 1894/99, diseases in livestock in organic production are to be prevented by means of:

- selection of appropriate breeds and strains
- the application of animal husbandry practices appropriate to the requirements of each species, encouraging strong resistance to disease and the prevention of infections
- use of high quality feed, together with regular exercise and access to pasturage, having the effect of encouraging the natural immunological defence of the animal
- ensuring an appropriate density of stocking, thus avoiding overstocking and any resulting animal health problems [3].

This does not preclude that animals sick or injured must be treated immediately, if necessary in isolation and in suitable housing. In order to provide further incentives to prevention, the withdrawal periods following medical treatment are extended. According to the EU livestock regulation No. 1894/99, a new conversion period (typically 3–6 months) is required if an individual animal with a production life longer than a year receives more than 3 courses of medical treatments within a 12 month period [3]. In my opinion these rules are not based on concern of food safety but rather a wish to minimize use of medicine and to promote means of disease prevention. It is also important to note that disease prevention through medication, formerly often used in conventional farming, is not compatible with organic practices. This may, however, create problems in certain areas e.g. ectoparasite control in extensively managed organic sheep (cf. [10]). In the USA., it is currently being proposed by the USDA that any antibiotic treatment will result in the animal losing its organic status permanently.

The *precautionary principle* has recently been declared a key tenet of risk management in e.g. protection of human, animal and plant health in EU [5]. According to the commission, the principle covers those specific circumstances where scientific evidence is insufficient, inconclusive or uncertain and that there are indications through preliminary objective scientific evaluation that there are reasonable grounds for concerns for potentially dangerous effects on the environment, human, animal or plant health. In organic farming, a precautionary principle has for a long time formed part of the background for the ban on genetically modified organisms (GMOs), pesticides, certain types of embryotechnology, and use of hormonal regulation of production. In a wider perspective, organic farming as a whole can be seen as the precautionary principle applied to agricultural production.

STATUS IN THE NORDIC COUNTRIES

Historical background

A small number of biodynamic farms established in the Nordic countries in the 1930-ties, later inspired the pioneers of organic farming in the 1970-ties. Increasing concern over environmental issues in intensive agriculture, in particular problems with the use of pesticides and commercial fertilizers, was the major driving force. Contributory factors were an increasing market for products perceived as healthy and concern for animal welfare. The development has thus been driven largely by consumer demand and visionary farmers and later supported by financial premiums or direct governmental subsidies (Table 1). Apart from NØLL formed in 1971, organic farmers organizations were formed in the 80-ties. Only in recent years has organic farming been supported by the research establishment with increasing numbers of Ph. D. students in the 90-ties, and later by large, government funded research programmes in e.g. Sweden and Denmark in the mid 90-ties and professorial chairs at universities.

The philosophical background for organic farming has been explained as a blend of some kind of deep ecology, placing humans with ethical responsibilities towards the soil, the plants, the entire living nature, and a more utilitarian and anthropocentric approach where organic farming is seen as a means of solving manmade environmental problems [9]. The latter approach is, however, not anthropocentric in a strict sense as improved animal welfare is an important element. The utilitarian point of view is hold by government authorities, as expressed in most legislative rules, whereas the organic farming movements share a more deep ecology view point [9]. In the process of generating knowledge, the development can also be seen as a move from learning from farmers experiences to a more, traditional expert knowledge based system e.g. the outcome of government funded research programmes.

Table 1: An overview of the institutional development of organic farming in the Nordic countries (modified from [17])(Govt.: governmental)

	Denmark	Finland	Iceland	Norway	Sweden
Organic farmers organizations	1981 Landsforeningen økologisk jordbrug (LØJ)	1985 Luomuliitto	1993 VOR	1971 NØLL 1987 Økoprodusentarne	1985 Alternativodlarnas riksförbund
National standards for organic farming	1981 LØJ 1987 Govt.	1986 1994 Govt.	1995 Gvt.	1986 and 1987 1995 Govt. 'connected'	1985 KRAV 1995 Govt. 'connected'
Organic standards for animal husbandry	1981 LØJ 1987 Govt.	1988	1995 Govt.	1989 1995 Govt. 'connected'	1987
Govt. economic support	1987: conversion 1993	1990: conversion	1999	1990	1989

First dissertation:					
plant production	1985 (Ph.D.)	1998	1995 (B.Sc.)	1993	1981
animal product.	1995 (Ph.D.)				

Brief statistics

Relatively few farms were certified organic before early 80-ties. With the introduction of governmental subsidies in the late 80-ties/early 90-ties in most Nordic countries, more farms converted. An EU-regulation in 1992 provided a framework for all EU member states to implement policies to support organic farming. The scheme was implemented in Denmark in 1994 resulting in higher subsidies and an almost exponential increase in number of farms converting. The scheme was implemented in Finland and Sweden in 1995 when they joined EU and both countries are now having a large organic agricultural production (Table 2). Organic livestock production is also influenced by the market. As an example, the premiums for organic milk in Denmark have decreased recently, and in 2000 less than 10 dairy herds started conversion compared to about 200 in 1999.

Table 2: Number of organic farms and total area of organic production in Nordic countries at end of 1999 (source: SÖL 2000,http://www.organic-europe.net) and examples of typical subsidies in 1997 (source: [16]) In most countries permanent grassland is excluded from subsidies.

	Denmark	Finland	Iceland	Norway	Sweden**
No. organic farms	3009	5197	33	1745	3252
As % of all farms	4.0*	6.1	0.8	2.6*	3.8
Total organic area (ha)	146 686	137 000*	2500	18 773	155 463
As % of total area	5.5	6.3	0.6	1.8*	5.5
Government support (ECU/ha/year):					
cereals/arable crops	114	162–312	n. a.	187	104/185
grass/forage	60	264–414	n. a.	137	173/254

* estimate

** only KRAV-certified farms. These farms together with organic farming supported under Council Regulation (EU) 2078/92 cover almost 10% of arable land in Sweden.

Standards

National standards for organic animal husbandry have been in effect in most Nordic countries since mid 1980ties (Table 1) and will be covered by the next lecturer. The many yes/no requirements in the standards are easy to practice and is perhaps one of the reasons for the success of organic farming: a clear concept which is simple to communicate. However, there is a trend of standards becoming more and more detailed, particularly in relation to animal housing and management. This may be seen as an obstacle to the continuing development of organic farming. A most recent development is the EU-legislation, only relevant to Sweden, Finland and Denmark of course. The first EU-directive [3] specifying so called minimum standards for organic livestock production took in effect August this year (2000) and is presently being implemented in the national standards.

VETERINARY CHALLENGES

Veterinary consultancy

Veterinarians play an important role in developing organic farming through work in clinical practice, as farm consultants, and in research. In Denmark, the involvement of veterinary practitioners in support and decision making on organic farms has been limited during the 90-ties. With the growing number of farms, this situation is gradually changing and a dialogue is now been established with organic farmers/organisations regarding e.g. revision of standards, development of herd health programmes and possibilities for use of alternative medicine. Organic farmers expect practitioners and farm consultants to have some understanding of aims in organic farming and knowledge of standards, particularly changes in standards. Furthermore, the advice should be tailored to the individual farmer's needs, and the practitioner should be willing to take up new ideas. On the other hand, the veterinary practitioner should be critical, particular in relation to animal welfare. Organic farms represent a good opportunity to carry out preventive medicine.

Research tasks

An important task for veterinary researchers in organic farming is to provide documentation of production and base line data on health and disease. Much data has been presented at meetings, in internal reports or in the local farming press but the access for outside researchers has been limited. However, there is a growing awareness of this problem and in association with the large scale research programmes and education of researchers in this field, more studies will be published internationally. Examples of Nordic base line studies include a Swedish survey of lesions recorded at abattoirs in 1997 in organic and conventional animals [7], studies on mastitis and somatic cell counts in organic dairy cattle in Denmark (e.g. [21]) and a long list of published recordings from case-studies on organic farms (e.g. [20, 11, 12, 14]).

Important base line data to be obtained include:

- risk assessment in organic pig production regarding food safety, particularly zoonotic agent like Salmonella, Campylobacter, parasites etc.
- use of antibiotics in pig and cattle enterprises and implications for development of antibiotic resistance
- profiling of nutritional quality of organic products for human consumption

Detailed data on many of these aspects will be provided in later sessions. Improved recording systems and recording of production method (e.g. organic or out-door conventional) in central databases are highly warranted and will facilitate future base line studies.

It is important alongside documentation of health on-farm to initiate basic research relevant to organic farming. Innovative research performed in universities and other research institutions may offer support to the continuing development of organic farming. It is often reductionistic and runs a high risk of becoming detached from the principles of organic farming. It is thus important to link this research to experiences of organic practitioners at some stage of the process.

An example of basic research leading to recognition of new biological relationships relevant and perhaps very important to organic farming is the effect of diet constituents and structure on bacterial and nematode infections in pigs. Several Danish studies have indicated that commercial feed triple the risk of establishment of Salmonella infections in individual pigs compared to home-grown feed (cf. [4]). Home-grown feed is in contrast to commercial feed not heat-treated, have a coarser structure after grinding and is not pelletted. It improves the condition of the gastrointestinal tract (e.g. less lesions in the stomach) and results in different fermentation pattern in the large intestine, but the production economy is generally affected negatively [8]. In organic farms, pigs receive a high proportion of home produced feed, and roughage (fresh grass or whole grain silage ad libitum) should be available at all times in all age groups. Although the influence of roughage per se on Salmonella infections has not been investigated, the feeding regime on organic farms is likely to decrease the risk of salmonellosis and perhaps counteract the potentially higher risk associated with a continuous production system and access to the outdoor. In contrast, high levels of insoluble dietary fibres and a non-pelleted diet have resulted in higher establishment rates and increased fecundity of the intestinal nematode *Oesophagostomum dentatum* compared to pelleted diets or diets of similar protein and energy levels but rich in digestible carbohydrates [19, 18]. *Ascaris*-infections were not affected in a similar way. These findings implicate that not only the high exposition outdoors but also the feeding regime, may affect the establishment of helminth infections in organic pig production systems. Also interactions between pathogens may be affected.

In summary, these studies clearly show that it may be difficult to predict the outcome at farm level when new pratices are introduced, or old practices re-adopted, in the organic standards. Access to roughage and out-door conditions in pig production may thus have wide implications for the epidemiology of different infections, behaviour, environment, economy and perhaps food safety. Basic research may play a significant role in the building of knowledge and understanding of the systems.

Other tasks where basic research is needed include methodology in welfare assessment (particularly welfare implications of subclinical disease), climatic and behavioural effects of new and near-natural housing conditions, methods for scientifical assessment of the effects of alternative medication e.g. herbal remedies and homoeopathy, relationships between welfare, immunity and disease resistance and the use of more holistic methods to assess food quality.

Future trends

In the last decade organic farming has become institutionalized [9]. From being a small grass-root movement in opposition to the established agricultural food industry, organic farming is now part of the 'establishment' with large production units, retail from supermarkets, a considerable market share and not least, substantial government funded research in several Nordic countries. From being able to market products solely by being alternative to conventional farming with its perceived environmental and animal welfare problems, the burden of proof now lies with the organic producers. Although this represents an oversimplification, it is a fact in Denmark for instance, that critical questions are now being raised by consumers, politicians, researchers and others: is organic produce healthier for humans considering both nutritional qualities and zoonotic aspects? is the level of animal welfare acceptable on organic farms? is it better to buy national/local conventional products than imported, organic products with a high energy expenditure for transportation etc. Consumers obviously focus on single issues and do not realize/debate the benefits of organic farming in a larger perspective. However, the concern of consumers and the critique from others should be recognized as an important factor in future development. More information is needed from the organic farming organizations (and researchers) to influence and re-direct the discussion, and the need for development of production methods and scientific documentation of product quality and animal welfare are further emphasized.

DECLARATIONS

Acknowledgements

I wish to thank Erik S. Jensen and Torben W. Bennedsgaard for their comments on the manuscript and The Hede Nielsen Foundation for financial support.

REFERENCES

1. Anonymous: Basic standards for organic production and processing. IFOAM. 1998, 57-
2. Anonymous: Udvalget til vurdering af de samlede konsekvenser af en afvikling af pesticidanvendelsen (Bicheludvalget) – rapport fra hovedudvalget. 1999, 144-
3. Anonymous: EU Regulation no. 1804/1999: Production of organic livestock (supplementing EU Regulation 2092/91). 1999
4. Dahl J: Foderrelaterede risikofaktorer for subklinisk salmonellainfektion. Veterinærinformation. 1997, 6:
5. Gremmen B, Van den Belt H: The precautionary principle, or the unbearable lightness of the burden of proof. Proc 2nd Congress EurSafe, Copenhagen. Edited by: Robinson. 2000, 125-128.
6. Hansen B, Kristensen ES: N-udvaksning og -balancer ved omlægning fra konventionelt til økologisk jordbrug. Kvælstofudvaskning og -balancer i konventionelle og økologiske produktionssystesmer. Edited by: Kristensen, Olesen. 1998, FØJO-report
7. Hansson I, Hamilton C, Forslund K, Ekman T: En jämförelse av slaktresultat mellem KRAV-uppfödda och konventionellt uppfödda djur. Ekologisk djurproduktion enligt KRAV, Supplementum 29, Svensk Veterinär tidning. Edited by: Hammarberg, Beck-Friis. 1999, 51 (2): 17-24.
8. Jørgensen L, Dahl J, Jensen BB, Poulsen HD: Effekt af ekspandering, pelletering og formalingsgrad på Salmonella, produktionsresultater og mavetarmsundhed

hos slagtesvin samt på fytaseaktivitet og vitaminstabilitet i foder. Info-svin, Danske Slagterier, Rapport no 426. 1999, 24-

9. Kaltoft P: What is the ethical foundation of organic farming in a reflexive modern society. Proc 2nd Congress EurSafe, Copenhagen. Edited by: Robinson. 2000, 165-168.
10. Keatinge R, Gray D, Thamsborg SM, Martini A, Plate P: EU-regulation 1804/1999 – the implications of limiting allopathic treatment. Diversity of livestock systems and definition of animal welfare. Proc 2nd workshop of the Network of Animal Health and Welfare in Organic Agriculture (NAHWOA). 2000, 92-98.
11. Kristensen T: Studier i økologiske jordbrugssystemer. Beretning fra Statens Husdyrbrugsforsøg, no.734. 1997, 159-
12. Kristensen T: Studier i økologiske jordbrugssystemer. DJF-rapport nr 1, Husdyrbrug. 1998, 178-
13. Kølster P: Økologisk jordbrug – fælles nordisk beskrivelse. Ecological agriculture in the Nordic countries. Report of 1989 meeting of Nordic researchers and advisors in ecological agriculture and Nordic IFOAM. 1989, KVL, København
14. Mogensen L, Kristensen T, Kristensen IS: Økologisk mælkeproduktion. Teknisk-økonomisk gårdresultater 1997–98. Typetal for økologisk mælke-produktion. DJF-Rapport nr 10, Husdyrbrug. 1999, 138-
15. Lampkin N: Organic Farming. 1990, Farming Press, 715
16. Lampkin N, Foster C, Padel S, Midmore P: The policy and regulatory environment for organic farming in Europe. 1999, Hohenheim Univ, 1: 166-
17. Lund V: What is ecological animal husbandry?. Ecological animal husbandry in the Nordic countries. Proc NJF-seminar No 3030. Edited by: Hermansen, Lund, Thuen. 2000, 9-12.
18. Petkevicius S, Nansen P, Knudsen KEB, Skjøth F: The effect of increasing levels of insoluble dietary fibre on the establishment and persistence of Oesophagostomum dentatum in pigs. Parasite. 1998, 6: 17-26.
19. Petkevicius S, Bjørn H, Roepstorff A, Nansen P, Knudsen KEB, Barnes EH, Jensen K: The effect of two types of diet on populations of *Ascaris suum* and *Oesophagostomum dentatum* in experimentally infected pigs. Parasitology. 1995, 11: 395-402.
20. Strømm T, Olesen I: Mjølkekvalitet, helse og hold-barhet på kyrne ved omlegging til økologisk mjølkeproduktion. Rapport Norsøk. 1997, 77-
21. Vaarst M, Enevoldsen C: Patterns of clinical mastitis manifestations in Danish organic dairy herds. Journal of Dairy Science. 1997, 64: 23-37. 10.1017/S002202999600194X.

8

CHAPTER

MICROBIAL DIVERSITY OF VERMICOMPOST BACTERIA THAT EXHIBIT USEFUL AGRICULTURAL TRAITS AND WASTE MANAGEMENT POTENTIAL

Jayakumar Pathma and Natarajan Sakthivel

Keywords: Vermicompost, Earthworms, Beneficial bacteria, Organic waste management, Pathogen suppression, Plant-growth promotion, Biofertilizer

INTRODUCTION

Soil, is the soul of infinite life that promotes diverse microflora. Soil bacteria viz., *Bacillus, Pseudomonas* and *Streptomyces* etc., are prolific producers of secondary metabolites which act against numerous co-existing phytopathogeic fungi and human pathogenic bacteria (Pathma et al. 2011b). Earthworms are popularly known as the "farmer's friend" or "nature's plowman". Earthworm influences microbial community, physical and chemical properties of soil. They breakdown large soil particles and leaf litter and thereby increase the availability of organic matter for microbial degradation and transforms organic wastes into valuable vermicomposts by grinding and digesting them with the help of aerobic and anaerobic microbes (Maboeta and Van Rensburg 2003). Earthworms activity is found to enhance the beneficial microflora and suppress harmful pathogenic microbes. Soil wormcasts are rich source of micro and macronutrients, and microbial enzymes (Lavelle and Martin 1992). Vermicomposting is an efficient nutrient recycling process that involves harnessing earthworms as versatile natural bioreactors for organic matter decomposition. Due to richness in nutrient availability and microbial activity vermicomposts increase soil fertility, enhance plant

Citation: Jayakumar Pathma and Natarajan Sakthivel, Microbial diversity of vermicompost bacteria that exhibit useful agricultural traits and waste management potential, DOI: 10.1186/2193-1801-1-26.

growth and suppress the population of plant pathogens and pests. This review paper describes the bacterial biodiversity and nutrient status of vermicomposts and their importance in agriculture and waste management.

EARTHWORMS

Earthworms are capable of transforming garbage into 'gold'. Charles Darwin described earthworms as the 'unheralded soldiers of mankind', and Aristotle called them as the 'intestine of earth', as they could digest a wide variety of organic materials (Darwin and Seward 1903; Martin 1976). Soil volume, microflora and fauna influenced by earthworms have been termed as "drilosphere" and the soil volume includes the external structures produced by earthworms such as surface and below ground casts, burrows, middens, diapause chambers as well as the earthworms body surface and internal gut associated structures in contact with the soil (Lavelle et al. 1989; Brown et al. 2000). Earthworms play an essential role in carbon turnover, soil formation, participates in cellulose degradation and humus accumulation. Earthworm activity profoundly affects the physical, chemical and biological properties of soil. Earthworms are voracious feeders of organic wastes and they utilize only a small portion of these wastes for their growth and excrete a large proportion of wastes consumed in a half digested form (Edwards and Lofty 1977; Kale and Bano 1986; Jambhekar 1992). Earthworms intestine contains a wide range of microorganisms, enzymes and hormones which aid in rapid decomposition of half-digested material transforming them into vermicompost in a short time (neary 4–8 weeks) (Ghosh et al. 1999; Nagavallemma et al. 2004) compared to traditional composting process which takes the advantage of microbes alone and thereby requires a prolonged period (nearly 20 weeks) for compost production (Bernal et al. 1998; Sánchez-Monedero et al. 2001). As the organic matter passes through the gizzard of the earthworm it is grounded into a fine powder after which the digestive enzymes, microorganisms and other fermenting substances act on them further aiding their breakdown within the gut, and finally passes out in the form of "casts" which are later acted upon by earthworm gut associated microbes converting them into mature product, the "vermicomposts" (Dominguez and Edwards 2004).

Earthworms, grouped under phylum annelida are long, narrow, cylindrical, bilaterally symmetrical, segmented soil dwelling invertebrates with a glistening dark brown body covered with delicate cuticle. They are hermaphrodites and weigh over 1,400–1,500 mg after 8–10 weeks. Their body contains 65% protein (70–80% high quality 'lysine rich protein' on a dry weight basis), 14% fats, 14% carbohydrates, and 3% ash. Their life span varies between 3–7 years depending upon the species and ecological situation. The gut of earthworm is a straight tube starting from mouth followed by a muscular pharynx, oesophagus, thin walled crop, muscular gizzard, foregut, midgut, hindgut, associated digestive glands, and ending with anus. The gut consisted of mucus containing protein and polysaccharides, organic and mineral matter, amino acids and microbial symbionts viz., bacteria, protozoa and microfungi. The increased organic carbon, total organic carbon and nitrogen and moisture content in the earthworm gut provide an optimal environment for the activation of dormant microbes and germination of endospores etc. A wide array of digestive enzymes such as amylase, cellulase, protease, lipase, chitinase and urease were reported from earthworm's alimentary canal. The gut microbes were found to be responsible for the cellulase and mannose activities (Munnoli et al. 2010). Earthworms comminutes the substrate, thereby increases the surface area for microbial degradation constituting to the active phase of vermicomposting. As this crushed organic matter passes through the gut it get mixed up with the gut associated microbes and the digestive enzymes and finally leaves the gut in partially digested form as "casts" after which the microbes takes up the process of decomposition contributing to the maturation phase (Lazcano et al. 2008).

Association of earthworms with microbes is found to be complex. Certain groups of microbes were found to be a part of earthworm's diet which is evidenced by the destruction of certain microbes as they pass through the earthworms digestive system. Few yeasts, protozoa and certain groups of fungi such as *Fusarium oxysporum, Alternaria solani*, and microfungi were digested by the earthworms, *Drawida calebi, Lumbricus terrestris* and *Eisenia foetida*. *Bacillus cereus* var *mycoides* were reported to decrease during gut passage while *Escherichia coli* and*Serratia marcessens* were completely eliminated during passage through earthworm gut (Edwards and Fletcher,1988).

Earthworms are classified into epigeic, anecic and endogeic species based on definite ecological and trophic functions (Brown 1995; Bhatnagar and Palta 1996) (Table 1). Epigeic earthworms are smaller in size, with uniformly pigmented body, short life cycle, high reproduction rate and regeneration. They dwell in superficial soil surface within litters, feeds on the surface litter and mineralize them. They are phytophagous and rarely ingest soil. They contain an active gizzard which aids in rapid conversion of organic matter into vermicomposts. In addition epigeic earthworms are efficient bio-degraders and nutrient releasers, tolerant to disturbances, aids in litter comminution and early decomposition and hence can be efficiently used for vermicomposting. Epigeic earthworms includes*Eisenia foetida, Lumbricus rubellus, L. castaneus, L. festivus, Eiseniella tetraedra, Bimastus minusculus, B. eiseni, Dendrodrilus rubidus, Dendrobaena veneta, D. octaedra.* Endogeics earthworms are small to large sized worms, with weakly pigmented body, life cycle of medium duration, moderately tolerant to disturbance, forms extensive horizontal burrows and they are geophagous feeding on particulate organic matter and soil. They bring about pronounced changes in soil physical structure and can efficiently utilize energy from poor soils, hence can be used for soil improvements. Endogeics include *Aporrectodea caliginosa, A. trapezoides, A. rosea, Millsonia anomala, Octolasion cyaneum, O. lacteum, Pontoscolex corethrurus, Allolobophora chlorotica* and *Aminthas* sp. They are further classified into polyhumic endogeic which are small sized, rich soil feeding earthworms, dwelling in top soil (A1); mesohumic endogeic which are medium sized worms, dwelling in A and B horizon, feeding on bulk (A1) soil; and oligohumic endogeic which are very large worms, dwelling in B and C horizons, feeding on poor, deep soil. Aneceics are larger, dorsally pigmented worms, with low reproductive rate, sensitive to disturbance, nocturnal, phytogeophagous, bury the surface litter, forms middens and extensive, deep, permanent vertical burrows, and live in them. Formation of vertical burrows affects airwater relationship and movement from deep layers to surface helps in efficient mixing of nutrients. *Lumbricus terrestris, L. polyphemus* and *Aporrectodea longa* are examples of aneceics earthworms (Kooch and Jalilvand 2008). Epigeics and aneceics are harnessed largely for vermicomposting (Asha et al. 2008). Epigeics namely *Eisenia foetida* (Hartenstein et al. 1979), *Eudrilus eugeniae*(Kale and Bano 1988), *Perionyx excavatus* (Sinha et al. 2002; Suthar and Singh 2008) and *Eisenia anderi* (Munnoli et al. 2010) have been used in converting organic wastes into vermicompost. These surface dwellers capable of working on litter layers converting them into manure are of no significant value in modifying the soil structure. In contrast, anecics such as *Lampito mauritii* are efficient creators of an effective drilosphere as well as excellent compost producers (Ismail 1997). Earthworms thus act as natural bio-reactors, altering the nature of the organic waste by fragmenting them.

Table 1: Ecological categories and niches of earthworms and their characteristic features and beneficial traits

Species	Ecological category	Ecological niche	Characteristic features	Beneficial trait
Eisenia foetida,	Epigeics	Superficial soil layers, leaf litter, compost	Smaller in size, body uniformly pigmented, active gizzard, short life cycle, high reproduction rate and regeneration, tolerant to disturbance, phytophagous	Efficient bio-degraders and nutrient releasers, efficient compost producers, aids in litter comminution and early decomposition
Lumbricus rubellus,				
L. castaneus, *L. festivus,*				
Eiseniella tetraedra,				
Bimastus minusculus, *B. eiseni,*				
Dendrodrilus rubidus,				
Dendrobaena veneta, *D. octaedra*				
Aporrectodea caliginosa,	Endogeics	Topsoil or subsoil	Small to large sized worms, weakly pigmented, life cycle of medium duration, moderately tolerant to disturbance, geophagous	Brings about pronounced changes in soil physical structure, can efficiently utilize energy from poor soils hence can be used for soil improvements
A. trapezoides,				
A. rosea, *Millsonia anomala,*				
Octolasion cyaneum,	Polyhumic endogeic	Top soil (A_1)	Small size, unpigmented, forms horizontal burrows, rich soil feeder	
O. lacteum, *Pontoscolex corethrurus,*	Mesohumic endogeic	A and B horizon	Medium size, unpigmented, forms extensive horizontal burrows, bulk (A_1) soil feeder	
Allolobophora chlorotica, *Aminthas sp*	Oligohumic endogeic	B and C horizon	Very large in size, unpigmented, forms extensive horizontal burrows, feeds on poor, deep soils	
L. terrestris	Anecics	Permanent deep burrows in soil	Large in size, dorsally pigmented, forms extensive, deep, vertical permanent burrows, low reproductive rate, sensitive to disturbance, phytogeophagous, noctumal	Forms vertical burrows affecting air-water relationship and movement from deep layers to surface helps in efficient mixing of nutrients
L. polyphemus,				
A. longa				

Earthworm activity engineers the soil by forming extensive burrows which loosen the soil and makes it porous. These pores improve aeration, water absorption, drainage and easy root penetration. Soil aggregates formed by earthworms and associated microbes, in the casts and burrow walls play an indispensible role in soil air ecosystem. These aggregates are mineral granules bonded in a way to resist erosion and to avoid soil compaction both in wet and dry condition. Earthworms speed up soil reclamation and make them productive by restoring beneficial microflora (Nakamura 1996). Thus degraded unproductive soils and land degraded by mining could be engineered physically, chemically and biologically and made productive by earthworms. Hence earthworms are termed as ecosystem engineers (Brown et al. 2000; Munnoli et al. 2010).

Vermicomposting

Vermicomposting is a non-thermophilic biological oxidation process in which organic material are converted into vermicompost which is a peat like material, exhibiting high porosity, aeration, drainage, water holding capacity and rich microbial activities (Edwards 1998; Atiyeh et al. 2000b; Arancon et al. 2004a), through the interactions between earthworms and associated microbes. Vermiculture is a cost-effective tool for environmentally sound waste management (Banu et al. 2001; Asha et al. 2008). Earthworms are the crucial drivers of the process, as they aerate, condition and fragment the substrate and thereby drastically alter the microbial activity and their biodegradation potential (Fracchia et al. 2006; Lazcano et al. 2008). Several enzymes, intestinal mucus and antibiotics in earthworm's intestinal tract play an important role in the breakdown of organic macromolecules. Biodegradable organic wastes such as crop residues, municipal, hospital and industrial wastes pose major problems in disposal and treatment. Release of unprocessed animal manures into agricultural fields contaminates ground water causing public health risk. Vermicomposting is the best alternative to conventional composting and differs from it in several ways (Gandhi et al. 1997). Vermicomposting hastens the decomposition process by 2–5 times, thereby quickens the conversion of wastes into valuable biofertilizer and produces much more homogenous materials compared to thermophilic composting (Bhatnagar and Palta 1996; Atiyeh et al. 2000a). Distinct differences exist between the microbial communities found in vermicomposts and composts and hence the nature of the microbial processes is quite different in vermicomposting and composting (Subler et al. 1998). The active phase of composting is the thermophilic stage characterized by thermophilic bacterial community where intensive decomposition takes place followed by a mesophilic maturation phase (Lazcano et al. 2008; Vivas et al. 2009). Vermicomposting is a mesophilic process characterized by mesophilic bacteria and fungi (Benitez et al. 1999). Vermicomposting comprises of an active stage during which earthworms and associated microbes jointly process the substrate and the maturation phase that involves the action of associated microbes and occurs once the worm's moves to the fresher layers of undigested waste or when the product is removed from the vermireacter. The duration of the active phase depends on the species and density of the earthworms involved (Ndegwa et al. 2000; Lazcano et al. 2008; Aira et al. 2011). A wide range of oganic wastes viz., horticultural residues from processed potatoes (Edwards1988); mushroom wastes (Edwards 1988; Tajbakhsh et al. 2008); horse wastes (Hartenstein et al. 1979; Edwards et al. 1998); pig wastes (Chan and Griffiths 1988; Reeh 1992); brewery wastes (Butt 1993); sericulture wastes (Gunathilagraj and Ravignanam 1996); municipal sewage sludge (Mitchell et al. 1980; Dominguez et al. 2000); agricultural residues (Bansal and Kapoor 2000); weeds (Gajalakshmi et al. 2001); cattle dung (Gunadi et al. 2002); industrial refuse such as paper wastes (Butt 1993; Elvira et al. 1995; Gajalakshmi et al. 2002); sludge from paper mills and dairy plants (Elvira et al. 1997; Banu et al. 2001); domestic kitchen wastes (Sinha et al. 2002); urban residues and animal wastes (Edwards et al. 1985; Edwards 1988) can be vermicomposted (Sharma et al. 2005).

Effects of vermicomposting on pH, electrical conductivity (EC), C:N ratio and other nutrients have been documented. Earthworm activity reduced pH and C:N ratio in manure (Gandhi et al. 1997; Atiyeh et al. 2000b). Chemical analysis showed vermicompost had a lower pH, EC, organic carbon (OC) (Nardi et al. 1983; Albanell et al. 1988; Mitchell1997), C:N ratio (Riffaldi and Levi-Minzi 1983; Albanell et al. 1988), nitrogen and potassium and higher amounts of total phosphorous and micronutrients compared to the parent material (Hashemimajd et al. 2004). Slightly decreased pH values of vermicompost compared to traditional compost might be attributed due to mineralization of N and P, microbial decomposition of organic materials into intermediate organic acids, fulvic acids, humic acids (Lazcano et al. 2008; Albanell et al. 1988; Chan and Griffiths 1988; Subler et al. 1998) and concomitant production of CO_2 (Elvira et al. 1998; Garg et al. 2006). Vermicomposting of paper mill and dairy sludge resulted in 1.2–1.7 fold loss of organic carbon as CO_2 (Elvira et al. 1998). In contrast to the parent material used, vermicomposts contain higher humic acid substances (Albanell et al. 1988). Humic acid substances occur naturally in mature animal manure, sewage sludge or paper-mill sludge, but vermicomposting drastically increases the rate of production and their amount from 40–60 percent compared to traditional composting. The enhancement in humification processes is by fragmentation and size reduction of organic matter, increased microbial activity within earthworm intestine and soil aeration by earthworm feeding and movement (Dominguez and Edwards, 2004). EC indicates the salinity of the organic amendment. Minor production of soluble metabolites such as ammonium and precipitation of dissolved salts during vermicomposting lead to lower EC values. Compared to the parent material used, vermicomposts contain less soluble salts and greater cation exchange capacity (Holtzclaw and Sposito 1979; Albanell et al. 1988). C:N ratio is an indicator of the degree of decomposition. During the process of biooxidation, CO_2 and N is lost and loss of N takes place at a comparatively lower rate. Comparison of compost and vermicompost showed that vermicompost had significantly less C:N ratios as they underwent intense decomposition (Lazcano et al. 2008).

Vermicomposting of cow manure using earthworm species *E. andrei* (Atiyeh et al. 2000b) and *E. foetida* (Hand et al.1988) favored nitrification, resulting in the rapid conversion of ammonium-nitrogen to nitrate-nitrogen. Vermicomposting increased the concentration of nitrate-nitrogen to 28 fold after 17 weeks, while in conventional compost there was only 3-fold increase (Subler et al. 1998; Atiyeh et al. 2000a). Increase in ash concentration during vermicomposting suggests that vermicomposting accelerates the rate of mineralization (Albanell et al. 1988). Mineralization is the process in which the chemical compounds in the organic matter decompose or oxidise into forms that could be easily assimilated by the plants. Increase in ash content increases the rate of mineralization. Ash is an alkaline substance which hinders the formation of H_2S as well as improves the availibility of O_2 and thereby renders composts odorless. Thus vermicomposting increases the ash content and accelerates the rate of mineralization which is essential to make nutrients available to plants. The observed increase of total phosphorous (TP) in vermicompost is probably due to mineralization and mobilization of phosphorus resulting from the enhanced phosphatase activity by microorganisms in the gut epithelium of the earthworms (Zhang et al. 2000; Garg et al.2006). Vermicomposts showed a significant increase in exchangeable Ca^{2+}, Mg^{2+} and K^{+} compared to fresh sludge indicating the conversion of nutrients to plant-available forms during passage through the earthworm gut (Garg et al.2006; Yasir et al. 2009a). Vermicomposts contain higher nutrient concentrations, but less likely to produce salinity, than composts. Additionally, vermicomposts possess outstanding biological properties and have microbial populations significantly larger and more diverse compared to conventional composts (Edwards 1998). Soil supplemented with vermicompost showed better plant growth compared to soil treated with inorganic fertilizers or cattle manure (Kalembasa 1996; Subler et al. 1998).

Diversity of bacteria associated with earthworms

Earthworm's ability to increase plant nutrient availability is likely to be dependent on the activity of earthworm gut microflora. Earthworms indirectly influence the dynamics of soil chemical processes, by comminuting the litter and affecting the activity of the soil micro-flora (Petersen and Luxton 1982; Lee 1985; Edwards and Bohlen 1996). Interactions between earthworms and microorganisms seem to be complex. Earthworms ingest plant growth-promoting rhizospheric bacteria such as *Pseudomonas, Rhizobium, Bacillus, Azosprillium, Azotobacter,* etc. along with rhizospheric soil, and they might get activated or increased due to the ideal micro-environment of the gut. Therefore earthworm activity increases the population of plant growth-promoting rhizobacteria (PGPR) (Sinha et al.2010). This specific group of bacteria stimulates plant growth directly by solubilization of nutrients (Ayyadurai et al.2007; Ravindra et al. 2008), production of growth hormone, 1-aminocyclopropane 1 carboxylate (ACC) deaminase (Correa et al. 2004), nitrogen fixation (Han et al. 2005), and indirectly by suppressing fungal pathogens. Antibiotics, fluorescent pigments, siderophores and fungal cell-wall degrading enzymes namely chitinases and glucanases (Han et al. 2005; Sunish et al. 2005; Ravindra et al. 2008; Jha et al. 2009; Pathma et al. 2010; Pathma et al. 2011a, b) produced by bacteria mediate the fungal growth-suppression. Earthworms are reported to have association with such free living soil bacteria and constitute the drilosphere (Ismail 1995). Earthworm microbes mineralize the organic matter and also facilitate the chelation of metal ions (Pizl and Novokova 1993; Canellas et al. 2002). Gut of earthworms *L. terrestris, Allolobophora caliginosa* and *Allolobophora terrestris* were reported to contain higher number of aerobes compared to soil (Parle 1963). Earthworms increased the number of microorganisms in soil as much as five times (Edwards and Lofty 1977) and the number of bacteria and 'actinomycetes' contained in the ingested material increased upto 1,000 fold while passing through their gut (Edwards and Fletcher 1988). Similar increase was observed in plate counts of total bacteria, proteolytic bacteria and actinomycetes by passage through earthworms gut (Parle 1963;

Daniel and Anderson 1992; Pedersen and Hendriksen 1993; Devliegher and Verstraete 1995). Similarly microbial biomass either decreased (Bohlen and Edwards 1995; Devliegher and Verstraete 1995), or increased (Scheu 1992) or remained unchanged (Daniel and Anderson 1992) after passage through the earthworm gut. An oxalate-degrading bacterium *Pseudomonas oxalaticus* was isolated from intestine of*Pheretima* species (Khambata and Bhat 1953) and an actinomycete *Streptomyces lipmanii* was identified in the gut of *Eisenia lucens* (Contreras 1980). Scanning electron micrographs provided evidence for endogenous microflora in guts of earthworms, *L. terrestris* and *Octolasion cyaneum* (Jolly et al. 1993). Gut of *E. foetida* contained various anaerobic N_2-fixing bacteria such as *Clostridium butyricum, C. beijerinckii* and *C. paraputrificum* (Citernesi et al.1977). Alimentary canal of *Lumbricus rubellus* and *Octolasium lacteum* were found to contain more numbers of aerobes and anerobes (Karsten and Drake 1995) and culturable denitrifiers (Karsten and Drake 1997). List of vermicompost bacteria and their beneficial traits is presented in Table 2.

Table 2: Biodiversity of vermicompost bacteria and their beneficial traits

Vermicompost earthworm	Names of bacteria	Beneficial traits	References
Pheretima sp.	*Pseudomonas oxalaticus*	Oxalate degradation	Khambata and Bhat, 1953
Unspecified	*Rhizobium trifolii*	Nitrogen fixation and growth of leguminous plants	Buckalew et al. 1982
Lumbricus rubellus	*R. japonicum, P. putida*	Plant growth promotion	Madsen and Alexander 1982
L. terrestris	*Bradyrhizobium japonicum*	Improved distribution of nodules on soybean roots	Rouelle, 1983
Aporrectodea trapezoids, *A. rosea*	*P. corrugata* 2 14OR	Suppress *Gaeumannomyces graminis* var. *Tritd* in wheat	Doube et al. 1994
A. trapezoids, *Microscolex dubius*	*R. meliloti* L5-30R	Increased root nodulation and nitrogen fixation in legumes	Stephens et al. 1994b
Eisenia foetida	*Bacillus* spp., *B. megaterium,*	Antimicrobial activity against *Enterococcus faecalis* DSM 2570, *Staphylococcus aureus* DSM 1104	Vaz-Moreira et al. 2008
	B. pumilus, B. subtilis		

L. terrestris	Fluorescent pseudomonads,	Suppress *Fusarium oxysporum* f. sp. *asparagi* and *F. proliferatum* in asparagus, *Verticillium dahlia* in eggplant and *F. oxysporum* f. sp. *lycopersici* Race 1 in tomato	Elmer, 2009
Eudrilus sp.	Filamentous actinomycetes Free-living N_2 fixers, *Azospirillum*, *Azotobacter*, Autotrophic *Nitrosomonas*, *Nitrobacter*, Ammonifying bacteria, Phosphate solubilizers,	Plant growth promotion by nitrification, phosphate solubilisation and plant disease suppression	Gopal et al. 2009
	Fluorescent pseudomonads		
E. foetida	Proteobacteria, Bacteroidetes,	Antifungal activity against *Colletotrichum coccodes, R. solani, P. ultimum, P. capsici* and *F. moliniforme*	Yasir et al. 2009a
	Verrucomicrobia, Actinobacteria,		
	Firmicutes		
Unspecified	*Eisenicola composti* YC06271^T	Antagonistic activity against *F. moniliforme*	Yasir et al. 2009b

Earthworms harbor 'nitrogen-fixing' and 'decomposer microbes' in their gut and excrete them along with nutrients in their excreta (Singleton et al. 2003). Earthworms stimulate and accelerate microbial activities by increasing the population of soil microorganisms (Binet et al. 1998), microbial numbers and biomass (Edwards and Bohlen 1996), by improving aeration through burrowing actions. Vermicomposting modified the original microbial community of the waste in a diverse way. Actinobacteria and Gammaproteobacteria were abundant in vermicompost, while conventional compost contained more Alphaproteobacteria and Bacteriodetes, the bacterial phylogenetic groups typical of non-cured compost (Vivas et al. 2009). Total bacterial counts exceeded 10^{-10}/ g of vermicompost and it included nitrobacter, azotobacter, rhizobium, phosphate solubilizers and actinomycetes (Suhane 2007). Molecular and culture-dependent analyses of bacterial community of vermicompost showed the presence of α-Proteobacteria, β-Proteobacteria, γ-Proteobacteria, Actinobacteria, Planctomycetes, Firmicutes and Bacteroidetes (Yasir et al.2009a). Several findings showed considerable increase in total viable counts of actinomycetes and bacteria in the worm treated compost (Parthasarathi and Ranganathan 1998; Haritha Devi et al. 2009). The increase of microbial population may be due to the congenial condition for the growth of microbes within the digestive tract of earthworm and by the ingestion of nutrient rich organic wastes which provide energy and also act as a substrate for the growth of microorganisms (Tiwari et al. 1989). The differences in microbial species, numbers and activity between the earthworm alimentary canal or burrow and bulk soil indirectly support the hypothesis that the bacterial community structures of these habitats are different from those of the soil. Specific phylogenetic groups of bacteria such as*Aeromonas hydrophila* in *E. foetida* (Toyota and Kimura 2000), fluorescent pseudomonads in *L. terrestris*(Devliegher and Verstraete 1997), and *Actinobacteria* in *L. rubellus* (Kristufek et al. 1993) have been found in higher numbers in earthworm guts, casts, or burrows.

Enzymatic activity characterization and quantification has a direct correlation with type and population of microbes and reflects the dynamics of the composting process in terms of the decomposition of organic matter and nitrogen transformations and provide information about the maturity of the compost (Tiquia 2005). Wormcasts contain higher activities of cellulase, amylase, invertase, protease, peroxidase, urease, phosphatase and dehydrogenase (Sharpley and Syers 1976; Edwards and Bohlen 1996). Dehydrogenase is an intracellular enzyme related to the oxidative phosphorylation process (Trevors 1984) and is an indicator of microbial activity in soil and other biological ecosystems (Garcia et al. 1997). The maximum enzyme activities (cellulase, amylase, invertase, protease and urease) were observed during 21–35 days in vermicomposting and on 42–49 days in conventional composting. Also, microbial numbers and their extracellular enzyme profiles were more abundant in vermicompost produced from fruitpulp, vegetable waste, groundnut husk and cowdung compared to the normal compost of the same parental origin (Haritha Devi et al. 2009). *Pseudomonas, Paenibacillus, Azoarcus, Burkholderia, Spiroplasm, Acaligenes,* and *Acidobacterium*, the potential degraders of several categories of organics are seen associated with the earthworm's intestine and vermicasts (Singleton et al. 2003). Firmicutes viz., *Bacillus benzoevorans, B. cereus,B. licheniformis, B. megaterium, B. pumilus, B. subtilis, B. macroides*; Actinobacteria namely *Cellulosimicrobium cellulans, Microbacterium* spp., *M. oxydans*; Proteobacteria such as *Pseudomonas* spp., *P. libaniensis*; ungrouped genotypes *Sphingomonas* sp., *Kocuria palustris* and yeasts namely *Geotrichum* spp. and *Williopsis californica* were reported from vermicomposts (Vaz-Moreira et al. 2008). Pinel et al. (2008) reported the presence of a novel nephridial symbiont, *Verminephrobacter eiseniae* from *E. foetida. Ochrobactrum* sp., *Massilia* sp., *Leifsonia* sp. and bacteria belonging to families Aeromonadaceae, Comamonadaceae, Enterobacteriaceae, Flavobacteriaceae, Moraxellaceae, Pseudomonadaceae, Sphingobacteriaceae, Actinobacteria and Microbacteriaceae were reported to occur in earthworms alimentary canal (Byzov et al. 2009). The microbial flora of earthworm gut and cast are potentially active and can digest a wide range of organic materials and polysaccharides including cellulose, sugars, chitin, lignin, starch and polylactic acids Zhang et al. (2000; Aira et al. 2007; Vivas et al. 2009). Single-strand conformation polymorphism (SSCP) profiles on the diversity of eight bacterial groups viz., Alphaproteobacteria, Betaproteobacteria, Bacteroidetes, Gammaproteobacteria, Deltaproteobacteria, Verrucomicrobia, Planctomycetes, and Firmicutes from fresh soil, gut, and casts of the earthworms *L. terrestris* and *Aporrectodea caliginosa* showed the presence of Bacteroidetes, Alphaproteobacteria, Betaproteobacteria and representatives of classes Flavobacteria, Sphingobacteria (Bacteroidetes) and *Pseudomonas* spp. in the worm casts in addition to unclassified Sphingomonadaceae (Alphaproteobacteria) and *Alcaligenes* spp. (Betaproteobacteria) (Nechitaylo et al. 2010).

ROLE OF VERMICOMPOST IN SOIL FERTILITY

Vermicomposts can significantly influence the growth and productivity of plants (Kale et al. 1992; Kalembasa 1996; Edwards 1988; Sinha et al. 2009) due to their micro and macro elements, vitamins, enzymes and hormones (Makulec 2002). Vermicomposts contain nutrients such as nitrates, exchangeable phosphorus, soluble potassium, calcium, and magnesium in plant available forms (Orozco et al. 1996; Edwards 1998) and have large particular surface area that provides many microsites for microbial activity and for the strong retention of nutrients (Shi-wei and Fu-zhen 1991). Uptake of nitrogen (N), phosphorus (P), potassium (K) and magnesium (Mg) by rice (*Oryza sativa*) plant was highest when fertilizer was applied in combination with vermicompost (Jadhav et al. 1997). N uptake by ridge gourd (*Luffa acutangula*) was higher when the fertilizer mix contained 50% vermicompost (Sreenivas et al. 2000). Apart from providing mineralogical nutrients, vermicomposts also contribute to the biological fertility by adding beneficial microbes to soil. Mucus, excreted through the earthworm`s digestive canal, stimulates antagonism and competition between diverse microbial populations resulting in the production of some antibiotics and hormone-like biochemicals, boosting plant growth (Edwards and Bohlen 1996). In addition, mucus accelerates and enhances decomposition of organic matter composing stabilized humic substances which embody water-soluble phytohormonal elements (Edwards and Arancon 2004) and plant-available nutrients at high levels (Atiyeh et al. 2000c). Adding vermicasts to soil improves soil structure, fertility, plant growth and suppresses diseases caused by soil-borne plant pathogens, increasing crop yield (Chaoui et al. 2002; Scheuerell et al. 2005; Singh et al. 2008). Kale (1995) reported the nutrient status of vermicomposts with organic carbon 9.15-17.98%, total nitrogen 0.5-1.5%, available phosphorus 0.1-0.3%, available potassium 0.15%, calcium and magnesium 22.70-70 mg/100 g, copper 2–9.3 ppm, zinc 5.7-11.5 ppm and available sulphur 128–548 ppm.

Effects of a variety of vermicomposts on a wide array of field crops (Chan and Griffiths 1988; Arancon et al. 2004b), vegetable plants (Edwards and Burrows 1988; Wilson and Carlile 1989;Subler et al. 1998; Atiyeh et al. 2000b), ornamental and flowering plants (Edwards and Burrows 1988; Atiyeh et al. 2000c) under greenhouse and field conditions have been documented. Vermicomposts are used as alternative potting media due to their low-cost, excellent nutrient status and physiochemical characters. Considerable improvements in plant growth recorded after amending soils with vermicomposts have been attributed to the physico-chemical and biological properties of vermicomposts.

Vermicompost addition favorably affects soil pH, microbial population and soil enzyme activities (Maheswarappa et al. 1999) and also reduces the proportion of water-soluble chemical, which cause possible environmental contamination (Mitchell and Edwards 1997). Vermicompost addition increases the macropore space ranging from 50–500 μm, resulting in improved air-water relationship in the soil, favourably affecting plant growth (Marinari et al.2000). Evaluation of various organic and inorganic amendments on growth of raspberry proves that vermicompost has beneficial buffering capability and ameliorate the damage caused by excess of nutrients which may otherwise cause phytotoxicity (Subler et al. 1998). Thus, vermicompost acts a soil conditioner (Albanell et al. 1988) and a slow-release fertilizer (Atiyeh et al. 2000a). During vermicomposting the heavy metals forms complex, aggregates with humic acids and other polymerized organic fractions resulting in lower availability of heavy metals to the plant, which are otherwise phytotoxic (Dominguez and Edwards 2004). Soil amended with vermicompost produced better quality fruits and vegetables with less content of heavy metals or nitrate, than soil fertilized with mineral fertilizers (Kolodziej and Kostecka 1994).

ROLE OF VERMICOMPOST BACTERIA IN BIOMEDICAL WASTE MANAGEMENT

The importance of sewage sludge, biosolids and biomedical waste management by safe, cheap and easy methods need no further emphasis. All these wastes are infectious and have to be disinfected before being disposed into the environment. Biosolids also contain an array of pathogenic microorganisms (Hassen et al. 2001). Biocomposting of wastes bring about biological transformation and stabilization of organic matter and effectively reduces potential risks of pathogens (Burge et al. 1987; Gliotti et al. 1997; Masciandaro et al. 2000). Vermicomposting does not involves a thermophilic phase which might increase the risk of using this technology for management of infectious wastes, but surprisingly vermicomposting resulted into a noticeable reduction in the pathogen indicators such as fecal coliform, *Salmonella* sp., enteric virus and helminth ova in the biosolids (Eastman 1999; Sidhu et al. 2001). Vermicomposting of biosolids resulted in reduction of faecal coliforms and *Salmonella* sp. from 39,000 MPN/g to 0 MPN/g and < 3 MPN to < 1MPN/g respectively (Dominguez and Edwards 2004). Vermicomposting of municipal sewage sludge with *L. mauritii* eliminated *Salmonella* and *Escherichia* sp., and the earthworm gut analysis also proved that *Salmonella* sp. ranging $15–17 \times 10^3$ CFU/g and *Escherichia* sp. ranging $10–14 \times 10^2$ CFU/g were completely eliminated in the gut after 70 days of vermicomposting period (Ganesh Kumar and Sekaran 2005). Activities by earthworms on sludge reduced levels of pathogens and odors of putrefaction and accelerated sludge stabilization (Mitchell 1978; Brown and Mitchell 1981; Hartenstein 1983). The reduction or removal of these enteric bacterial populations at the end of vermicomposting period, correlates with the findings that earthworm's diet include microorganisms and earthworms ability to selectively digest them (Bohlen and Edwards 1995; Edwards and Bohlen1996). Apart from solid waste management, earthworms are also used in sewage water treatment. Earthworms promote the growth of 'beneficial decomposer bacteria' in wastewater and acts as aerators, grinders, crushers, chemical degraders, and biological stimulators (Dash 1978; Sinha et al. 2002). Earthworms also granulate the clay particles and increase the hydraulic conductivity and natural aeration and further grind the silt and sand particles and increase the total specific surface area and thereby enhance adsorption of the organic and inorganic matter from the wastewater. In addition, earthworms body acts as a 'biofilter' and remove the biological oxygen demand (BOD), chemical oxygen demand (COD), total dissolved solids (TDS) and total suspended solids (TSS) from wastewater by 90%, 80–90%, 90–92% and 90–95% respectively by 'ingestion' and biodegradation of organic wastes, heavy metals, and solids from wastewater and by their 'absorption' through body walls (Sinha et al. 2008). Reports reveal that vermicomposting converts the infected biomedical waste containing various pathogens viz.,*Staphylococcus aureus*, *Proteus vulgaris*, *Pseudomonas pyocyaneae* and *Escherichia coli* to an innocuous waste containing commensals like *Citrobactor freundii* and aerobic spore bearing microorganism usually found in the soil and alimentary canal of earthworms (Umesh et al. 2006). Vermicomposting plays a vital role for safe management of biomedical wastes and solid wastes generated from wastewater treatment plants and its bioconversion into valuable composts free from enteric bacterial populations. Depending on the earthworm species, vermicomposting was known to reduce the level of different pathogens such as *Salmonella enteriditis, Escherichia coli,* total and faecal coliforms, helminth ova and human viruses in different types of waste. Direct means of reduction in these microbial numbers during gut passage might be due to the digestive enzymes and mechanical grinding, while indirect means of pathogen removal might be due to promotion of aerobic conditions which could bring down the load of coliforms (Monroy et al. 2009; Edwards 2011; Aira et al. 2011).

ROLE OF VERMICOMPOST IN PLANT GROWTH PROMOTION

Use of vermicomposts as biofertilizers has been increasing recently due to its extraordinary nutrient status, and enhanced microbial and antagonistic activity. Vermicompost produced from different parent material such as food waste, cattle manure, pig manure, etc., when used as a media supplement, enhanced seedling growth and development, and increased productivity of a wide variety of crops (Edwards and Burrows 1988; Wilson and Carlile1989; Buckerfield and Webster 1998; Edwards 1998; Subler et al. 1998; Atiyeh et al. 2000c). Vermicompost addition to soil-less bedding plant media enhanced germination, growth, flowering and fruiting of a wide range of green house vegetables and ornamentals (Atiyeh et al. 2000a, b, c), marigolds (Atiyeh et al. 2001), pepper (Arancon et al.2003a), strawberries (Arancon et al. 2004b) and petunias (Chamani et al. 2008). Vermicompost application in the ratio of 20:1 resulted in a significant and consistent increase in plant growth in both field and greenhouse conditions (Edwards et al. 2004), thus providing a substantial evidence that biological growth promoting factors play a key role in seed germination and plant growth (Edwards and Burrows 1988; Edwards 1998). Investigations revealed that plant hormones and plant-growth regulating substances (PGRs) such as auxins, gibberellins, cytokinins, ethylene and abscisic acid are produced by microorganisms (Barea et al. 1976; Arshad and Frankenberger 1993).

Several researchers have documented the presence of plant growth regulators such as auxins, gibberellins, cytokinins of microbial origin (Krishnamoorthy and Vajranabhiah 1986; Grappelli et al. 1987; Tomati et al. 1988; Muscolo et al. 1999) and humic acids (Senesi et al. X1992; Masciandaro et al. 1997; Atiyeh et al. 2002) in vermicompost in appreciable quantities. Cytokinins produced by *Bacillus* and *Arthrobacter* spp. in soils increase the vigour of seedlings (Inbal and Feldman 1982; Jagnow 1987). Microbially produced gibberellins influence plant growth and development (Mahmoud et al. 1984; Arshad and Frankenberger 1993) and auxins produced by*Azospirillum brasilense* affects the growth of plants belonging to paoceae (Barbieri et al. 1988).

Extensive investigations on the biological activities of humic substances showed that they also posses plant growth stimulating property (Chen and Aviad 1990). Humic substances increased the dry matter yields of corn and oat seedlings (Lee and Bartlett 1976; Albuzio et al. 1994); number and length of tobacco roots (Mylonas and Mccants 1980); dry weights of roots, shoots and number of nodules of groundnut, soyabean and clover plants (Tan and Tantiwiramanond 1983) and vegetative growth of chicory plants (Valdrighi et al. 1996) and induced root and shoot formation in plant tissue culture (Goenadi and Sudharama 1995). High levels of humus have been reported from vermicomposts originating from food wastes, animal manure, sewage and paper mill sludges (Atiyeh et al. 2002; Canellas et al. 2002; Arancon et al. 2003c). The humic and fulvic acid in the humus dissolves insoluble minerals in the organic matter and makes them readily available to plants and in addition they also help plants to overcome stress and stimulates plant growth (Sinha et al. 2010). Studies on biological activities of vermicompost derived humic substances, revealed that they had similar growth-promoting hormonal effect (Dell'Agnola and Nardi 1987; Nardi et al. 1988; Muscolo et al. 1993). The humic materials extracted from vermicomposts have been reported to produce auxin-like cell growth and nitrate metabolism in carrots (*Daucus carota*) (Muscolo et al. 1996). Humates obtained from pig manure vermicompost increased growth of tomato (Atiyeh et al. 2002) and those obtained from cattle, food and paper waste vermicompost increased the growth of strawberries and peppers (Arancon et al.2003a).

Earthworms produce plant growth regulators (Gavrilov 1963). Since earthworms increase the microbial activity by several folds they are considered as important agents which enhance the production of plant growth regulators (Nielson 1965; Graff and Makeschin 1980; Dell'Agnola and Nardi 1987; Grappelli et al. 1987; Tomati et al. 1987,1988; Edwards and Burrows 1988; Nardi et al. 1988; Edwards 1998). Plant growth stimulating substances of microbial origin were isolated from tissues of *Aporrectodea longa, L. terrestris* and *Dendrobaena rubidus* and indole like substances were detected from the tissue extracts of *A. caliginosa, L. rubellus* and *E. foetida* which increased the growth of peas (Nielson 1965) and dry matter production of rye grass (Graff and Makeschin 1980). *A. trapezoids*aided in the dispersal of *Rhizobium* through soil resulting in increased root colonization and nodulation of leguminous plants (Bernard et al. 1994). Use of earthworm casts in plant propagation promoted root initiation, increased root numbers and biomass. The hormone-like effect produced by earthworm casts on plant metabolism, growth and development causing dwarfing, stimulation of rooting, internode elongation and precociousness of flowering was attributed to the fact of presence of microbial metabolites (Tomati et al. 1987; Edwards 1998). Earthworm casts stimulated growth of ornamental plants and carpophore formation in *Agaricus bisporus* when used as casing layer in mushroom cultivation (Tomati et al. 1987). Aqueous extracts of vermicompost produced growth comparable to the use of hormones such as auxins, gibberellins and cytokinins on Petunia, Begonia and Coleus, providing solid evidence that vermicompost is a rich source of plant growth regulating substances (Grappelli et al.1987; Tomati et al. 1987, 1988). Addition of vermicompost at very low levels to the growth media dramatically increased the growth of hardy ornamentals *Chamaecyparis lawsonian, Elaeagnus pungens, Pyracantha* spp.,*Viburnum bodnantense, Cotoneaster conspicus* and *Cupressocyparis leylandi*. Cucumber (Hahn and Bopp 1968), dwarf maize (Sembdner et al. 1976) and coleus bioassays (Edwards et al. 2004) evidenced that vermicompost contained appreciable amounts of cytokinins, gibberellins and auxins respectively. Maize seedlings dipped in vermicompost water showed marked difference in plumule length compared to normal water indicating that plant growth promoting hormones are present in vermicompost (Nagavallemma et al. 2004). Comparative studies on the impact of vermiwash and urea solution on seed germination, root and shoot length in *Cyamopsis tertagonoloba*proved that vermiwash contained hormone like substances (Suthar 2010). High performance liquid chromatography (HPLC) and gas chromatography-mass spectroscopy (GC-MS) analyses of aqueous extracts of cattle waste derived vermicompost showed presence of significant amounts indole-acetic-acid (IAA), gibberellins and cytokinins (Edwards et al. 2004).

Earthworm gut associated microbes enrich vermicomposts with highly water-soluble and light-sensitive plant growth hormones, which gets absorbed on humic acid substances in vermicompost making them extremely stable and helps them persist longer in soils thereby influencing plant growth (Atiyeh et al. 2002; Arancon et al. 2003c). This is confirmed by presence of exchangeable auxin group in the macrostructure of humic acid extract from vermicompost (Canellas et al. 2002). Apart from the rich nutritional status and ready nutrient availability, presence of humic acids and plant growth regulating substances makes vermicompost a biofertilizer which increases germination, growth, flowering and fruiting in a wide range of crops. Vermicompost substitution in a relatively small proportion (10–20%) to the potting mixture increased dry matter production and tomato growth significantly (Subler et al. 1998). Soil amended with 20% vermicompost was more suitable for tomato seedling production (Valenzuela et al. 1997). Similarly vermicompost addition upto 50% in the medium resulted in enhanced growth of *Chamaecyparis lawsoniana* (Lawson's Cypress), *Juniperus communis* (Juniper) and *Elaeagnus pungens* (Silverberry) rooted liners (Bachman and Edgar Davice 2000).

Vermicompost application increased plant spread (10.7%), leaf area (23.1%), dry matter (20.7%) and increased total strawberry fruit yield (32.7%) (Singh et al. 2008). Substitution of vermicompost drastically reduced the incidence of physiological disorders like albinism (16.1–4.5%), fruit malformation (11.5–4.0%) and occurrence of grey mould (10.4–2.1%) in strawberry indicating its significance in reducing nutrient-related disorders and *Botrytis* rot, thereby increasing the marketable fruit yield upto 58.6% with better quality parameters. Fruit harvested from plant receiving vermicompost were firmer, had higher total soluble solids (TSS), ascorbic acid content and attractive colour. All these parameters appeared to be dose dependent and best results were achieved at 7.5 t ha^{-1} (Singh et al. 2008). Vermicompost application showed significant increase in germination percent (93%), growth and yield of mung bean (*Vigna radiata*) compared to the control (Karmegam et al. 1999). Similarly, the fresh and dry matter yields of cowpea (*Vigna unguiculata*) were higher in soil amended with vermicompost than with biodigested slurry, (Karmegam and Daniel 2000). Combined application of vermicompost with N fertilizer gave higher dry matter (16.2 g $plant^{-1}$) and grain yield (3.6 t ha^{-1}) of wheat (*Triticum aestivum*) and higher dry matter yield (0.66 g plant^{-1}) of the following coriander (*Coriandrum sativum*) crop in wheat-coriander cropping system (Desai et al. 1999). Vermicompost application produced herbage yields of coriander cultivars comparable to those obtained with chemical fertilizers (Vadiraj et al. 1998). Yield of pea (*Pisum sativum*) increased with the application of vermicompost (10 t ha^{-1}) along with recommended NPK (Meena et al. 2007). Vermicompost application to sorghum (*Sorghum bicolor*) (Patil and Sheelavantar 2000), sunflower (*Helianthus annuus*) (Devi et al. 1998), tomato (*Lycopersicon esculentum*) (Nagavallemma et al. 2004), eggplant (*Solanum melangona*) (Guerrero and Guerrero, 2006), okra (*Abelmoschus esculentus*) (Gupta et al. 2008), hyacinth bean (*Lablab purpureas*) (Karmegam and Daniel 2008), grapes (Buckerfield and Webster 1998) and cherry (Webster 2005) showed a positive result. Vermicompost amendment at the rate of 10 t ha^{-1} along with 50% of recommended dose of NPK fertilizer increased the number and fresh weight of flowers per plant, flower diameter and yield, while at the rate of 15 t ha^{-1} along with 50% of recommended dose of NPK increased vase life of *Chrysanthemum chinensis* (Nethra et al. 1999). Red Clover and cucumber grown in soil amended with vermicompost showed an increase in mineral contents viz., Ca, Mg, Cu, Mn and Zn in their shoot tissues (Sainz et al. 1998). Vermicomposted cow manure stimulated the growth of lettuce and tomato plants while the unprocessed parent material did not (Atiyeh et al. 2000b). Similarly, vermicomposted duck wastes resulted in better growth of tomatoes, lettuce, and peppers than the unprocessed wastes (Wilson and Carlile 1989). Theenhancement in plant growth might be attributed to the fact that processed waste had improved physicochemical characteristics and nutrients, in forms readily available to the plant as well as the presence of plant growth promoting and antagonistic disease suppressing beneficial bacteria.

ROLE OF VERMICOMPOST IN PLANT DISEASE MANAGEMENT

Plant pathogen control

Soils with low organic matter and microbial activity are conducive to plant root diseases (Stone et al. 2004) and addition of organic amendments can effectively suppress plant disease (Raguchander et al. 1998; Blok et al. 2000; Lazarovits et al. 2000). Several researchers reported the disease suppressive properties of thermophilic compost (Hoitink et al. 1997; Goldstein 1998; Pitt et al. 1998) on a wide range of phytopathogens viz., *Rhizoctonia* (Kuter et al. 1983), *Phytopthora* (Hoitink and Kuter 1986; Pitt et al. 1998), *Plasmidiophora brassicae* and *Gaeumannomyces graminis* (Pitt et al. 1998) and *Fusarium* (Kannangowa et al. 2000; Cotxarrera et al. 2002). Microbial antagonism might be one of the possible reasons for disease suppression as organic amendments enhances the microbial population and diversity. Traditional thermophilic composts promote only selected microbes while non-thermophilic vermicomposts are rich sources of microbial diversity and activity and harbour a wide variety of antagonistic bacteria thus acts as effective biocontrol agents aiding in suppression of diseases caused by soil-borne phytopathogenic fungi (Chaoui et al. 2002; Scheuerell et al. 2005; Singh et al. 2008). Earthworm feeding reduces the survival of plant pathogens such as *Fusarium* sp. and *Verticillium dahliae* (Yeates1981; Moody et al. 1996) and increases the densities of antagonistic fluorescent pseudomonads and filamentous actinomycetes while population densities of *Bacilli* and *Trichoderma* spp. remains unaltered (Elmer 2009). Earthworm activities reduce root diseases of cereals caused by *Rhizoctonia* (Doube et al. 1994). It has been proved that earthworms decreased the incidence of field diseases of clover, grains, and grapes incited by *Rhizoctonia* spp. (Stephens et al. 1994a; Stephens and Davoren 1997) and *Gaeumannomyces* spp. (Clapperton et al. 2001). Earthworms *Aporrectodea trapezoides* and *Aporrectodea rosea* act as vectors of *Pseudomonas corrugata 2* 14OR, a biocontrol agent for wheat take-all caused by *G. graminis* var. *tritd* (Doube et al. 1994). Greenhouse studies on augmentation of pathogen infested soils with *L. terrestris* showed a significant reduction of disease caused by*Fusarium oxysporum* f. sp. *asparagi* and *F. proliferatum* on susceptible cultivars of asparagus (*Asparagus officinalis*), *Verticillium dahliae* on eggplant (*Solanum melongena*) and *F. oxysporum* f. sp. *lycopersici* race 1 on tomato. Plant weights increased by 60-80% and disease severity reduced by 50-70% when soils were augmented with earthworms. Incorporation of soil with vermicompost effectively suppressed *R. solani* in wheat (Stephens et al.1993), *Phytophthora nicotianae* (Nakamura 1996; Szczech 1999; Szczech and Smolinska 2001) and *Fusarium* in tomatoes (Nakamura 1996; Szczech 1999), *Plasmodiophora brassicae* in tomatoes and cabbage (Nakamura 1996),*Pythium* and *Rhizoctonia* (root rot) in cucumber and radish (Simsek Ersahin et al. 2009), *Botrytis cineria* (Singh et al.2008) and *Verticillium* (Chaoui et al. 2002) in strawberry and *Sphaerotheca fulginae* in grapes (Edwards et al. 2004). Vermicompost application drastically reduced the incidence of 'Powdery Mildew', 'Color Rot' and 'Yellow Vein Mosaic' in Lady's finger (*Abelmoschus esculentus*) (Agarwal et al. 2010). Substitution of vermicompost in the growth media reduced the fungal diseases caused by *R. solani, P. drechsleri and F. oxysporum* in gerbera (Rodriguez et al.2000). Amendment of vermicompost at low rates (10-30%) in horticulture bedding media resulted in significant suppression of *Pythium* and *Rhizoctonia* under green house conditions (Edwards et al. 2004). Research findings proved that vermicompost when added to container media significantly reduced the infection of tomato plants by *P. nicotianae* var. nicotianae and *F. oxysporum* sp. *lycopersici* (Szcech et al. 1993; Szczech 1999). Club-rot of cabbage caused by *P. brassicae* was inhibited by dipping cabbage roots into a mixture of clay and vermicompost (Szcech et al. 1993). Potato plants treated with vermicompost were less susceptible to *P. infestans* than plants treated with inorganic fertilizers (Kostecka et al. 1996a). Aqueous extracts of vermicompost inhibited mycelial growth of *B. cineria, Sclerotinia sclerotiorum, Corticium rolfsii*, *R. solani* and *F. oxysporum* (Nakasone et al. 1999), effectively controlled powdery mildew of barley (Weltzien 1989) and affected the development of powdery mildews on balsam (*Impatiens balsamina*) and pea (*Pisum sativum*) caused by *Erysiphe cichoracearum* and *E. pisi,* respectively in field conditions (Singh et al. 2003).

Mechanisms that mediate pathogen suppression

Two possible mechanisms of pathogen suppression have been described, one depends on systemic plant resistance and the other is mediated by microbial competition, antibiosis and hyperparasitism (Hoitink and Grebus1997). The microbially mediated suppression is again classified into two mechanisms viz., 'general suppression' where a wide range of microbes suppress the pathogens such as *Pythium* and *Phytopthora* (Chen et al. 1987) and 'specific suppression' where a narrow range of organisms facilitates suppression, for instance disease caused by*Rhizoctonia* (Hoitink et al. 1997). The disease suppressive effect of vermicompost against fusarium wilt of tomato clearly depicted that fungus inhibition was purely biotic and no chemical factors played any role, since the experiments with heat-sterilized vermicompost failed to control the disease (Szczech 1999). Experiments on suppression of damping-off caused by *R. solani,* in vermicompost amended nurseries of white pumpkin proved that vermicompost suppressed the disease in a dosage and temperature dependent manner (Rivera et al. 2004). Earthworm castings are rich in nutrients (Lunt and Jacobson 1944; Parle 1963) and calcium humate, a binding agent (Edwards 1998) that reduces desiccation of individual castings and favors the incubation and proliferation of beneficial microbes, such as *Trichoderma* spp. (Tiunov and Scheu 2000), *Pseudomonas* spp. (Schmidt et al. 1997), and mycorrhizal spores (Gange 1993; Doube et al. 1995). Earthworm activity increased the communities of Gram-negative bacteria (Clapperton et al. 2001; Elmer 2009). Vermicompost associated chitinolytic bacterial communities viz., *Nocardioides oleivorans*, several species of *Streptomyces* and *Staphylococcus epidermidis* showed inhibitory effects against plant phytopathogens such as, *R. solani, Colletotrichum coccodes, Pythium ultimum, P. capsici* and*Fusarium moniliforme* (Yasir et al. 2009a).

ROLE OF VERMICOMPOST IN ARTHROPOD PEST CONTROL

Addition of organic amendments helped in suppression of various insect pests such as European corn borer (Phelan et al. 1996), other corn insect pests (Biradar et al. 1998), aphids and scale insects (Culliney and Pimentel 1986; Costello and Altiei 1995; Huelsman et al. 2000) and brinjal shoot and fruit borer (Sudhakar et al. 1998). Several reports also evidenced that vermicompost addition decreased the incidence of *Spodoptera litura, Helicoverpa armigera*, leaf miner (*Apoaerema modicella)*, jassids (*Empoasca kerri*), aphids (*Aphis craccivora)* and spider mites on groundnuts (Rao et al. 2001; Rao 2002, 2003) and psyllids (*Heteropsylla cubana*) on a tropical leguminous tree (*Leucaena leucocephala*) (Biradar et al. 1998). Vermicompost amendment decreased the incidence of sucking pests under field conditions (Ramesh 2000) and suppressed the damage caused by of two-spotted spider mite (*Tetranychus* spp.), aphid (*Myzus persicae*) (Edwards et al. 2007) and mealy bug (*Pseudococcus* spp.) under green house conditions (Arancon et al. 2007). Vermicompost substitution to soil less plant growth medium MetroMix 360 (MM360) at a rate less then 50% reduced the damage caused by infestation of pepper seedlings by *M. persicae* and*Pseudococcus* spp. and tomato seedlings by *Pseudococcus* spp., cabbage seedlings by *M. persicae* and cabbage white caterpillars (*Pieris brassicae* L.) (Arancon et al. 2005). Greenhouse cage experiments conducted on tomatoes and cucumber seedlings infested with *M. persicae*, citrus mealybug (*Planococcus citri*), two spotted spider mite (*Tetranychus urticae*); striped cucumber beetles (*Acalymna vittatum*) attacking cucumbers and tobacco hornworms (*Manduca sexta*) attacking tomatoes proved that treatment of infested plants with aqueous extracts of vermicompost suppressed pest establishment, and their rates of reproduction. Vermicompost teas at higher dose also brought about pest mortality

(Edwards et al. 2010b). Suppression of aphid population gains importance since they are key vectors in transmission of plant viruses. Addition of solid vermicompost reduced damage by *A. vittatum* and spotted cucumber beetles (*Diabotrica undecimpunctata*) on cucumbers and larval hornworms (*Manduca quinquemaculata)*on tomatoes in both greenhouse and field experiments (Yardim et al. 2006). Combined application of vermicompost and vermiwash spray to chilli (*Capiscum annum*) significantly reduced the incidence of 'Thrips' (*Scirtothrips dorsalis*) and 'Mites' (*Polyphagotarsonemus latus*) (Saumaya et al. 2007).

Mechanisms that mediate pest control

Plants grown in inorganic fertilizers are more prone to pest attack than those grown on organic fertilizers (Culliney and Pimentel 1986; Yardim and Edwards 2003; Phelan 2004). Inorganic nitrogen fertilization improves the nutritional quality and palatability of the host plants, inhibits the raise of secondary metabolite concentrations (Fragoyiannis et al. 2001; Herms 2002), enhances the fecundity of insects dieting on them, attracts more individuals for oviposition (Bentz et al. 1995) and increases the population growth rates of insects (Culliney and Pimentel 1986; Jannsson and Smilowitz 1986). Though organic fertilizer has an enhanced nutritional composition they release nutrients at a slower rate (Patriquin et al. 1995) hence plants grown with organic fertilizers possess decreased N levels (Steffen et al. 1995) and have higher phenol content (Asami et al. 2003) resulting in resistance of these plants to pest attack. Similarly vermicomposts exhibit a slow, balanced nutritional release pattern, particularly in release of plant available N, soluble K, exchangeable Ca, Mg and P (Edwards and Fletcher 1988; Edwards 1998). Vermicomposts are rich in humic acid and phenolic compounds. Phenolic compounds act as feeding deterrents and hence significantly affect pest attacks (Kurowska et al. 1990; Summers and Felton 1994; QiTian 2004; Hawida et al. 2007; Koul 2008; Mahanil et al. 2008; Bhonwong et al. 2009). Soil containing earthworms contained polychlorinated phenols and their metabolites (Knuutinen et al. 1990). An endogenous phenoloxidase present in *L. rubellus* bioactivate compounds to form toxic phenols viz., *p*-nitrophenol (Park et al. 1996). Monomeric phenols could be absorbed by humic acids in the gut of earthworms (Vinken et al. 2005). Uptake of soluble phenolic compounds from vermicompost, by the plant tissues makes them unpalatable thereby affecting pest rates of reproduction and survival (Edwards et al. 2010a; Edwards et al. 2010b).

ROLE OF VERMICOMPOST IN NEMATODE CONTROL

It has been well documented that addition of organic amendments decreases the populations of plant parasitic nematodes (Addabdo 1995; Sipes et al. 1999; Akhtar and Malik 2000). Vermicompost amendments appreciably suppress plant parasitic nematodes under field conditions (Arancon et al. 2003b). Vermicomposts also suppressed the attack of *Meloidogyne incognita* on tobacco, pepper, strawberry and tomato (Swathi et al. 1998; Edwards et al.2007; Arancon et al. 2002; Morra et al. 1998) and decreased the numbers of galls and egg masses of *Meloidogyne javanica* (Ribeiro et al. 1998).

Mechanisms that mediate nematode control

There are several feasible mechanisms that attribute to the suppression of plant parasitic nematodes by vermicompost application and it involves both biotic and abiotic factors. Organic matter addition to the soil stimulates the population of bacterial and fungal antagonists of nematodes (*e.g.*, *Pasteuria penetrans*, *Pseudomonas* spp. and chitinolytic bacteria, *Trichoderma* spp.), and other typical nematode predators including nematophagous mites viz.,*Hypoaspis calcuttaensis* (Bilgrami 1996), Collembola and other arthropods which selectively feeds on plant parasitic nematodes. (Thoden et al. 2011). Vermicompost amendment promoted fungi capable of trapping nematode and destroying nematode cysts (Kerry 1988) and increased the population of plant growth-promoting rhizobacteria which produce enzymes toxic to plant parasitic nematodes (Siddiqui and Mahmood 1999). Vermicompost addition to soils planted with tomatoes, peppers, strawberry and grapes showed a significant reduction of plant parasitic nematodes and increased the population of fungivorous and bacterivorous nematodes compared to inorganic fertilizer treated plots (Arancon et al. 2002). In addition, few abiotic factors viz., nematicidal compounds such as hydrogen sulphide, ammonia, nitrates, and organic acids released during vermicomposting, as well as low C/N ratios of the compost cause direct adverse effects while changes in soil physiochemical characterists viz., bulk density, porosity, water holding capacity, pH, EC, CEC and nutrition posses indirect adverse effects on plant parasitic nematodes (Rodriguez-Kabana 1986; Thoden et al. 2011).

CONCLUSION

Vermicomposting is a cost-effective and eco-friendly waste management technology which takes the previlige of both earthworms and the associated microbes and has many advantages over traditional thermophilic composting. Vermicomposts are excellent sources of biofertilizers and their addition improves the physiochemical and biological properties of agricultural soil. Vermicomposting amplifies the diversity and population of beneficial microbial communities. Although there are some reports indicating that few harmful microbes such as spores of *Pythium* and*Fusarium* are dispersed by earthworms (Edwards and Fletcher 1988), the presence and amplification of antagonistic disease-suppressing and other plant growth-promoting beneficial bacteria during vermicomposting out weigh these harmful effects (Edwards and Fletcher 1988; Gammack et al. 1992; Brown 1995). Vermicomposts with excellent physio-chemical properties and buffering ability, fortified with all nutrients in plant available forms, antagonistic and plant growth-promoting bacteria are fantabulous organic amendments that act as a panacea for soil reclamation, enhancement of soil fertility, plant growth, and control of pathogens, pests and nematodes for sustainable agriculture.

DECLARATIONS

Acknowledgement

The financial support from Department of Biotechnology (DBT), New Delhi, India, and Department of Science and Technology (DST), New Delhi, India, through Fund for Improvement of Science and Technology Infrastructure in Higher Educational Institutions (FIST), is gratefully acknowledged.

Competing interests

The authors declare that they have no competing interests.

AUTHORS' CONTRIBUTIONS

JP: Collected and reviewed the literature and drafted the manuscript. NS: Formulated the objectives, provided guidance and improved the quality of the manuscript. Both authors read and approved the final manuscript.

REFERENCES

1. Addabdo TD: The nematicidal effect of organic amendments: a review of the literature 1982–1994. *Nematol Mediterranea* 1995, 23: 299-305.
2. Aira M, Monroy F, Dominguez J: Earthworms strongly modify microbial biomass and activity triggering enzymatic activities during vermicomposting independently of the application rates of pig slurry. *Sci Total Environ* 2007, 385:252-261.
3. Aira M, Gómez-Brandón M, González-Porto P, Domínguez J: Selective reduction of the pathogenic load of cow manure in an industrial-scale continuous-feeding vermireactor. *Bioresource Technol* 2011, 102: 9633-9637.
4. Agarwal S, Sinha RK, Sharma J, *et al.*: Ver-miculture for sustainable horticulture: Agronomic impact studies of earthworms, cow dung compost and vermi-compost vis-à-vis chemical fertilizers on growth and yield of lady's finger (Abelmoschus esculentus). In *Special Issue on 'Vermiculture Technology', International Journal of Environmental Engineering*. Edited by: Sinha RK. Inderscience Publishers, Geneva, Switzerland; 2010.
5. Akhtar M, Malik A: Role of organic amendments and soil organisms in the biological control of plant parasitic nematodes: a review. *Bioresour Technol* 2000, 74: 35-47.
6. Albanell E, Plaixats J, Cabrero T: Chemical changes during vermicomposting (Eisenia fetida) of sheep manure mixed with cotton industrial wastes. *Biol Fertil Soils* 1988, 6: 266-269.
7. Albuzio A, Concheri G, Nardi S, Dell'Agnola G: Effect of humic fractions of different molecular size on the development of oat seedlings grown in varied nutritional conditions. In *Humic substances in the Global Environment and Implications on Human Health*. Edited by: Senesi N, Miano TM. Elsevier, Amsterdam, Netherlands; 1994:199-204.
8. Arancon NQ, Edwards CA, Atiyeh R, Metzger JD: Effects of vermicomposts produced from food waste on the growth and yields of greenhouse peppers. *Bioresour Technol* 2004, 93: 139-144.

9. Arancon NQ, Edwards CA, Bierman P, Metzger JD, Lee S, Welch C: Effects of vermicomposts to tomatoes and peppers grown in the field and strawberries under high plastic tunnels. *Pedobiologia* 2003, 47: 731-735.
10. Arancon NQ, Edwards CA, Bierman P, Welch C, Metzger JD: The influence of vermicompost applications to strawberries: Part 1. Effects on growth and yield. *Bioresour Technol* 2004, 93: 145-153.
11. Arancon NQ, Edwards CA, Lee S: Management of plant parasitic nematode populations by use of vermicomposts.*Proceedings Brighton Crop Protection Conference – Pests and Diseases* 2002, 705-716.
12. Arancon NQ, Edwards CA, Yardim EN, Oliver TJ, Byrne RJ, Keeney G: Suppression of two-spotted spider mite (Tetranychus urticae), mealy bug (Pseudococcus sp) and aphid (Myzus persicae) populations and damage by vermicomposts. *Crop Prot* 2007, 26: 29-39.
13. Arancon NQ, Galvis PA, Edwards CA: Suppression of insect pest populations and damage to plants by vermicomposts. *Bioresour Technol* 2005, 96: 1137-1142.
14. Arancon NQ, Galvis P, Edwards CA, Yardim E: The trophic diversity of nematode communities in soils treated with vermicomposts. *Pedobiologia* 2003, 47: 736-740.
15. Arancon NQ, Lee S, Edwards CA, Atiyeh RM: Effects of humic acids and aqueous extracts derived from cattle, food and paper-waste vermicomposts on growth of greenhouse plants. *Pedobiologia* 2003, 47: 744-781.
16. Arshad M, Frankenberger WT Jr: Microbial production of plant growth regulators. In *Soil Microbial Ecology: Applications in Agricultural and Environmental Management.* Edited by: Metting FB Jr. Marcell Dekker, New York; 1993:307-347.
17. Asami DK, Hang YJ, Barnett DM, Mitchell AE: Comparison of the total phenolic and ascorbic acid content of freeze-dried and air-dried marionberry, strawberry and corn grown using conventional organic and sustainable agricultural practices. *J Agric Food Chem* 2003, 51: 1237-1241.
18. Asha A, Tripathi AK, Soni P: Vermicomposting: A Better Option for Organic Solid Waste Management. *J Hum Ecol*2008, 24: 59-64.
19. Atiyeh RM, Arancon NQ, Edwards CA, Metzger JD: The influence of earthworm-processed pig manure on the growth and productivity of marigolds. *Bioresour Technol* 2001, 81: 103-108.
20. Atiyeh RM, Dominguez J, Subler S, Edwards CA: Changes in biochemical properties of cow manure during processing by earthworms (Eisenia andrei, Bouché) and the effects on seedling growth. *Pedobiologia* 2000, 44:709-724.
21. Atiyeh RM, Lee S, Edwards CA, Arancon NQ, Metzger JD: The influence of humic acids derived from earthworm-processed organic wastes on plant growth. *Bioresour Technol* 2002, 84: 7-14.
22. Atiyeh RM, Subler S, Edwards CA, Bachman G, Metzger JD, Shuster W: Effects of vermicomposts and composts on plant growth in horticulture container media

and soil. *Pedobiologia* 2000, 44: 579-590.

23. Atiyeh RM, Arancon NQ, Edwards CA, Metzger JD: Influence of earthworm-processed pig manure on the growth and yield of green house tomatoes. *Bioresour Technol* 2000, 75: 175-180.
24. Ayyadurai N, Ravindra Naik P, Sakthivel N: Functional characterization of antagonistic fluorescent pseudomonads associated with rhizospheric soil of rice (Oryza sativa L.). *J Microbiol Biotechnol* 2007, 17: 919-927.
25. Bachman GR, Edgar Davice W: *Growth of magnolia virginiana liners in vermicompost-amended media*. Proceeding of SNA Research Conference, Southern Nursery Association, Atlanta; 2000:65-67.
26. Bansal S, Kapoor KK: Vermicomposting of crop residues and cattle dung with Eisenia foetida. *Bioresour Technol*2000, 73: 95-98.
27. Banu JR, Logakanthi S, Vijayalakshmi GS: Biomanagement of paper mill sludge using an indigenous (Lampito mauritii) and two exotic (Eudrilus eugineae and Eisenia foetida) earthworms. *J Environ Biol* 2001, 22: 181-185.
28. Barbieri P, Bernardi A, Galli E, Zanetti G: Effects of inoculation with different strains of Azospirillum brasilense on wheat roots development. In *Azospirillum IV, Genetics, Physiology, Ecology*. Edited by: Klingmüller W. SpringerVerlag, Berlin; 1988:181-188.
29. Barea JM, Navarro E, Montana E: Production of plant growth regulators by rhizosphere phosphate solubilizing bacteria. *J Appl Bacteriol* 1976, 40: 129-134.
30. Benitez E, Nogales R, Elvira C, Masciandaro G, Ceccanti B: Enzymes activities as indicators of the stabilization of sewage sludges composting by Eisenia foetida. *Bioresour Technol* 1999, 67: 297-303.
31. Bentz JA, Reeves J, Barbosa P, Francis B: Nitrogen fertilizer effect on selection, acceptance and suitability of Euphorbia pulcherrima (Euphorbiaceae) as a host plant to Bemisia tabaci (Homoptera: Aleyrodidae). *Environ Entomol* 1995, 24: 40-45.
32. Bernal MP, Faredes C, Sanchez-Monedero MA, Cegarra J: Maturity and stability parameters of composts prepared with a, wide range of organic wastes. *Bioresour Technol* 1998, 63: 91-99.
33. Bernard MD, Peter MS, Christopher WD, Maarten HR: Interactions between earthworms, beneficial soil microorganisms and root pathogens. *Appl Soil Ecol* 1994, 1: 3-10.
34. Bhatnagar RK, Palta RK: *Earthworm-Vermiculture and Vermicomposting*. Kalyani Publishers, New Delhi; 1996.
35. Bhonwong A, Stout MJ, Attajarusit J, Tantasawat P: Defensive role of tomato polyphenoloxidases against cotton bollworm (Helicoverpa armigera) and beet army worm (Spodoptera exigua). *J Chem Ecol* 2009, 35: 28-38.
36. Bilgrami AL: Evaluation of the predation abilities of the mite Hypoaspis calcuttaensis, predaceous on plant and soil nematodes. *Fund Appl Nematol* 1996, 20: 96-98.

37. Binet F, Fayolle L, Pussard M: Significance of earthworms in stimulating soil microbial activity. *Biol Fertil Soils* 1998,27: 79-84.

38. Biradar AP, Sunita ND, Teggel RG, Devaradavadgi SB: Effect of vermicompost on the incidence of subabul psyllid.*Insect- Environ* 1998, 4: 55-56.

39. Blok WJ, Lamers JG, Termoshuizen AJ, Bollen GJ: Control of soil-borne plant pathogens by incorporating fresh organic amendments followed by tarping. *Phytopathology* 2000, 90: 253-259.

40. Bohlen PJ, Edwards CA: Earthworm effects on N dynamics and soil respiration in microcosms receiving organic and inorganic nutrients. *Soil Biol Biochem* 1995, 27: 341-348.

41. Brown BA, Mitchell MJ: Role of the earthworm, Eisenia fetida, in affecting survival of Salmonella enteritidis ser. Typhimurium. *Pedobiologia* 1981, 22: 434-438.

42. Brown GG: How do earthworms affect microfloral and faunal community diversity? *Plant Soil* 1995, 170: 209-231.

43. Brown GG, Barois I, Lavelle P: Regulation of soil organic matter dynamics and microbial activity in the drilosphere and the role of interactions with other edaphic functional domains. *Eur J Soil Biol* 2000, 36: 177-198.

44. Buckalew DW, Riley RK, Yoder WA, Vail WJ: Invertebrates as vectors of endomycorrhizal fungi and Rhizobium upon surface mine soils. *West Virginia Acad Sci Proc* 1982, 54: 1.

45. Buckerfield JC, Webster KA: Worm-worked waste boosts grape yields: prospects for vermicompost use in vineyards.*Australian and New Zealand Wine Industry Journal* 1998, 13: 73-76.

46. Burge WD, Enkiri NK, Hussong D: Salmonella regrowth in compost as influenced by substrate. *Microbial Ecol* 1987,14: 243-253.

47. Butt KR: Utilization of solid paper mill sludge and spent brewery yeast as a feed for soil dwelling earthworms.*Bioresour Technol* 1993, 44: 105-107.

48. Byzov BA, Nechitaylo TY, Bumazhkin BK, Kurakov AV, Golyshin PN, Zvyagintsev DG: Culturable microorganisms from the earthworm digestive tract. *Microbiology* 2009, 78: 360-368.

49. Canellas LP, Olivares FL, Okorokova FAR: Humic acids isolated from earthworm compost enhance root elongation, lateral root emergence and plasma membrane H+– ATPase activity in maize roots. *Plant Physiol* 2002, 130: 1951-1957.

50. Chamani E, Joyce DC, Reihanytabar A: Vermicompost Effects on the Growth and Flowering of Petunia hybrida 'Dream Neon Rose'. *American-Eurasian J Agric And Environ Sci* 2008, 3: 506-512.

51. Chan LPS, Griffiths DA: The vermicomposting of pretreated pig manure. *Biol Wastes* 1988, 24: 57-69.

52. Chaoui H, Edwards CA, Brickner M, Lee S, Arancon N: Suppression of the plant diseases, Pythium (damping off), Rhizoctonia (root rot) and Verticillum (wilt) by vermicomposts. *Proceedings of Brighton Crop Protection Conference – Pests*

and Diseases 2002,II(8B-3):711-716.

53. Chen W, Hoitink HA, Schmitthenner AF, Touvinen O: The role of microbial activity in suppression of damping off caused by Pythium ultimum. *Phytopathology* 1987, 78: 314-322.

54. Chen Y, Aviad T: Effects of humic substances on plant growth. In *Humic Substances in Soil and Crop Sciences*. Edited by: MacCarthy P, Clapp CE, Malcolm RL, Bloom PR. Selected Reading ASA and SSSA, Madison; 1990:161-186.

55. Citernesi U, Neglia R, Seritti A, Lepidi AA, Filippi C, Bagnoli G, Nuti MP, Galluzzi R: Nitrogen fixation in the gastro-enteric cavity of soil animals. *Soil Biol Biochem* 1977, 9: 71-72.

56. Clapperton MJ, Lee NO, Binet F, Conner RL: Earthworms indirectly reduce the effect of take-all (Gaeumannomyces graminis var. tritici) on soft white spring wheat (Triticium aestivum cv. Fielder). *Soil Biol Biochem* 2001, 33: 1531-1538.

57. Contreras E: Studies on the intestinal actinomycete flora of Eisenia lucens (Annelida, Oligochaeta). *Pedobiologia*1980, 20: 411-416.

58. Correa JD, Barrios ML, Galdona RP: Screening for plant growth promoting rhizobacteria in Chamaecytisus proliferus (tagasaste), a forage tree-shrub legume endemic to the Canary Islands. *Plant Soil* 2004, 266: 75-84.

59. Costello MJ, Altiei MA: Abundance, growth rate and parasitism of Brevicoryne brassicae and Myzus persicae (Homoptera: Aphididae) on broccoli grown in living mulches. *Agric Ecosyst Environ* 1995, 52: 187-196.

60. Cotxarrera L, Trillas-Gayl MI, Steinberg C, Alabouvette C: Use of sewage sludge compost and Trichoderma asperellum isolates to suppress Fusarium wilt of tomato. *Soil Biol Biochem* 2002, 34: 467-476.

61. Culliney TW, Pimentel D: Ecological effects of organic agricultural practices on insect populations. *Agric Ecosyst Environ* 1986, 15: 253-256.

62. Daniel O, Anderson JM: Microbial biomass and activity in contrasting soil material after passage through the gut of earthworm Lumbricus rubellus Hoffmeister. *Soil Biol Biochem* 1992, 24: 465-470.

63. Darwin F, Seward AC: More letters of Charles Darwin. In *A record of his work in series of hitherto unpublished letters*. Edited by: John M. London; 1903:508.

64. Dash MC: Role of earthworms in the decomposer system. In *Glimpses of ecology*. Edited by: Singh JS, Gopal B. India International Scientific Publication, New Delhi; 1978:399-406.

65. Dell'Agnola G, Nardi S: Hormone-like effect and enhanced nitrate uptake induced by depolyconder humic fractions obtained from Allolobophora rosea and A. caliginosa faeces. *Biol Fertil Soils* 1987, 4: 111-118.

66. Desai VR, Sabale RN, Raundal PV: Integrated nitrogen management in wheat-coriander cropping system. *J Maharasthra Agric Univ* 1999, 24: 273-275.

67. Devi D, Agarwal SK, Dayal D: Response of sunflower (Helianthus annuus) to organic manures and fertilizers. *Indian J Agron* 1998, 43: 469-473.

68. Devliegher W, Verstraete W: Lumbricus terrestris in a soil core experiment: nutrient-enrichment processes (NEP) and gut-associated processes (GAP) and their effect on microbial biomass and microbial activity. *Soil Biol Biochem* 1995, 27: 1573-1580.

69. Devliegher W, Verstraete W: Microorganisms and soil physicochemical conditions in the drilosphere of Lumbricus terrestris. *Soil Biol Biochem* 1997, 29: 1721-1729.

70. Dominguez J, Edwards CA, Webster M: Vermicomposting of sewage sludge: effects of bulking materials on the growth and reproduction of the earthworm Eisenia andrei. *Pedobiologia* 2000, 44: 24-32.

71. Dominguez J, Edwards CA: Vermicomposting organic wastes: A review. In *Soil Zoology for sustainable Development in the 21st century*. Edited by: Shakir Hanna SH, Mikhail WZA. Cairo; 2004:369-395.

72. Doube BM, Ryder MH, Davoren CW, Meyer T: Earthworms: a down under delivery system service for biocontrol agents of root disease. *Acta Zool Fennica* 1995, 196: 219-223.

73. Doube BM, Stephens PM, Davorena CW, Ryderb MH: Interactions between earthworms, beneficial soil microorganisms and root pathogens. *Appl Soil Ecol* 1994, 1: 3-10.

74. Eastman BR: Achieving pathogen stabilization using vermicomposting. *BioCycle* 1999, 40: 62-64.

75. Edwards CA: Breakdown of animal, vegetable and industrial organic wastes by earthworms. In *Earthworms in Waste and Environmental Management SPB*. Edited by: Edwards CA, Neuhauser EF. The Hague, Netherlands; 1988:21-31.

76. Edwards CA: The use of earthworms in the breakdown and management of organic wastes. In *Earthworm Ecology*. Edited by: Edwards CA. CRC Press, Boca Raton; 1998:327-354.

77. Edwards CA: Human pathogen reduction during vermicomposting. In *Vermiculture technology: earthworms, organic wastes and environmental management*. Edited by: Edwards CA, Arancon NQ, Sherman R. CRC Press, Boca Raton; 2011:249-261.

78. Edwards CA, Arancon NQ, Bennett MV, Askar A, Keeney G, Little B: Suppression of green peach aphid (Myzus persicae) (Sulz.), citrus mealybug (Planococcus citri) (Risso), and two spotted spider mite (Tetranychus urticae) (Koch.) attacks on tomatoes and cucumbers by aqueous extracts from vermicomposts. *Crop Prot* 2010, 29: 80-93.

79. Edwards CA, Arancon NQ, Bennett MV, Askar A, Keeney G: Effect of aqueous extracts from vermicomposts on attacks by cucumber beetles (Acalymna vittatum) (Fabr.) on cucumbers and tobacco hornworm (Manduca sexta) (L.) on tomatoes. *Pedobiologia* 2010, 53: 141-148.

80. Edwards CA, Arancon NQ, Emerson E, Pulliam R: Supressing plant parasitic nematodes and arthropod pests with vermicompost teas. *Biocycle* 2007, 38-39.

81. Edwards CA, Arancon NQ: Vermicomposts suppress plant pest and disease

attacks. *BioCycle* 2004, 45: 51-53.

82. Edwards CA, Bohlen PJ: *Biology and Ecology of earthworms*. Chapman and Hall, London; 1996:426.
83. Edwards CA, Burrows I, Fletcher KE, Jones BA: The use of earthworms for composting farm wastes. In *Composting Agricultural and Other Wastes*. Edited by: Gasser JKR. Elsevier, London; 1985:229-241.
84. Edwards CA, Burrows I: The potential of earthworm composts as plant growth media. In *Earthworms in Waste and Environmental Management*. Edited by: Edwards CA, Neuhauser E. SPB Academic Press, The Hague; 1988:21-32.
85. Edwards CA, Dominguez J, Arancon NQ: The influence of vermicomposts on pest and diseases. In *Soil Zoology for Sustainable Development in the 21st centuary*. Edited by: Shakir Hanna SH, Mikhail WZA. Cairo; 2004:397-418.
86. Edwards CA, Dominguez J, Neuhauser EF: Growth and reproduction of Perionyx excavatus (Perr.) (Megascolecidae) as factors in organic waste management. *Biol Fertil Soils* 1998, 27: 155-161.
87. Edwards CA, Fletcher KE: Interaction between earthworms and microorganisms in organic matter breakdown. *Agric Ecosyst Environ* 1988, 20: 235-249.
88. Edwards CA, Lofty R: *The Biology of Earthworms*. Chapmann and Hall, London; 1977.
89. Elmer WH: Influence of earthworm activity on soil microbes and soilborne diseases of vegetables. *Plant Dis* 2009,93: 175-179.
90. Elvira C, Dominguez J, Sampedro L, Mato S: Vermicomposting for the pulp industry. *Biocycle* 1995, 36: 62-63.
91. Elvira C, Sampedro L, Benítez E, Nogales R: Vermicomposting of sludges from paper mill and dairy industries with Eisenia andrei: a pilot-scale study. *Bioresour Technol* 1998, 63: 205-211.
92. Elvira C, Sampedro L, Dominguez J, Mato S: Vermicomposting of wastewater sludge from paper-pulp industry with nitrogen rich materials. *Soil Biol Biochem* 1997, 9: 759-762.
93. Fracchia L, Dohrmann AB, Martinotti MG, Tebbe CC: Bacterial diversity in a finished compost and vermicompost: differences revealed by cultivation independent analyses of PCR-amplified 16S rRNA genes. *Appl Microbiol Biotechnol* 2006, 71: 942-952.
94. Fragoyiannis DA, McKinlay RG, D'Mello JPF: Interactions of aphids herbivory and nitrogen availability on the total foliar glycoalkoloid content of potato plants. *J Chem Ecol* 2001, 27: 1749-1762.
95. Gajalakshmi S, Ramasamy EV, Abbasi SA: Assessment of sustainable vermiconversion of water hyacinth at different reactor efficiencies employing Eudrilus engeniae Kingburg. *Bioresour Technol* 2001, 80: 131-135.
96. Gajalakshmi S, Ramasamy EV, Abbasi SA: Vermicomposting of paper waste with the anecic earthworm Lampito mauritii Kingburg. *Indian J Chem Technol* 2002, 9: 306-311.

97. Gammack SM, Paterson E, Kemp JS, Cresser MS, Killham K: Factors affecting movement of microorganisms in soils.In *Soil Biochemistry, 7*. Edited by: Stotzky G, Bolla LM. Marcel Dekker, New York; 1992:263-305.

98. Gandhi M, Sangwan V, Kapoor KK, Dilbaghi N: Composting of household wastes with and without earthworms.*Environ Ecol* 1997, 15: 432-434.

99. Ganesh kumar A, Sekaran G: Enteric pathogen modification by anaecic earthworm, Lampito Mauritii. *J Appl Sci Environ Mgt* 2005, 9: 15-17.

100. Gange AC: Translocation of mycorrhizal fungi by earthworms during early succession. *Soil Biol Biochem* 1993, 25:1021-1026.

101. Garcia C, Hernandez T, Costa F: Potential use of dehydrogenase activity as an index of microbial activity in degraded soils. *Commun Soil Sci Plant Anal* 1997, 28: 123-134.

102. Garg P, Gupta A, Satya S: Vermicomposting of different types of waste using Eisenia foetida: a comparative study.*Bioresour Technol* 2006, 97: 391-395.

103. Gavrilov K: Earthworms, producers of biologically active substances. *Zh Obshch Biol* 1963, 24: 149-154.

104. Ghosh M, Chattopadhyay GN, Baral K: Transformation of phosphorus during vermicomposting. *Bioresour Technol*1999, 69: 149-154.

105. Gliotti C, Giusquiani PL, Businelli D, Machioni A: Composition changes of dissolved organic matter in a soil amended with municipal waste compost. *Soil Sci* 1997, 162: 919-926.

106. Goenadi DH, Sudharama IM: Shoot initiation by humic acids of selected tropical crops grown in tissue culture. *Plant Cell Rep* 1995, 15: 59-62.

107. Goldstein J: Compost suppresses diseases in the lab and fields. *BioCycle* 1998, 39: 62-64.

108. Gopal M, Gupta A, Sunil E, Thomas VG: Amplification of plant beneficial microbial communities during conversion of coconut leaf substrate to vermicompost by Eudrilus sp. *Curr Microbiol* 2009, 59: 15-20.

109. Graff O, Makeschin F: Beeinlussung des Ertrags von Weidelgrass (Lolium muttiflorum) Ausscheidungen von Regenwurmen dreier verschiedener Arten. *Pedobiologia* 1980, 20: 176-180.

110. Grappelli A, Galli E, Tomati U: Earthworm casting effect on Agaricus bisporus fructification. *Agrochimica* 1987, 2: 457-462.

111. Guerrero RD, Guerrero LA: Response of eggplant (Solanum melongena) grown in plastic containers to vermicompost and chemical fertilizer. *Asia Life Sciences* 2006, 15: 199-204.

112. Gunadi B, Blount C, Edward CA: The growth and fecundity of Eisenia foetida (Savigny) in cattle solids pre-composted for different periods. *Pedobiologia* 2002, 46: 15-23.

113. Gunathilagraj K, Ravignanam T: Vermicomposting of sericulture wastes. *Madras Agric J* 1996, 83: 455-457.

114. Gupta AK, Pankaj PK, Upadhyava V: Effect of vermicompost, farm yard manure, biofertilizer and chemical fertilizers (N, P, K) on growth, yield and quality of lady's finger (Abelmoschus esculentus). *Pollution Research* 2008, 27: 65-68.

115. Hahn H, Bopp M: A cytokinin test with high specificity. *Planta* 1968, 83: 115-118.

116. Han J, Sun L, Dong X, Cai Z, Yang H, Wang Y, Song W: Characterization of a novel plant growth-promoting bacteria strain Delftia tsuruhatensis HR4 both as a diazotroph and a potential biocontrol agent against various pathogens.*Syst Appl Microbiol* 2005, 28: 66-76.

117. Hand P, Hayes WA, Frankland JC, Satchell JE: Vermicomposting of cow slurry. *Pedobiologia* 1988, 31: 199-209.

118. Haritha Devi S, Vijayalakshmi K, Pavana Jyotsna K, Shaheen SK, Jyothi K, Surekha Rani M: Comparative assessment in enzyme activities and microbial populations during normal and vermicomposting. *J Environ Biol* 2009, 30: 1013-1017.

119. Hartenstein R, Neuhauser EF, Kaplan DL: Reproductive potential of the earthworm Eisenia foetida. *Oecologia* 1979,43: 329-340.

120. Hartenstein R: Assimilation by earthworm Eisenia fetida. In *Earthworm ecology. From Darwin to vermiculture*. Edited by: Satchell JE. Chapman and Hall, London; 1983:297-308.

121. Hashemimajd K, Kalbasi M, Golchin A, Shariatmadari H: Comparison of vermicompost and composts as potting media for growth of tomatoes. *J Plant Nutr* 2004, 27: 1107-1123.

122. Hassen A, Belguith K, Jedidi N, Cherif A, Cherif M, Boudabous A: Microbial characterization during composting of municipal solid waste. *Bioresour Technol* 2001, 80: 217-225.

123. Hawida S, Kapari L, Ossipov V, Ramtala MJ, Ruuhola T, Haukioja E: Foliar phenolics are differently associated with Epirrita autmnata growth and immuno competence. *J Chem Ecol* 2007, 33: 1013-1023.

124. Herms DA: Effects of fertilization on insect resistance of woody ornamental plants. *Environ Entomol* 2002, 31: 923-933.

125. Hoitink HA, Kuter GA: Efects of composts in growth media on soil-bome pathogens. In *The role of organic matter in modern agriculture*. Edited by: Chen Y, Avnimelech Y. Martinus Nijhoff Publishers, Dordrecht; 1986:289-306.

126. Hoitink HA, Stone AG, Han DY: Suppression of plant diseases by compost. *Hort Sci* 1997, 32: 184-187.

127. Hoitink HA, Grebus ME: Composts and Control of Plant Diseases. In *Humic Substances Peats and Sludges Health and Environmental Aspects*. Edited by: Hayes MHB, Wilson WS. Royal Society of Chemistry, Cambridge; 1997:359-366.

128. Holtzclaw KM, Sposito G: Analytical properties of the soluble metal-complexing fractions in sludge-soil mixtures. IV. Determination of carboxyl groups in fulvic

acid. *Soil Sci Soc Am J* 1979, 43: 318-323.

129. Huelsman MF, Edwards CA, Lawrence JL, Clarke-Harris DO: A study of the effect of soil nitrogen levels on the incidence of insect pests and predators in Jamaican sweet potato (Ipomoea batatus) and Callaloo (Amaranthus).*Proc Brighton Pest Control Conference: Pests and Diseases* 2000, 8D–13: 895-900.
130. Inbal E, Feldman M: The response of a hormonal mutant of common wheat to bacteria of the Azospirillium. *Israel J Bot* 1982, 31: 257-263.
131. Ismail SA: Earthworms in soil fertility management. In *Organic Agriculture* Edited by: Thampan PK. 1995, 77-100.
132. Ismail SA: *Vermicology: The biology of Earthworms*. Orient Longman Limited, Chennai; 1997.
133. Jadhav AD, Talashilkar SC, Pawar AG: Influence of the conjunctive use of FYM, vermicompost andurea on growth and nutrient uptake in rice. *J Maharashtra Agric Univ* 1997, 22: 249-250.
134. Jagnow G: Inoculation of cereal crops and forage grasses with nitrogen-fixing rhizosphere bacteria: a possible cause of success and failure with regard to yield response - a review. *Z Pflanzenernaehr Dueng Bodenkde* 1987, 150:361-368.
135. Jambhekar H: *Use of earthworm as a potential source of decompose organic wastes*. Proc Nat Sem Org Fmg, Coimbatore; 1992:52-53.
136. Jannsson RK, Smilowitz Z: Influence of nitrogen on population parameters of potato insects: abundance, population growth and within-plant distribution of the green peach aphid, Myzus persicae (Homoptera: Aphididae). *Environ Entomol* 1986, 15: 49-55.
137. Jha BK, Gandhi Pragash M, Cletus J, Raman G, Sakthivel N: Simultaneous phosphate solubilization potential and antifungal activity of new fluorescent pseudomonad strains, Pseudomonas aeruginosa, P. plecoglossicida and P. mosselii. *W J Microbiol Biotech* 2009, 25: 573-581.
138. Jolly JM, Lappin-Scott HM, Anderson JM, Clegg CD: Scanning electron microscopy of the gut microflora of two earthworms: Lumbricus terrestris and Octolasion cyaneum. *Microbial Ecol* 1993, 26: 235-245.
139. Kale RD, Bano K: Field Trials with vermicompost (vee comp. E. 8. UAS) on organic fertilizers. In *Proceedings of the national seminar on organic waste utilization*. Edited by: Dass MC, Senapati BK, Mishra PC. Sri Artatrana Ront, Burla; 1986:151-157.
140. Kale RD, Bano K: Earthworm cultivation and culturing techniques for the production of vee COMP83E UAS. *Mysore J Agric Sci* 1988, 2: 339-344.
141. Kale RD, Mallesh BC, Bano K, Bagyaray DJ: Influence of vermicompost application on the available macronutrients and selected microbial populations in paddy field. *Soil Biol Biochem* 1992, 24: 1317-1320.
142. Kale RD: Vermicomposting has a bright scope. *Indian Silk* 1995, 34: 6-9.
143. Kalembasa D: The influence of vermicomposts on yield and chemical composition of tomato. *Zesz Probl Post Nauk Roln* 1996, 437: 249-252.

144. Kannangowa T, Utkhede RS, Paul JW, Punja ZK: Effect of mesophilic and thermophilic composts on suppression of Fusarium root and stem rot of greenhouse cucumber. *Canad J Microbiol* 2000, 46: 1021-1022.

145. Karmegam N, Alagermalai K, Daniel T: Effect of vermicompost on the growth and yield of greengram (Phaseolus aureus Rob.). *Trop Agric* 1999, 76: 143-146.

146. Karmegam N, Daniel T: Effect of biodigested slurry and vermicompost on the growth and yield of cowpea (Vigna unguiculata (L.). *Environ Ecol* 2000, 18: 367-370.

147. Karmegam N, Daniel T: Effect of vermi-compost and chemical fertilizer on growth and yield of Hyacinth Bean (Lablab purpureas). *Dynamic Soil, Dynamic Plant, Global Science Books* 2008, 2: 77-81.

148. Karsten GR, Drake HL: Comparative assessment of the aerobic and anaerobic microfloras of earthworm guts and forest soils. *Appl Environ Microbiol* 1995, 61: 1039-1044.

149. Karsten GR, Drake HL: Denitrifying bacteria in the earthworm gastrointestinal tract and in vivo emission of nitrous oxide (N2O) by earthworms. *Appl Environ Microbiol* 1997, 63: 1878-1882.

150. Kerry B: Fungal parasites of cyst nematodes. In *Biological Interactions in Soil.* Edited by: Edwards CA, Stinner BR, Stinner D, Rabatin S. Elsevier, Amsterdam; 1988:293-306.

151. Khambata SR, Bhat JV: Studies on a new oxalate-decomposing bacterium, Pseudomonas oxalaticus. *J Bacteriol*1953, 66: 505-507.

152. Knuutinen J, Palm H, Hakala H, Haimi J, Huhta V, Salminen J: Polychlorinated phenols and their metabolites in soil and earthworms of a saw mill environment. *Chemosphere* 1990, 20: 609-623.

153. Kolodziej M, Kostecka J: Some qualitative features of the cucumbers and carrots cultivated on the vermicompost.*Zeszyty Naukowe Akademii Rolniczej W Krakowie* 1994, 292: 89-94.

154. Kooch Y, Jalilvand H: Earthworm as ecosystem engineers and the most important detritivors in forst soils. *Pak J Boil Sci* 2008, 11: 819-825.

155. Kostecka J, Blazej JB, Kolodziej M: Investigations on application of vermicompost in potatoes farming in second year of experiment. *Zeszyty Naukowe Akademii Rolniczej W Krakowie* 1996a, 310: 69-77.

156. Koul O: Phytochemicals and insect control: an antifeedant approach. *Crit Rev Plant Sci* 2008, 27: 1-24.

157. Krishnamoorthy RV, Vajranabhiah SN: Biological activity of earthworm casts: An assessment of plant growth promoter levels in casts. *Proc Indian Acad Sci (Anim Sci)* 1986, 95: 341-35.

158. Kristufek V, Ravasz K, Pizl V: Actinomycete communities in earthworm guts and surrounding soil. *Pedobiologia* 1993,37: 379-384.

159. Kurowska A, Gora J, Kalemba D: Effects of plant phenols on insects. *Pol Wiad Chem* 1990, 44: 399-409.

160. Kuter GA, Nelson GB, Hoitink HA, Madden LV: Fungal population in container media amended with composted hardwood bark suppressive and conductive to Rhizoctonia damping-off. *Phytopathology* 1983, 73: 1450-1456.

161. Lavelle E, Barois I, Martin A, Zaidi Z, Schaefer R: Management of earthworm populations in agro-ecosystems: A possible way to maintain soil quality? In *Ecology of Arable Land: Perspectives and Challenges*. Edited by: Clarholm M, Bergstrom L. Kluwer Academic Publishers, London; 1989:109-122.

162. Lavelle P, Martin A: Small-scale and large-scale effects of endogeic earthworms on soil organic matter dynamics in soils of the humid tropics. *Soil Biol Biochem* 1992, 12: 1491-1498.

163. Lazarovits G, Tenuta M, Conn KL, Gullino ML, Katan J, Matta A: Utilization of high nitrogen and swine manure amendments for control of soil-bome diseases: efficacy and mode of action. *Acta Hortic* 2000, 5: 559-564.

164. Lazcano C, Gomez-Brandon M, Dominguez J: Comparison of the effectiveness of composting and vermicomposting for the biological stabilization of cattle manure. *Chemosphere* 2008, 72: 1013-1019.

165. Lee KE: *Earthworms: Their Ecology and Relationships with Soils and Land Use*. Academic Press, Sydney; 1985.

166. Lee YS, Bartlett RJ: Stimulation of plant growth by humic substances. *Soil Sci Soc Am J* 1976, 40: 876-879.

167. Lunt HA, Jacobson HGM: The chemical composition of earthworm casts. *Soil Sci* 1944, 58: 367-375.

168. Maboeta MS, Van Rensburg L: Vermicomposting of industrially produced wood chips and sewage sludge utilizing Eisenia foetida. *Ecotoxicol Environ Saf* 2003, 56: 265-270.

169. Madsen EL, Alexander M: Transport of Rhizobium and Pseudomonas through soil. *Soil Sci Soc Am J* 1982, 46: 557-560.

170. Mahanil S, Attajarusit J, Stout MJ, Thipayong P: Over expression of tomato phenol oxidase increases resistance to the common cutworm. *Plant Sci* 2008, 174: 456-466.

171. Maheswarappa HP, Nanjappa HV, Hegde MR: Influence of organic manures on yield of arrowroot, soil physico-chemical and biological properties when grown as intercrop in coconut garden. *Ann Agr Res* 1999, 20: 318-323.

172. Mahmoud SA, Ramadan Z, Thabet EM, Khater T: Production of plant growth promoting substance rhizosphere organisms. *Zentrbl Mikrobiol* 1984, 139: 227-232.

173. Makulec G: The role of Lumbricus rubellus Hoffm. In determining biotic and abiotic properties of peat soils. *Pol J Ecol* 2002, 50: 301-339.

174. Marinari S, Masciandaro G, Ceccanti B, Grego S: Influence of organic and mineral fertilisers on soil biological and physical properties. *Bioresour Technol* 2000, 72: 9-17.

175. Martin JP: *Darwin on earthworms: the formation of vegetable moulds*. Bookworm

Publishing, Ontario; 1976.

176. Masciandaro G, Ceccanti B, Gracia C: Soil agro-ecological management: fertigation and vermicompost treatments.*Bioresour Technol* 1997, 59: 199-206.

177. Masciandaro G, Ceccanti B, Gracia C: `In situ' vermicomposting of biological sludges and impacts on soil quality.*Soil Biol Biochem* 2000, 32: 1015-1024.

178. Meena RN, Singh Y, Singh SP, Singh JP, Singh K: Effect of sources and level of organic manure on yield, quality and economics of garden pea (Pisum sativam L.) in eastern uttar pradesh. *Vegetable Science* 2007, 34: 60-63.

179. Mitchell A, Edwards CA: The production of vermicompost using Eisenia fetida from cattle manure. *Soil Biol Biochem*1997, 29: 3-4.

180. Mitchell A: Production of Eisenia fetida and vermicompost from feed-lot cattle manure. *Soil Biol Biochem* 1997, 29:763-766.

181. Mitchell MJ, Hornor SG, Abrams BI: Decomposition of sewage sludge in drying beds and the potential role of the earthworm, Eisenia foetida. *J Environ Qual* 1980, 9: 373-378.

182. Mitchell MJ: Role of invertebrates and microorganisms in sludge deomposition. In *Utilisation of soil organisms in sludge management. natural technology information services*. Edited by: Hartenstein R. Springfield, Virginia; 1978:35-50.

183. Monroy F, Aira M, Domínguez J: Reduction of total coliform numbers during vermicomposting is caused by short-term direct effects of earthworms on microorganisms and depends on the dose of application of pig slurry. *Sci Tot Environ* 2009, 407: 5411-5416.

184. Moody SA, Piearce TG, Dighton J: Fate of some fungal spores associated with wheat straw decomposition on passage through the guts of Lumbricus terrestris and Aporrectodea longa. *Soil Biol Biochem* 1996, 28: 533-537.

185. Morra L, Palumbo AD, Bilotto M, Ovieno P, Ptcascia S: Soil solarization: organic fertilization grafting contribute to build an integrated production system in a tomato-zucchini sequence. *Colture-Protte* 1998, 27: 63-70.

186. Munnoli PM, Da Silva JAT, Saroj B: Dynamics of the soil-earthworm-plant relationship: a review. *Dynamic soil, dynamic plant* 2010, 1-21.

187. Muscolo A, Bovalo F, Gionfriddo F, Nardi S: Earthworm humic matter produces auxin-like effect on Daucus carota cell growth and nitrate metabolism. *Soil Biol Biochem* 1999, 31: 1303-1311.

188. Muscolo A, Felici M, Concheri G, Nardi S: Effect of earthworm humic substances on esterase and peroxidase activity during growth of leaf explants of Nicotiana plumbaginifolia. *Biol Fertil Soils* 1993, 15: 127-131.

189. Muscolo A, Panuccio MR, Abenavoli MR, Concheri G, Nardi S: Effect of molecular complexity acidity of earthworm faeces humic fractions on glutamale dehydrogenase, glutamine synthetase, and phosphoenolpyruvate carboxylase in Daucus carota II cells. *Biol Fertil Soils* 1996, 22: 83-88.

190. Mylonas VA, Mccants CB: Effects of humic and fulvic acids on growth of tobacco. I. Root initiation and elongation.*Plant Soil* 1980, 54: 485-490.

191. Nagavallemma KP, Wani SP, Stephane L, Padmaja VV, Vineela C, Babu Rao M, Sahrawat KL: *Vermicomposting: Recycling wastes into valuable organic fertilizer. Global Theme on Agrecosystems Report no.8. Patancheru 502324*. International Crops Research Institute for the Semi-Arid Tropics, Andhra Pradesh; 2004:20.

192. Nakamura Y: Interactions between earthworms and microorganisms in biological control of plant root pathogens.*Farming Jpn* 1996, 30: 37-43.

193. Nakasone AK, Bettiol W, de Souza RM: The effect of water extracts of organic matter on plant pathogens. *Summa Phytopathology* 1999, 25: 330-335.

194. Nardi S, Arnoldi G, Dell'Agnola G: Release of hormone-like activities from Alloborophora rosea and Alloborophora caliginosa feces. *J Soil Sci* 1988, 68: 563-657.

195. Nardi S, Dell'Agnola G, Nuti PM: *Humus production from farmyard wastes by vermicomposting. Proc. Int. Symp. On Agricultural and Environmental Prospects in Earthworm Farming*. Rome; 1983:87-94.

196. Ndegwa PM, Thompson SA, Das KC: Effects of stocking density and feeding rate on vermicomposting of biosolids.*Bioresour Technol* 2000, 71: 5-12.

197. Nechitaylo TY, Yakimov MM, Godinho M, Timmis KN, Belogolova E, Byzov BA, *et al.*: Effect of the earthworms Lumbricus terrestris and Aporrectodea caliginosa on bacterial diversity in soil. *Microbial Ecol* 2010, 59: 574-587.

198. Nethra NN, Jayaprasad KV, Kale RD: China aster (Callistephus chinensis (L)) cultivation using vermicompost as organic amendment. *Crop Research, Hisar* 1999, 17: 209-215.

199. Orozco FH, Cegarra J, Trujillo LM, Roig A: Vermicomposting of coffee pulp using the earthworm Eisenia fetida: effects on C and N contents and the availability of nutrients. *Biol Fertil Soils* 1996, 22: 162-166.

200. Park SR, Cho EJ, Yu KH, Kim YS, Suh JJ, Chang CS: Endogenous phenoloxidase from an earthworm Lumbricus rubellus. *Tongmul Hakoehi* 1996, 39: 36-46.

201. Parle JN: A Microbiological Study of Earthworm Casts. *J Gen Microbiol* 1963, 31: 13-22.

202. Parthasarathi K, Ranganathan LS: Pressmud vermicast are hot spots of fungi and bacteria. *Ecol Environ Cons* 1998, 4:81-86.

203. Pathma J, Ayyadurai N, Sakthivel N: Assessment of Genetic and Functional Relationship of Antagonistic Fluorescent Pseudomonads of Rice Rhizosphere by Repetitive Sequence, Protein Coding Sequence and Functional Gene Analyses. *J Microbiol* 2010, 48: 715-727.

204. Pathma J, Kamaraj Kennedy R, Sakthivel N: Mechanisms of fluorescent pseudomonads that mediate biological control of phytopathogens and plant growth promotion of crop plants. In *Bacteria in Agrobiology: Plant Growth Responses*. Edited by: Maheswari DK. SpringerVerlag, Berlin; 2011a:77-105.

205. Pathma J, Rahul GR, Kamaraj Kennedy R, Subashri R, Sakthivel N: Secondary metabolite production by bacterial antagonists. *Journal of Biological Control* 2011, 25: 165-181.

206. Patil SL, Sheelavantar MN: Effect of moisture conservation practices, organic sources and nitrogen levels on yield, water use and root development of rabi sorghum (Sorghum bicolor (L.)) in the vertisols of semiarid tropics. *Ann Agric Res* 2000, 21: 32-36.

207. Patriquin DG, Baines D, Abboud A: Diseases, pests and soil fertility. In *Soil Management in Sustainable Agriculture*. Edited by: Cook HF, Lee HC. Wye College Press, Wye; 1995:161-174.

208. Pedersen JC, Hendriksen NB: Effect of passage through the intestinal tract of detritivore earthworms (Lumbricus spp.) on the number of selected gram-negative and total bacteria. *Biol Fertil Soils* 1993, 16: 227-232.

209. Petersen H, Luxton MA: A comparative analysis of soil fauna populations and their role in decomposition process.*Oikos* 1982, 39: 287-388.

210. Phelan PL, Norris KH, Mason JF: Soil management history and host preference by Ostrinia nubilatis: evidence for plant mineral balance mediating insect-plant interactions. *Environ Entom* 1996, 25: 1329-1336.

211. Phelan PL: Connecting below-ground and above-ground food webs: the role of organic matter in biological buffering. In *Soil Organic Matter in sustainable agriculture*. Edited by: Magadoff F, Well RR. CRC Press, Boca Raton; 2004:199-226.

212. Pinel N, Davidson SK, Stahl DA: Verminephrobacter eiseniae gen. nov., sp. nov., a nephridial symbiont of the earthworm Eisenia foetida (Savigny). *Int J Syst Evol Microbiol* 2008, 58: 2147-2157.

213. Pitt D, Tilston EL, Groenhof AC, Szmidt RA: Recycled organic materials (ROM) in the control of plant disease. *Acta Hortic* 1998, 469: 391-403.

214. Pizl V, Novokova A: Interactions between microfungi and Eisenia andrei (Oligochaeta) during cattle manure vermicomposting. *Pedobiologia* 1993, 47: 895-899.

215. QiTian S: Research on prevention and elimination of agricultural pests by ginkgo phenols phenolic acids. *Chem Ind For Prod* 2004, 24: 83.

216. Raguchander T, Rajappan K, Samiyappan R: Influence of biocontrol agents and organic amendments on soybean root rot. *Int J Trop Agri* 1998, 16: 247-252.

217. Ramesh P: Effects of vermicomposts and vermicornposting on damage by sucking pests to ground nut (Arachis hypogea). *Indian J Agri Sci* 2000, 70: 334.

218. Rao KR, Rao PA, Rao KT: Influence of fertilizers and manures on the population of coccinellid beetles and spiders in groundnut ecosystem. *Ann Plant Protect Sci* 2001, 9: 43-46.

219. Rao KR: Induce host plant resistance in the management sucking pests of groundnut. *Ann Plant Protect Sci* 2002,10: 45-50.

220. Rao KR: Influence of host plant nutrition on the incidence of Spodoptera litura and Helicoverpa armigera on groundnuts. *Indian J Entomol* 2003, 65: 386-392.

221. Ravindra NP, Raman G, Badri Narayanan K, Sakthivel N: Assessment of genetic and functional diversity of phosphate solubilizing fluorescent pseudomonads

isolated from rhizospheric soil. *BMC Microbiol* 2008, 8: 230.

222. Reeh U: Influence of population densities on growth and reproduction of the earthworm Eisenia andrei on pig manure. *Soil Biol Biochem* 1992, 24: 1327-1331.
223. Ribeiro CF, Mizobutsi EH, Silva DG, Pereira JCR, Zambolim L: Control of Meloidognye javanica on lettuce with organic amendments. *Fitopatol Brasileira* 1998, 23: 42-44.
224. Riffaldi R, Levi-Minzi R: Osservazioni preliminari sul ruolo dell Eisenia foetida nell›umificazione del letame.*Agrochimica* 1983, 27: 271-274.
225. Rivera AMC, Wright ER, López MV, Fabrizio MC: Temperature and dosage dependent suppression of damping-off caused by Rhizoctonia solani in vermicompost amended nurseries of white pumpkin. *Phyton* 2004, 53: 131-136.
226. Rodriguez JA, Zavaleta E, Sanchez P, Gonzalez H: The effect of vermicomposts on plant nutrition, yield and incidence of root and crown rot of gerbera (Gerbera jamesonii H. Bolus). *Fitopatol* 2000, 35: 66-79.
227. Rodriguez-Kabana R: Organic and inorganic amendments to soil as nematode suppressants. *J Nematol* 1986, 18:129-135.
228. Rouelle J: Introduction of an amoeba and Rhizobium Japonicum into the gut of Eisenia fetida (Sav.) and Lumbricus terrestris L. In *Earthworm Ecology: From Darwin to Vermiculture*. Edited by: Satchel JE. Chapman and Hall, New York; 1983:375-381.
229. Sánchez-Monedero MA, Roig A, Paredes C, Bernal MP: Nitrogen transformation during organic waste composting by the Rutgers system and its effects on ph, EC and maturity of the composting mixtures. *Bioresour Technol* 2001, 78:301-308.
230. Sainz MJ, Taboada-Castro MT, Vilariño A: Growth, mineral nutrition and mycorrhizal colonization of red clover and cucumber plants grown in a soil amended with composted urban wastes. *Plant Soil* 1998, 205: 85-92.
231. Saumaya G, Giraddi RS, Patil RH: Utility of vermiwash for the management of thrips and mites on chilli (Capiscum annum) amended with soil organics. *Karnataka J Agric Sci* 2007, 20: 657-659.
232. Scheu S: Automated measurement of the respiratory response of soil micro-compartments: active microbial biomass in earthworm faeces. *Soil Biol Biochem* 1992, 24: 1113-1118.
233. Scheuerell SJ, Sullivan DM, Mahaffee WF: Suppression of seedling damping-off caused by Pythium ultimum, and Rhizoctonia solani in container media amended with a diverse range of Pacific Northwest compost sources. *Phytopathology* 2005, 95: 306-315.
234. Schmidt O, Doubre BM, Ryder MH, Killman K: Population dynamics of Pseudomonas corrugata 2140R LUX8 in earthworm food and in earthworm cast. *Soil Biol Biochem* 1997, 29: 523-528.
235. Sembdner G, Borgman E, Schneider G, Liebisch HW, Miersch O, Adam G, Lischewski M, Schieber K: Biological activity of some conjugated

gibberellins. *Planta* 1976, 132: 249-257.

236. Senesi N, Saiz-Jimenez C, Miano TM: Spectroscopic characterization of metal-humic acid-like complexes of earthworm-composted organic wastes. *Sci Total Environ* 1992, 117–118: 111-120.

237. Sharma S, Pradhan K, Satya S, Vasudevan P: Potentiality of earthworms for waste management and in other uses - A Review. *The Journal of American Science* 2005, 1: 4-16.

238. Sharpley AN, Syers JK: Potential role of earthworm casts for the phosphorous enrichment of runoff waters. *Soil Biol Biochem* 1976, 8: 341-346.

239. Shi-wei Z, Fu-zhen H: The nitrogen uptake efficiency from 15N labeled chemical fertilizer in the presence of earthworm manure (cast). In *Advances in Management and Conservation of Soil Fauna*. Edited by: Veeresh GK, Rajagopal D, Viraktamath CA. Oxford and IBH publishing Co, New Delhi; 1991:539-542.

240. Siddiqui ZA, Mahmood I: Role of bacteria in the management of plant parasitic nematodes: a review. *Bioresour Technol* 1999, 69: 167-179.

241. Sidhu J, Gibbs RA, Ho GE, Unkovich I: The role of indigenous microorganisms in suppression of Salmonella regrowth in composted biosolids. *Water Res* 2001, 35: 913-920.

242. Simsek Ersahin Y, Haktanir K, Yanar Y: Vermicompost suppresses Rhizoctonia solani Kühn in cucumber seedlings. *J Plant Dis Protect* 2009, 9: 15-17.

243. Singh R, Sharma RR, Kumar S, Gupta RK, Patil RT: Vermicompost substitution influences growth, physiological disorders, fruit yield and quality of strawberry (Fragaria x ananassa Duch.). *Bioresour Technol* 2008, 99: 8507-8511.

244. Singh UP, Maurya S, Singh DP: Antifungal activity and induced resistance in pea by aqueous extract of vermicompost and for control of powdery mildew of pea and balsam. *J Plant Dis Protect* 2003, 110: 544-553.

245. Singleton DR, Hendrixb PF, Colemanb DC, Whitmana WB: Identification of uncultured bacteria tightly associated with the intestine of the earthworm Lumbricus rubellus (Lumbricidae; Oligochaeta). *Soil Biol Biochem* 2003, 35: 1547-1555.

246. Sinha RK, Agarwal S, Chauhan K, Valani D: The wonders of earthworms and its vermicompost in farm production: Charles Darwin's 'friends of farmers', with potential to replace destructive chemical fertilizers from agriculture.*Agricultural sciences* 2010, 1: 76-94.

247. Sinha RK, Bharambe G, Chaudhari U: Sewage treatment by vermifiltration with synchronous treatment of sludge by earthworms: a low-cost sustainable technology over conventional systems with potential for decentralization. *The Environmentalist* 2008, 28: 409-420.

248. Sinha RK, Heart S, Agarwal S, Asadi R, Carretero E: Vermiculture technology for environmental management: study of the action of the earthworms Eisenia foetida, Eudrilus euginae and Perionyx excavatus on biodegradation of some community wastes in India and Australia. *The Environmentalist* 2002, 22: 261-268.

249. Sinha RK, Herat S, Valani D, Chauhan K: Vermiculture and sustainable agriculture. *Am-Euras J Agric and Environ Sci, IDOSI Publication* 2009, 5: 1-55.

250. Sipes BS, Arakaki AS, Schmitt DP, Hamasaki RT: Root-knot nematode management in tropical cropping systems with organic products. *J Sustain Agr* 1999, 15: 69-76.

251. Sreenivas C, Muralidhar S, Rao MS: Vermicompost, a viable component of IPNSS in nitrogen nutrition of ridge gourd. *Ann Agr Res* 2000, 21: 108-113.

252. Steffen KL, Dan MS, Harper JK, Fleischer SJ, Mkhize SS, Grenoble DW, MacNab AA, Fager K: Evaluation of the initial season for implementation of four tomato production systems. *J Am Soc Hort Sci* 1995, 120: 148-156.

253. Stephens PM, Davoren CW, Doube BM, Ryder MH: Reduced superiority of Rhizoctonia solani disease on wheat seedlings associated with the presence of the earthworm Aporrectodea trapezoids. *Soil Biol Biochem* 1993, 11:1477-1484.

254. Stephens PM, Davoren CW, Ryder MH, Doube BM, Correll RL: Field evidence for reduced severity of Rhizoctonia bare-patch disease of wheat, due to the presence of the earthworms Aporrectodea rosea and Aporrectodea trapezoides. *Soil Biol Biochem* 1994, 26: 1495-1500.

255. Stephens PM, Davoren CW, Ryder MH, Doube BM: Influence of the earthworm Aporrectodea trapezoides (Lumbricidae) on the colonization of alfalfa (Medicago sativa L.) roots by Rhizobium melilotti strain LS-30R and the survival of L5-30R in soil. *Biol Fertil Soils* 1994, 18: 63-70.

256. Stephens PM, Davoren CW: Influence of the earthworms Aporrectodea trapezoides and A. rosea on the disease severity of Rhizoctonia solani on subterranean cloves and ryegrass. *Soil Biol Biochem* 1997, 29: 511-516.

257. Stone AG, Scheurell SJ, Darby HM: Suppression of soilborne diseases in field agricultural systems: organic matter management, cover cropping and other cultural practices. In *Soil Organic Matter in Sustainable Agriculture*. Edited by: Magdoff F, Weil . CRC Press LLC, Boca Raton; 2004:131-177.

258. Subler S, Edwards CA, Metzger PJ: Comparing vermicomposts and composts. *Biocycle* 1998, 39: 63-66.

259. Sudhakar K, Punnaiah KC, Krishnayya PV: Influence of organic and inorganic fertilizers and certain insecticides on the incidence of shoot and fruit borer, Leucinodes orbonalis Guen, infesting brinjal. *J Entomol Res* 1998, 22: 283-286.

260. Suhane RK: *Vermicompost*. Publication of Rajendra Agriculture University, Pusa; 2007. 88

261. Summers G, Felton GW: Prooxidation effects of phenolic acids on the generalist herbivore Helicoverpa zea: potential mode of action of phenolic compounds on plant anti-herbivory chemistry. *Insect Biochem Mol Biol* 1994,24: 943-953.

262. Sunish KR, Ayyadurai N, Pandiaraja P, Reddy AV, Venkateshwarlu Y, Prakash O, Sakthivel N: Characterization of antifungal metabolite produced by a new strain Pseudomonas aeruginosa PuPa3 that exhibits broad-spectrum antifungal activity and biofertility traits. *J Appl Microbiol* 2005, 98: 145-154.

263. Suthar S: Evidence of plant hormone like sub-stances in vermiwash: An ecologically safe option of synthetic chemicals for sustainable farming. *J Ecol Eng* 2010, 36: 1089-1092.

264. Suthar S, Singh S: Vermicomposting of domestic waste by using two epigeic earthworms (Perionyx excavatus and Perionyx sansibaricus). *Int J Evniron Sci and Technol* 2008, 5: 99-106.

265. Swathi P, Rao KT, Rao PA: Studies on control of root-knot nematode Meloidogyne incognita in tobacco miniseries.*Tobacco Res* 1998, 1: 26-30.

266. Szcech M, Rondomanski W, Brzeski MW, Smolinska U, Kotowski JF: Suppressive effect of a commercial earthworm compost on some root infecting pathogens of cabbage and tomato. *Biol Agric and Hortic* 1993, 10: 47-52.

267. Szczech M, Smolinska U: Comparison of suppressiveness of vermicomposts produced from animal manures and sewage sludge against Phytophthora nicotianae Breda de Haan var. nicotiannae. *J Phytopathology* 2001, 149: 77-82.

268. Szczech MM: Suppressiveness of vermicomposts against fusarium wilt of tomato. *J Phytopathology* 1999, 147: 155-161.

269. Tajbakhsh J, Abdoli MA, Mohammadi Goltapeh E, Alahdadi I, Malakouti MJ: Trend of physico-chemical properties change in recycling spent mushroom compost through vermicomposting by epigeic earthworms Eisenia foetida and E. andrei. *J Agric Technol* 2008, 4: 185-198.

270. Tan KH, Tantiwiramanond D: Effect of humic acids on nodulation and dry matter production of soybean, peanut, and clover. *Soil Sci Soc Am J* 1983, 47: 1121-1124.

271. Thoden TC, Korthals GW, Termorshuizen : Organic amendments and their influences on plant-parasitic and free living nematodes: a promosing method for nematode management. *Nematology* 2011, 13: 133-153.

272. Tiquia SM: Microbiological parameters as indicators of compost maturity. *J Appl Microbiol* 2005, 99: 816-828.

273. Tiunov AV, Scheu S: Microfungal communities in soil litter and casts of Lumbricus terrestris (Lumbricidae): a laboratory experiment. *Appl Soil Ecol* 2000, 14: 17-26.

274. Tiwari SC, Tiwari BK, Mishra RR: Microbial populations, enzyme activities and nitrogen, phosphorous, potassium enrichment in earthworm casts and in the surrounding soil of pine apple plantation. *Biol Fertil Soils* 1989, 8: 178-182.

275. Tomati U, Grapppelli A, Galli E: The presence of growth regulators in earthworm worked waste. In *On Earthworms. Proceedings of International Symposium on Earthworms, Selected Symposia and Monographs, Union Zoologica Italian, 2.* Edited by: Bonvicini Paglioi AM, Omodeo P. Modena, Mucchi; 1987:423-435.

276. Tomati U, Grapppelli A, Galli E: The hormone-like effect of earthworm casts on plant growth. *Biol Fertil Soils* 1988, 5:288-294.

277. Toyota K, Kimura M: Microbial community indigenous to the earthworm Eisenia foetida. *Biol Fertil Soils* 2000, 31: 187-190.

278. Trevors JT: Dehydrogenase activity in soil. A comparison between the INT and TTC assay. *Soil Biol Biochem* 1984,16: 673-674.

279. Umesh B, Mathur LK, Verma JN, Srivastava : Effects of vermicomposting on microbiological flora of infected biomedical waste. *ISHWM Journal* 2006, 5: 28-33.

280. Vadiraj BA, Siddagangaiah D, Potty SN: Response of coriander (Coriandrum sativum L.) cultivars to graded levels of vermicompost. *J Spices Aromatic Crops* 1998, 7: 141-143.

281. Valdrighi MM, Pera A, Agnolucci M, Frassinetti S, Lunardi D, Vallini G: Effects of compost-derived humic acids on vegetable biomass production and microbial growth within a plant (Cichorium intybus) soil system: a comparative study. *Agric Ecosyst Environ* 1996, 58: 133-144.

282. Valenzuela O, Gluadia Y, Gallardo S: Use of vermicompost as a growing medium for tomato seedlings (cv. Pltense).*Revista Cientifica Agropecuaria* 1997, 1: 15-21.

283. Vaz-Moreira I, Maria E, Silva CM, Manaia Olga C, Nunes : Diversity of Bacterial Isolates from Commercial and Homemade Composts. *Microbial Ecol* 2008, 55: 714-722.

284. Vinken R, Schaeffer A, Ji R: Abiotic association of soil-borne monomeric phenols with humic acids. *Org Geochem*2005, 36: 583-593.

285. Vivas A, Moreno B, Garcia-Rodriguez S, Benitez E: Assessing the impact of composting and vermicomposting on bacterial community size and structure, and functional diversity of an olive-mill waste. *Bioresour Technol* 2009, 100:1319-1326.

286. Webster KA: *Vermicompost increases yield of cherries for three years after a single application*. EcoResearch, South Australia; 2005.

287. Weltzien HC: Some effects of composted organic materials on plant health. *Agric Ecosyst Environ* 1989, 27: 439-446.

288. Wilson DP, Carlile WR: Plant growth in potting media containing worm-worked duck waste. *Acta Hortic* 1989, 238:205-220.

289. Yardim EN, Arancon NQ, Edwards CA, OliverTJ BRJ: Suppression of tomato hornworm (Manduca quinquemaculata) and cucumber beetles (Acalymma vittatum and Diabotrica undecimpunctata) populations and damage by vermicomposts. *Pedobiologia* 2006, 50: 23-29.

290. Yardim EN, Edwards CA: Effects of organic and synthetic fertilizer sources on pest and predatory insects associated with tomatoes. *Phytoparasitica* 2003, 31: 324-329.

291. Yasir M, Aslam Z, Kim SW, Lee SW, Jeon CO, Chung YR: Bacterial community composition and chitinase gene diversity of vermicompost with antifungal activity. *Bioresour Technol* 2009, 100: 4396-4403.

292. Yasir M, Aslam Z, Song GC, Jeon CO, Chung YR: Eiseniicola composti gen. nov., sp. nov., with antifungal activity against plant pathogenic fungi. *Int J Sys*

Evol Microbiol 2009b, 60: 268.

293. Yeates GW: Soil nematode populations depressed in the presence of earthworms. *Pedobiologiaogia* 1981, 22: 191-202.
294. Zhang BG, Li GT, Shen TS, Wang JK, Sun Z: Changes in microbial biomass C, N, and P and enzyme activities in soil incubated with the earthworms Metaphire guillelmi or Eisenia foetida. *Soil Biol Biochem* 2000, 32: 2055-2062.

9

CHAPTER

ANAEROBIC BIODIGESTION OF MANURE FROM FINISHING PIG SUPPLEMENTED WITH RACTOPAMINE OVER DIFFERENT PERIODS

Tânia M. B. dos Santos, Paula S. F. Trevizan , Cristiane A. N. Xavier, Charles Kiefer, André L. J. Ferraz

Keywords: β-adrenergic, biogas, ractopamine hydrochloride, environmental impact and nutrient partitioning

INTRODUCTION

Ractopamine hydrochloride, a beta-adrenergic agonist with nutrient partitioning function, has been used in pig farming due to its ability to redistribute nutrients that would be destined for lipids synthesis and deposition for muscular tissue deposition (Ferreira et al., 2013).

It is a synthetic drug and its use as a growth promoter is approved in countries with significant pig production, as the United States, Brazil and Canada, which currently occupy respectively the third, fourth and seventh positions in the ranking of world production. However, its use is prohibited in countries such as China and the European Union (Alemanno and Capodieci, 2012; Almeida et al., 2012).

Among the factors that affect the ractopamine efficiency in pig diets are the supplementation level, which can vary from 5 to 20 ppm, and administration time, which is generally performed in the finishing phase for periods ranging from 7 to 41 days pre-slaughter (Neill et al., 2010).

The benefits of ractopamine as a feed additive for pigs are well known in the

Citation: Santos, Tânia M. B. dos, Trevizan, Paula S. F., Xavier, Cristiane A. N., Kiefer, Charles, & Ferraz, André L. J.. (2016). Anaerobic Biodigestion of Manure from Finishing Pig Supplemented With Ractopamine Over Different Periods. Engenharia Agrícola, 36(3), 399-407. https://dx.doi.org/10.1590/1809-4430-Eng.Agric.v36n3p399-407/2016.

scientific literature regarding the improvements in the performance and quantitative characteristics of carcass (Kutzler et al., 2011; Pereira et al., 2011; Marçal et al., 2015). However, it is scarce the information about the effects of this additive on the manure characteristics and resulting effects on the environment and/or manure treatment systems.

In the United States, studies have reported environmental benefits brought by ractopamine in pig feeds, since a same amount of meat could be produced with reduction of 5.3 to 6.3% in slaughtered animals; hence, reducing the amounts of consumed soybean and corn. This may indirectly decrease the use of fertilizers, pesticides and water in those crops, besides of decreasing manure generation and nutrient concentration (Woods et al., 2011).

According to Decamp et al. (2001), pigs fed with ractopamine had a decrease in the total N excretion of 43% via urine and of 33% via feces. According to the authors, the amount of retained N increases by 38% with the use of ractopamine due to increasing lean tissue deposition.

When evaluating four levels (0, 5, 10 and 15 ppm) and four periods of supplementation with ractopamine (1 to 4 weeks) in female finishing, Watanabe et al. (2013) observed that there was no effect on the contents of dry matter (DM) and organic matter (OM) of feces. However, the authors observed quadratic effect for the weekly production of DM and OM, with higher values in the second and third week, respectively, therefore, manure quality depends on the period of ractopamine supply.

Since the use of ractopamine may interfere in the manure composition, it has been hypothesized the possible interference of it in the anaerobic biodigestion process, currently the main way of manure treatment and/or recycling generated in the Brazilian pig farming. Thus, the aim of this study was to evaluate the physical and chemical manure characteristics, anaerobic biodigestion process efficiency and biogas production by manure from finishing pig supplemented with ractopamine over different periods.

MATERIAL AND METHODS

The experiment was conducted at the Laboratory of Animal Waste Management in the *Universidade Estadual de Mato Grosso do Sul – UEMS* (State University of Mato Grosso do Sul), campus in Aquidauana - MS, Brazil. The local geographic coordinates are 20° 28'S and 55° 48'W, with an altitude of 184 meters. The average air temperature in the laboratory during the experimental period was 29.44 °C.

Six continuous bench biodigesters were used for evaluations. Each equipment was composed of one fermentation chamber and one gasometer, being built in polyvinyl chloride (PVC) and with a volume of 7.8 L of fermentation substrate.

The manure was derived from 50 barrows with initial and final weight respectively of 65.85 ± 4.34 and 96.55 ± 5.17 kg, housed in stalls at the Swine Sector of the Experimental Farm located at the Universidade Federal de Mato Grosso do Sul (UFMS), Campo Grande, MS, Brazil.

The animals were submitted to experimental corn and soybean meal-based diets supplemented with minerals and vitamins, formulated as in Rostagno et al., (2011), with 20%

adjustment for amino acids. Diets formulation was based on the ideal protein concept, since the nutritional requirements of amino acids in diets are higher for supplemented animals (Table 1).

Table 1: Compositions and calculated values of the experimental diets.

Ingredients	Ractopamine (ppm)	
	0	20
Corn	537.97	537.97
Soybean meal 45%	213.25	213.25
Millet	200.00	200.00
Soybean oil	17.92	17.92
Dicalcium phosphate	8.15	8.15
Calcitic limestone	6.29	6.29
Regular salt	1.00	1.00
Vitamin supplement[1]	4.00	4.00
Mineral supplement[2]	3.60	3.60
L-lysine HCl	4.13	4.13
L-Threonine	1.44	1.44
DL-Methionine	1.17	1.17
L-Tryptophan	0.09	0.09
Inert (kaolin)	1.00	0.00
Ractopamine	0.00	1.00
Calculated composition*		
Crude protein (g kg^{-1})	168.6	168.6
Metabolizable energy (kcal, kg)	3.230	3.230
Net energy (kcal, kg)	2.447	2.447
Total lysine (g kg^{-1})	11.09	11.09
Digestible lysine (g kg^{-1})	10.16	10.16
Digestible methionine + cystine (g kg^{-1})	6.10	6.10
Digestible threonine (g kg^{-1})	6.81	6.81
Digestible tryptophan (g kg^{-1})	1.83	1.83
Digestible valine (g kg^{-1})	7.01	7.01
Calcium (g kg^{-1})	5.10	5.10

Available phosphorus (g kg^{-1})	2.50	2.50
Total phosphorus (g kg^{-1})	4.68	4.68
Sodium (g kg^{-1})	1.60	1.60

\[1]Content kg^{-1} of product: iron, 100 g; copper, 10 g; cobalt, 0.2 g; manganese, 30 g; zinc, 100 g; iodine, 1.0 g; selenium, 0.3 g; excipient QSP, 1,000 g.

[2]Content kg^{-1} of feed: Vitamin A, 6,000,000 IU; Vitamin D3, 1,000,000 IU; Vitamin E, 12,000 IU; Vitamin B1, 0.5 g; Vitamin B2, 2.6 g; Vitamin B6, 0.7 g; pantothenic acid, 10 g; Vitamin K3, 1.5 g; nicotinic acid, 22 g; Vitamin B12, 0.015 g; folic acid, 0.2 g; biotin, 0.05 g; choline, 100 g; excipient QSP 1,000 g.

*Calculated composition according to the recommendations proposed by Rostagno et al. (2011), increased by 20% for the amino acid levels.

The animals received food and water ad libitum. Daily feed intake was determined and carried out initial and final weight animal measurements to obtain the weight gain and feed conversion. In addition, the animals with diet supplemented with 20-ppm ractopamine received it for periods of 7, 14, 21, 28 and 35 days pre-slaughter. The animals that took the control diet received the same diet (without ractopamine supplementation) throughout the experimental period.

The manure was collected in the last two days of the experimental phase (from 8:00 to 16:00 h), after excretion, avoiding to collect it near the drinking trough and feed trough. Samples were taken to determine the levels of total and volatile solids and total nitrogen and phosphorus. Subsequently, the manure were placed in labeled plastic bags and stored in a freezer until the experiment of anaerobic biodigestion, when it was defrosted at ambient temperature for 12 hours.

Starting affluents were formulated to contain approximately 4% of TS. Each biodigester received 1.10 kg of manure and 6.40 kg of water and the starting-up time was 28 days. After the starting-up time, the biodigesters were operated with hydraulic retention time of 30 days, with daily charges of 65 days. Within this period, in the first 30 days, the biodigesters were supplied daily with control affluent with 0.03 kg of manure, 0.22 kg of water and 0.5% (mass/volume) of sodium bicarbonate for pH correction, according to pretests conducted.

In the following 35 days, assessments were carried out, with differentiated supply of the biodigesters according to animal supplementation with ractopamine, as follows: 0 day - control affluent throughout the experimental period (without ractopamine); 7 days - control affluent for 28 days and supplementation affluent for 7 days; 14 days - control affluent for 21 days and supplementation affluent for 14 days; 21 days - control affluent for 14 days and supplementation affluent for 21 days; 28 days - control affluent for 7 days and supplementation affluent for 28 days; and 35 days - supplementation affluent for 35 days.

Affluent and effluent samples were collected daily for TS and VS content determination and weekly for pH and ammoniacal nitrogen (ammoniacal N) determination, following the methodologies described by APHA, AWWA and WPCF (2012). Fortnightly affluent and effluent analyses were performed for the quantification of total nitrogen (N) content using the semi-micro Kjeldahl method, and total phosphorus content (P) by the colorimetric method, according to APHA, AWWA and WPCF (2012).

Biogas production were calculated based on the gasometer displacement and biodigester area (0.00785 m²). Biogas volume correction to the conditions of 1 atm and 25 °C was made through the expression that results from a combination of Boyle and Gay-Lussac laws:

$$\frac{V_0 P_0}{T_0} = \frac{V_1 P_1}{T_1} \qquad (1)$$

where,

V_0 is the corrected biogas volume (m^3);

P_0 is the corrected biogas pressure (10,332.72 mm H_2O);

T_0 is the corrected biogas temperature (298.15 K);

V_1 is the biogas volume in the gasometer;

P_1 is the biogas pressure at the time of reading (10,315 mm H_2O), and

T_1 is the biogas temperature at the time of reading (K).

Considering the average atmospheric pressure in Aquidauana in the period equal to 10,293 mm H_2O, it has as result the following expression for the biogas volume correction:

$$V_0 = \frac{V_1}{T_1} \times 297.6515 \qquad (2)$$

For the variance analysis of pH, ammoniacal nitrogen, total nitrogen, total phosphorus and TS and VS reduction, during all the experimental period, we adopted a randomized block design, in which weeks (nine blocks) were considered as a cofactor. The means were compared by the Tukey's test at 5% probability.

The data of accumulated biogas production obtained during the last 35 days were submitted to regression analysis as a function of the animal supplementation period (in days) with ractopamine. After this analysis, the GLM procedure of SAS was used to compare the estimated means by linear equations generated for each supplementation period using the test at 5% probability.

RESULTS AND DISCUSSION

Daily animal weight gain was not affected ($p > 0.05$) either by ractopamine inclusion in the diet or by supplementation period. However, the increased supplementation period provided a linear improvement ($P < 0.05$) of feed conversion (equation: $Y = 3.08461 - 0.01144x$, $R^2 = 0.72$) (Table 2).

Table 2: Average daily weight gain (WG), feed conversion (FC) and average values of total solids (TS), volatile solids (VS), total nitrogen (total N) and total phosphorus (total P) in the manure and reductions of total (TS) and volatile solids (VS), in percentage, obtained from anaerobic biodigestion of pig manure supplemented with ractopamine over different periods.

Supplementation periods	0 day	7 days	14 days	21 days	28 days	35 days	P	CV (%)
Variables	Animal performance							
WG, g dia^{-1}	810	823	878	900	938	914	0.006	12.59
FC	3.13	2.93	3.04	2.77	2.65	2.77	0.028	10.37
	Physicochemical characteristics of manure							
TS^1, mg	39.07^{ab}	39.49^{a}	35.46^{d}	37.08^{c}	38.64^{ab}	37.81^{bc}	<0.001	1.39
VS^2 %	52.80^{c}	58.46^{b}	59.21^{b}	65.79^{a}	59.94^{b}	56.95^{bc}	<0.001	1.94
Total N^2, %	2.02^{b}	2.18^{ab}	2.40^{a}	2.40^{a}	2.03^{b}	2.09^{ab}	0.0082	3.69
Total P^2, %	5.10^{b}	5.78^{ab}	5.59^{ab}	5.16^{b}	6.04^{a}	5.85^{a}	0.0073	3.06
	Solids reduction post anaerobic biodigestion							
TS, %	47.41	49.24	42.55	52.44	55.32	53.26	0.339	18.84
VS, %	54.49	54.02	48.73	59.23	62.23	60.37	0.386	18.83

Means followed by the same letter in the row do not differ by the Tukey's test at 5% probability. [1] In percentage of fresh matter. [2]In percentage of dry matter. Tukey's test at 5% probability. P: statistical probability. CV: coefficient of variation.

Although not significant, the improvement in weight gain for animals supplemented for 28 days pre-slaughter, combined with low feed conversion, may result in benefit to the producer when considering the total of animals slaughtered, feed cost savings and carcass quality, especially if there is a bonus for lean carcasses.

Supplementation periods with ractopamine influenced the composition of the animals' fresh manure (Table 2). There was difference ($P < 0.05$) in the total (TS) and volatile solids (VS) percentage in the different supplementation periods with ractopamine.

Manure from animals supplemented with ractopamine during 0, 7, 28 and 35 days showed lower total N concentrations ($P < 0.05$). Increased protein synthesis may be associated with improved dietary protein utilization efficiency and may justify the reduction of the total N concentrations in the manure.

The highest concentrations of organic matter (VS) in the manure were observed when the animals were supplemented during 21 days. There was no effect ($p > 0.05$) of treatments in TS and VS reductions, obtaining averages of 50.04 and 56.51%, respectively (Table 2), which is possibly due to the high coefficient of variation (18.83%). The reduction in the TS and VS contents after anaerobic biodigestion indicates the process efficiency, both in removing solids and producing biogas.

Although the same amount of sodium bicarbonate has been added to all affluents, there was difference ($P < 0.05$) among the pH values of affluents that contained manure of animals supplemented with ractopamine during 35 days and those non-supplemented or supplemented during only 7 days (Table 3). Lower pH values in manure of animals supplemented for a long period indicate the presence of higher amounts of acids from organic matter degradation.

Table 3: Values of pH, ammoniacal nitrogen (ammoniacal N), total nitrogen (total N) and total phosphorus (total P) in the affluents (Aff.) and effluents (Eff.) from continuous biodigesters operated with manure from pigs supplemented with ractopamine over different periods

	pH		Ammoniacal N mg L^{-1}		Total N[1] %		Total P[1] %	
	Aff.	Eff.	Aff.	Eff.	Aff.	Eff.	Aff.	Eff.
0 day	7.67[a]	7.68	245[a]	822[a]	2.82	3.91	3.33	5.12
7 days	7.66[a]	7.69	235[ab]	824[a]	2.53	3.65	2.83	5.01
14 days	7.60[ab]	7.68	227[ab]	804[ab]	2.47	3.67	2.87	5.09
21 days	7.60[ab]	7.62	222[ab]	805[ab]	2.43	3.56	3.25	5.08
28 days	7.62[ab]	7.66	226[ab]	784[ab]	2.25	3.21	2.34	4.26
35 days	7.53[b]	7.65	207[b]	761[b]	2.26	3.30	2.46	3.67

P	0.017	0.367	0.029	0.002	0.173	0.477	0.322	0.093
CV (%)	1.26	0.89	10.84	4.18	15.96	16.67	30.75	19.24

Means followed by the same letter in the column do not differ by the Tukey's test at 5% probability. [1]In percentage of dry matter. P: statistical probability. CV: coefficient of variation.

However, despite the difference between the affluents pH values, there was no difference among the pH values of effluents from biodigesters. In stable biodigesters, the production and consumption of volatile fatty acids are balanced, resulting in an environment with adequate tamponade capacity. Thus, affluents slightly acids can result in neutral or alkaline effluents depending on the biodigesters operating conditions. The main components responsible for the tamponade power, and thus for the pH control in manure, are the inorganic carbon, ammoniacal N and volatile fatty acids (Christensen and Sommer, 2013).

The pH values of effluents were higher than the pH range recommended by Yadvika et al. (2004) (between 6.8 and 7.2) probably due to nitrogenous compounds degradation and free ammonia formation during the anaerobic biodigestion process, as can be verified by the difference between the initial and final ammoniacal N concentrations. The ammoniacal N is beneficial for the anaerobic biodigestion process, serving as nitrogen source and as a tamponade, avoiding pH changes. However, high concentrations (above 8.5) inhibit the process to be toxic to methanogenic archaea (Yenigün and Demirel, 2013).

The concentration of ammoniacal N in the effluent ranged from 761 to 824 mg L^{-1}, which is below the levels of inhibition of the process, which according to Rajagopal et al. (2013) it is above than 1,500 mg L^{-1}. When the manure was from animals that received ractopamine for a long period (35 days), lower concentrations of ammoniacal N were observed in affluents and effluents compared to animals non-supplemented or supplemented for only 7 days.

Considering that ractopamine may alter the animal metabolism, improving their feed conversion, changes in the composition of manure can be expected since feed polymers were better used. The lower ammoniacal N concentrations may be explained by the manure composition, which depends on the use of nutrients in the feed consumed by the animals. The longer period of animals supplementation with ractopamine could lead to the production of manure with lower nutrient concentrations, such as nitrogen compounds which, when degraded, may form ammonium ions or free ammonia. Thus, prolonged animal supplementation may contribute to lower ammoniacal N concentration in the biodigester effluent.

Thus, supplementation with ractopamine can be seen not only as a nutritional tool to increase animal performance, but also as a mitigation tool of the manure environmental impact since it provides the decrease of N content excreted in the manure, and consequently its volatilization, and in biofertilizers, when the manure is treated in biodigesters.

In this study, the results of pH and ammoniacal N as indicator parameters of the process balance and stability were satisfactory, offering no collapse risks for the anaerobic biodigestion process.

The different periods of animal supplementation with ractopamine did not interfere in the average concentrations of total N and P of the affluents and effluents from the biodigesters, being 2.46 and 3.55% of N and 2.80 and 4.70% of P, respectively.

Although the amount of N was not different in the affluents and effluents, the ammoniacal N concentrations differed. This fact is due to the ammoniacal N formation from different N sources in the manure (organic and inorganic N). Ammoniacal N formation occurs mainly in an anaerobic environment and remains dissolved in the manure liquid phase; in pig manure, the ammoniacal N concentrations are high since the animals receive high amounts of organic N in their diets, being more readily converted into ammoniacal N (Christensen and Sommer, 2013).

Probably the differences among the ammoniacal N concentrations in the manure were due to differences in the relations ammoniacal N: total N according to different periods in which the animals were supplemented with ractopamine, which may be related to the use of N from the diet.

There was difference (P <0.05) in the daily cumulative production of biogas according to different periods of supplementation (Table 4). In addition, although the highest VS levels in the manure have been observed for the 21-day supplementation period, the animals supplemented for a longer period (28 and 35 days) showed higher biogas production; the lowest productions were obtained from manure of animals non-supplemented or supplemented with 7 days before slaughtering, with no difference between these periods.

Once the volatile solids include two fractions, the biodegradable and refractory (Kangle et al., 2012), being the biodegradable fraction related to the organic matter contained in the manure and that can be transformed into biogas by the methanogenic bacteria, biogas production were consistent with the VS reductions, despite they were not significant. The results suggest that by improving use-efficiency of nutrients, the manure quality is also improved, resulting in higher biogas production.

In addition to the benefits of animal performance reported in the literature, ractopamine supply for long periods can add other benefits related to lower nutrient excretion and greater biogas production when the anaerobic biodigestion process is adopted as a manure treatment system.

Table 4: Regression and statistics equations of the daily accumulated biogas production (y) from manure of pigs supplemented with ractopamine over different periods (x).

Supplementation period	Equation	P value	R^2	Average estimation (m^3)	CV (%)
0 day	Y= 0.00063+0.00120x	<0.0001	0.99	0.0006 d	6.92
7 days	Y= 0.00001+0.00122x	<0.0001	0.98	0.0086 d	7.79
14 days	Y= 0.00070+0.00093x	<0.0001	0.98	0.0138 c	7.96
21 days	Y= 0.00074+0.00114x	<0.0001	0.99	0.0247 b	6.27
28 days	Y= -0.00015+0.00216x	<0.0001	0.99	0.0603 a	6.27
35 days	Y= 0.00077+0.00204x	<0.0001	0.99	0.0722 a	5.15

P: statistical probability. R^2: coefficient of determination. Means followed by the same letter in the column do not differ by the t test at 5% probability. CV: coefficient of variation.

CONCLUSIONS

The period of 28 days of supplementation with 20-ppm ractopamine is the most recommended since it combines good animal performance, manure production with lower N content, and increased biogas production by the manure.

The anaerobic biodigestion efficiency, in terms of pH, ammoniacal N and solids reduction, is not affected by the pig diet supplementation with 20-ppm ractopamine.

ACKNOWLEDGEMENTS

The authors want to thank the *Coordenação de Aperfeiçoamento de Pessoal de Nível Superior - CAPES*(Coordination for the Improvement of Higher Education Personnel - CAPES) for granting a Master's scholarship and to the *Fundação de Apoio ao Desenvolvimento do Ensino e a Ciência e Tecnologia do Estado de Mato Grosso do Sul – FUNDECT* (Foundation of Support for Education, Science and Technology of the Mato Grosso do Sul State - FUNDECT) for financial support.

ETHICS COMMISSION

Proceeding No. 425/2012, filed with the Ethics Committee on Animal Use of the Universidade Federal de Mato Grosso do Sul (Federal University of Mato Grosso do Sul) - CEUA/ UFMS.

REFERENCES

1. ALEMANNO, A.; CAPODIECI, G. Testing the Limits of Global Food Governance: The Case of Ractopamine.European Journal of Risk Regulation, Berlin, v.3, n. 1, p.1-2, 2012.
2. ALMEIDA, V.V.; NUÑEZ, A.J.C; MIYADA, V.S. Ractopamine as a Metabolic Modifier Feed Additive for Finishing Pigs: A Review. Brazilian Archives of Biology and Technology, Curitiba, v.55, n3, p. 445-456, 2012.
3. APHA, AWWA, WPCF. Standard methods for the examination of water and wastewater. 21th. ed. Washington, 2012. 1496 p.
4. CHRISTENSEN, M.L.; SOMMER, S.G. Manure characterization and inorganic chemistry. In: SOMMER, S.G.; CHRISTENSEN, M.L.; SCHIMIDT, T.; JENSEN, L.S. Animal manure recycling. West Sussex: Willey, 2013. p. 41-65.
5. DECAMP, S.A.; HANKINS, S.L.; CARROLL, A.L.; IVERS, D.J.; RICHERT, B.T.; SUTTON, A.L.; ANDERSON, D.B.Effects of ractopamine and level of dietary crude protein on nitrogen and phosphorus excretion from finishing pigs. Swine (Research Report). West Lafayette: Purdue University, 2001. p.97-102,
6. FERREIRA, M.S.S.; GARBOSSA, C.A.P.; OBERLENDER, G.; PEREIRA, L.J.; ZANGERONIMO, M.G.; SOUSA, R. V.; CANTARELLI, V.S. Effect of ractopamine on lipid metabolism in vivo - a systematic review. Brazilian Archives of Biology and Technology, Curitiba, v. 56, n. 1, p. 35-43, 2013.
7. KANGLE, K.M.; KORE, S.V.; KORE, V.S.; KULKARNI, G.S. Recent trends

in anaerobic codigestion. Universal Journal of Environmental Research and Technology, Pune, v.2, n. 4, p. 210-219, 2012.

8. KUTZLER, L. W.; HOLMER, S. F.; BOLER, D. D.; CARR, S. N.; RITTER, M. J.; PARKS, C. W.; MCKEITH, F. K.; KILLEFER, J. Comparison of varying doses and durations of ractopamine hydrochloride on late-finishing pig carcass characteristics and meat quality. American Society of Animal Science, Champaign, v.89, p.:2176–2188, 2011.
9. MARCAL, D.A.; KIEFER, C.; SOUZA, K.M.R.; ABREU, R.C.; ROSA, R.A.; ROSA, L.S. Ractopamina em dietas sem ajustes aminoacídicos para suínos machos castrados em terminação. Revista Ceres, Viçosa, MG, v. 62, n. 3, p. 259-264, 2015.
10. NEILL, C.R; GOODBAND, R.D.; TOKACH, M.D.; NELSSEN, J.L.; DRITZ, S.S.; DEROUCHEY, J.M, et al. Effects of intermittent ractopamine hydrochloride use on pig growth performance in late finishing. The Professional Animal Scientist, Sarvy, v.26, n.2, p.239-242, 2010.
11. PEREIRA, F.A.; FONTES, D.O.; VASCONCELLOS, C.H.F.; SILVA, F.C.O.; SILVA, M.A.; MARINHO, P.C.; AROUCA, C.L.C.; SALUM, G.M. Efeito da ractopamina e de métodos de formulação de ração sobre o desempenho e as características de carcaça de leitoas em terminação. Arquivos Brasileiros de Medicina Veterinária e Zootecnia, Belo Horizonte, v.63, n.1, p.203-213, 2011.
12. RAJAGOPAL, R.; MASSÉ, D.I.; SINGH, G. A critical review on inhibition of anaerobic digestion process by excess ammonia. Bioresource Technology, New York, v. 143, p. 632-641, 2013.
13. ROSTAGNO, H.S.; ALBINO, L.F.T.; DONZELE, J.L. GOMES. P.C.; OLIVEIRA, R.F.; LOPES, D.C.; FERREIRA, A.S.; BARRETO, S.L.T; EUCLIDES, R.F. Tabelas brasileiras para aves e suínos: composição de alimentos e exigências nutricionais. 3.ed. Viçosa: Universidade Federal de Viçosa, 2011. 252 p.
14. WATANABE, P.H.; THOMAZ, M.C.; PASCOAL, L.A.F.; RUIZ, U.S.; DANIEL, E.; AMORIM, A.B. Manure production and mineral excretion in feces of gilts fed ractopamine. Acta Scientiarum Agronomy, Maringá, v. 35, n. 3, p. 267-272, 2013.
15. WOODS, A.L.; ARMSTRONG, T.A.; ANDERSON, D.B.; ELAM, T.E.; SUTTON, A.L. Case Study: Environmental benefits of ractopamine use in United States finisher swine. The Professional Animal Scientist, Sarvey, v. 27, p.492–499, 2011.
16. YADVIKA, T.R.; SANTOSH, S.; SREEKRISHNAN, T. R.; KOHLI, S.; RANA, V. Enhancement of biogas production from solid substrates using different techniques – a review. Bioresource Technology, New York, v. 95, n. 1, p. 1-10, 2004.
17. YENIGÜN, O.; DEMIREL, B. Ammonia inhibition in anaerobic digestion. Process Biochemistry, New York, v.48, p.901–911, 2013.

10

CHAPTER

DETECTION OF HEPATITIS E VIRUS (HEV) THROUGH THE DIFFERENT STAGES OF PIG MANURE COMPOSTING PLANTS

M. García, S. Fernández-Barredo and M. T. Pérez-Gracia

INTRODUCTION

Hepatitis E virus (HEV) is the main causative agent of enterically transmitted non-A non-B hepatitis (Purcell and Emerson, 2000; Perez-Gracia and Rodriguez-Iglesias, 2003). Hepatitis E is considered an infectious disease endemic in developing areas such as India, Africa and South-east Asia, because of poor sanitary conditions in drinking water. When it was first reported in developed countries, HEV was related to travel to endemic areas. However, the epidemiology of HEV in industrialized countries like Spain have changed in the last years, with an increasing number of non-travel associated sporadic cases (Perez-Gracia *et al.*, 2004). In industrialized countries, hepatitis E actually represents less than 3% of acute viral hepatitis cases (Purdy and Khudyakov, 2011). However, the overall relevance of HEV infection has been underestimated and the disease may to be considered like a global health problem at the present time. The WHO estimates that more than 3 million individuals suffer symptomatic acute hepatitis E and the disease causes around of 70 000 deaths at year (Pischke and Wedemeyer, 2013). The overall mortality of hepatitis E in the general population is less than 1%; but it can reach up to 28% in infected pregnant women (Khuroo and Kamili, 2003).

The high phylogenetic homology observed between pig and humans strains of genotype 3 in the same geographical area suggests that hepatitis E is a zoonotic disease. This statement has been reinforced by the results of several studies in Japan (Takahashi *et al.*, 2003), Korea (Ahn *et al.*, 2005), the UK (Banks *et al.*, 2004; Ijaz *et al.*, 2005; Tolari *et al.*, 2006) and also in Spain (Pina *et al.*, 2000; Perez-Gracia *et al.*, 2007), where they

Citation: García, M., Fernández-Barredo, S. and Pérez-Gracia, M. T. (2014), Detection of hepatitis E virus (HEV) through the different stages of pig manure composting plants. Microbial Biotechnology, 7: 26–31. doi:10.1111/1751-7915.12064.

reported a nucleotide similarity of 94% between human and pig strains. Besides the high nucleotide homology observed, HEV has been demonstrated to be able of crossing the species barrier, raising concern for the potential ways of zoonotic transmission from swine to humans (Meng, 2011).

The direct application of the slurries in agricultural areas is the more economic and ecological method in the management and recycling of the same ones (Ziemer *et al.*, 2010; Pardo *et al.*, 2011), following the parameters established by the law. Nevertheless, the appearance in the last years of zones with high density of cattle intensive developments generates a large amount of wastes. In this sense, Spain is the second country in the business of pig livestock in European Union (EU), with 25 million of units. The area studied has 170 farms distributed in an approximated surface of 1000 square miles, generating a total amount of 1 million of tones of slurries at year (source: EUROSTAT DATABASE, 2009) of 41 million of tones at year generated in EU. The management of these products turns out to be complicated and inefficient, constituting an environmental and sanitary problem because of zoonotic pathogens like HEV, may be transported to drinking water resources (Bolado-Rodriguez *et al.*, 2010; Hazam *et al.*, 2010; Meng, 2011) and other problems like excessive nitrification of soil and presence of excessive concentration of heavy metals. These problems worry both the livestock sector and Public Health authorities.

In this sense, the pig manure composting plants constitutes an innovative purification system in Spain, and these are the only plants that are currently being used. These procedures performed in these plants are safe alternative methods in the management of the slurries.

The aim of this study was to evaluate the presence of HEV in the different stages of pig manure composting plants.

RESULTS

Two different types of composting plants (type A and type B), based on the type of manure treatment, were studied. Plants performing a total treatment of the slurry (type A plants), include an initial separation of liquid and solid phases of the slurry by centrifugation and posterior purification by fermentation procedures.

The purification process in type A plants (Fig. 1A) starts with an initial centrifuge. Liquid fraction of the slurries is carried to the anaerobic lagoon, where the slurries will remain 100 days. In this time; methanogenics, acetogenics and hydrolysis processes lead to a reduction of the smell, decrease of pathogens and stabilization of organic matter and hence reduction of BOD (biological oxygen demand).

Liquid from the anaerobic pond goes on later to the aerobic ponds 1 and 2, where the slurries will remain 10 days in each one. In this stage, offensive odours decrease and reduction of pathogen continues, conversion of available nitrogen to ammoniacal nitrogen and a decrease of C : N ratio is performed also.

The final product of this step is a depurated liquid fraction that will be mixed with the solid fraction and other vegetables, and carried to the fermentation channels where the composting process is performed. Composting is an aerobic process that requires a continuous supply of air with mechanical mixing during 21 days approximately. The temperature in the pile can rise in the 10 first days to as high as 65°C.

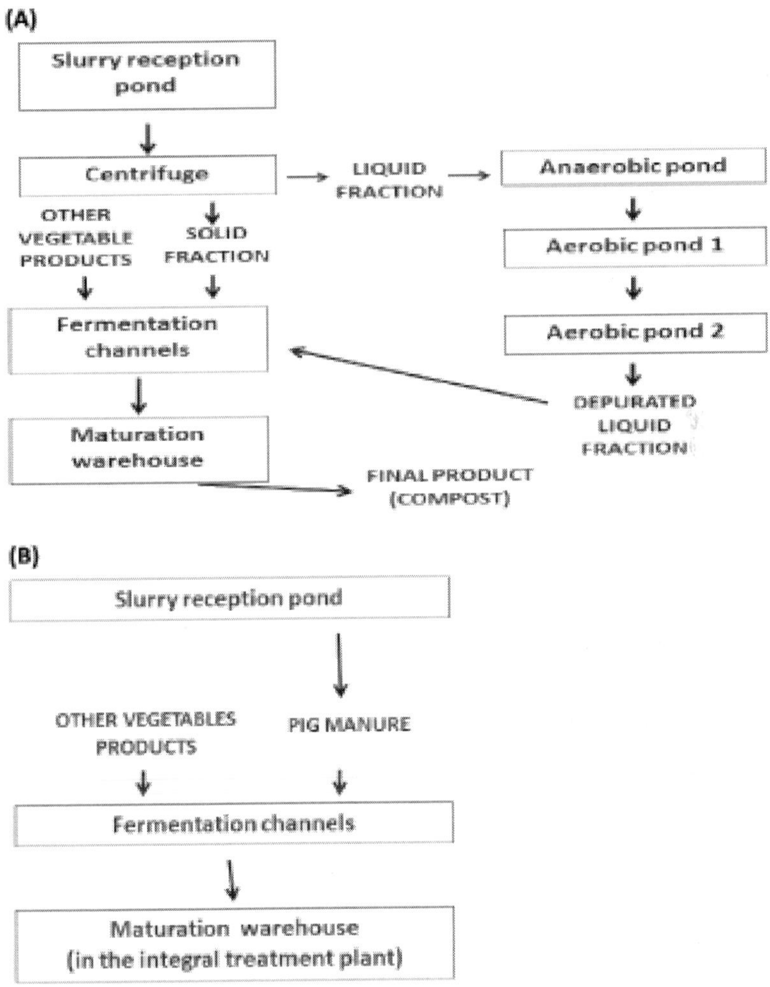

Figure 1: A. Plants type A flow-chart. B. Plants type B flow-chart.

As opposed to type A plants, the type B plants does not make a fractionated treatment of the slurry (Fig. 1B) and directly use it in the composting process.

The final product is the compost, an inoffensive, depurated and a better soil fertilizer than raw slurries.

We have studied two type A plants (A1and A2) and three type B plants (B1, B2 and B3). A total number of four (80%) of five plants studied were positive to the presence of RNA-HEV. According to the type of plant, HEV was detected in two (100%) of the two type A plants and in two (66.66%) of the three type B plants (Table 1).

Table 1: Nested-PCR results according to the number of sampling sessions in the different stages of treatment in type A and B plants

Plant	Reception pond	Anaerobic pond	Solid product from centrifuge	Aerobic pond 1	Aerobic pond 2	Fermenta channel (Start)
A1	24/27 [a] (88.89%) [b]	18/27 (66.66%)	3/27 (11.11%)	3/27 (11.11%)	0/27 (0%)	0/27 (0%)
A2	9/9 (100%)	6/9 (66.66%)	6/9 (66.66%)	0/9 (0%)	0/9 (0%)	0/9 (0%)
B1	0/3 (0%)	-	-	-	-	0/3 (0%)
B2	3/6 (50%)	-	-	-	-	0/6 (0%)
B3	3/9 (33.33%)	-	-	-	-	0/9 (0%)

a Sampling sessions HEV positive/total number of sampling sessions.
b Percentage of positive sampling sessions.

According to the different stage of treatment, in plant A1, HEV was detected in 24 (88.89%) of 27 sampling sessions in the reception pond. In 18 (66.67%) of 27 sampling sessions, HEV was detected in the anaerobic pond; which is the first step of purification process. Only in three (11.11%) of the 27 sampling sessions, HEV was detected in the aerobic pond number 1; from this stage forward to the end of the composting process, HEV was not detected (Fig. 2).

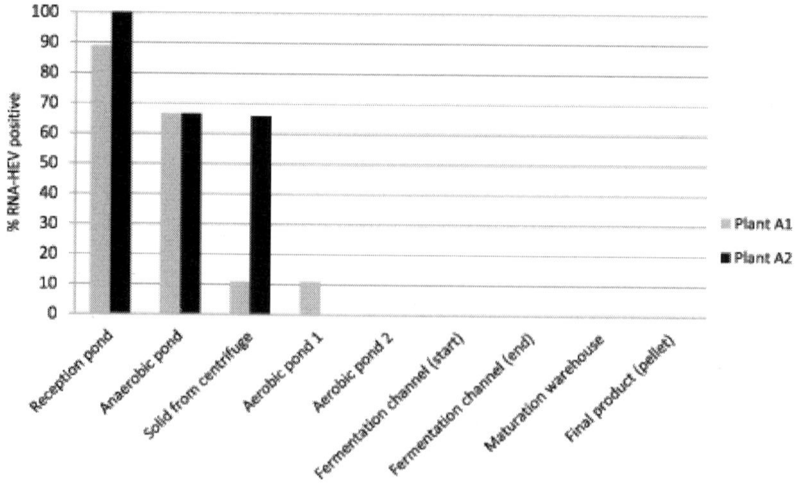

Figure 2: RNA-HEV detection through the different stage of pig manure treatment in type A plants.

In plant A2, HEV was detected in nine (100%) of nine sampling sessions in the reception pond. In six (66.66%) of nine sampling sessions, HEV was detected in the anaerobic pond; after this stage of treatment, HEV was not detected in either solid or liquid samples (Fig. 2).

In type B plants, HEV was not detected in plant B1 and it was detected in three (50%) of six sampling sessions in plant B2, and in three (33.33%) of nine sampling sessions in plant B3 in the respective reception ponds. HEV was not detected in any solid samples in these plants.

DISCUSSION

In recent years, the increase of animal farming has generated an intensive and megascale livestock operations that produce a large amount of animal wastes worldwide (Cole *et al.*, 2000; Spencer and Guan, 2004). A lot of causative agents of many infectious diseases have been identified in slurries, including HEV. The storage and treatment before land application are commonly performed in farms; however, it does not result in a total removal of pathogens from the manure (Ziemer *et al.*, 2010). Pig manure slurry is an emergent health and environmental problem and a potential source of infectious pathogens like HEV (Kasorndorkbua *et al.*, 2005). If the slurries do not undergo a process of purification, and are spread without control over crop fields, population can be infected by the virus by consuming contaminated water from aquifers or vegetables irrigated with HEV contaminated water. Moreover, veterinarians, farmers and swine workers have been observed to be at a higher risk (Perez-Gracia *et al.*, 2007; Galiana *et al.*, 2008), as well as the contact with pig manure constitutes a factor of higher risk. To reduce the impact of animal waste production several technologies have been developed around the world. One of them has been performed in a high swine production zone in Castellon (Spain), with the installation of five pig manure composting plants (A1, A2, B1, B2, B3) with a pioneer treatment system.

Previous works evaluated the survival of pathogens after the treatment of the sludge (Sobsey *et al.*, 2003; Costantini *et al.*, 2007; Wagner *et al.*, 2008), but this is the first study that evaluates the effectiveness of swine manure composting plants in the elimination of HEV.

The HEV is widely widespread in the zone of action of the plants studied (Fernández-Barredo *et al.*, 2007). In fact, in this study HEV was detected in 80% of the plants. These data coincide with other works in the same geographical zone (Seminati *et al.*, 2008) that reported in some cases 76.19% of HEV prevalence in farms closer to these plants (Fernández-Barredo *et al.*, 2006). In plants with total treatment of the slurries (type A plants), HEV was not detected in any sample from the liquid fraction after the stage of treatment corresponding to the aerobiosis 1 pond (Fig. 1A), suggesting that a high percentage of HEV had been inactivated to this point in anaerobic and aerobic ponds, and reduced its presence at levels not detectable by PCR. There are no data reporting the resistance of HEV to anaerobic purification processes; however, a work about the resistance of porcine enteric virus of the*Caliciviridae* family at different purification treatments of slurry (Costantini *et al.*, 2007) confirmed the reduction of the number of viral particles after subjecting slurry to anaerobic digestion. Additionally, this study confirmed that the aerobic treatment was able to reduce the concentration of more resistant virus, such as rotavirus type A, B and C, which themselves were capable to resist anaerobic processes, to undetectable levels by PCR. Additionally, the fact of HEV in type A plants was detected in a higher percentage in the reception pond compared with the detection of HEV in solid products from centrifuge, could be due to the lesser resistance of the virus in solid matrices (Pesaro *et al.*, 1995).

HEV was not detected in any solid samples of processed product by fermentation. The high temperatures achieved in the fermentation channels (approximately 65°C), could explain the lack of HEV amplification. In this sense, a study performed by Emerson and colleagues in 2005, conducted with human and swine strains, showed that HEV genotypes 1 and 2, were 50% inactivated at temperatures between 45°C and 50°C, and totally inactivated at 60°C. Only Mex 14 (genotype 2) strain was observed to be 100% resistant to 56°C and 20% to temperatures of 60°C (Emerson *et al.*, 2005). In addition, a recent study (Barnaud *et al.*, 2012), shows an efficient inactivation of the virus when food products from infected by HEV pork livers, were heated at least 20 min to an internal temperature of 71°C.

In this way, the low percentage of HEV positive samples after anaerobic digestion, suggests that large amounts of virus do not exceed this stage of treatment. The lack of detection of HEV in solid samples at the end phases of the composting treatment and in the final product, suggests that composting process is effective to eliminate this virus from the slurries, and thus to reduce the transmission of HEV from pigs to humans.

EXPERIMENTAL PROCEDURES

Collection of samples of composting plants

A total number of 594 liquid and solid samples were collected in 54 different sampling sessions from March 2006 to December 2011 in five plants distributed as follows (Table 1): two type A plants (A1 and A2) and three type B plants (B1, B2 and B3). The higher number of samples collected from type A plants, 504 (87.5%) is due to the fact that these plants assemble the majority of purification processes and therefore receive more manure than type B plants. In the same way, the higher number of liquid samples, 360 (60.6%) corresponds to the higher number of purification processes carried out in the liquid phase of the slurry.

The plants are located in Eastern Spain, in a zone endemic for porcine HEV (Fernández-Barredo *et al.*, 2006). The plants were sampled in 54 different sampling sessions, in which a variable number of samples were taken in each stage taking into account the volume of each pond. Thus, in the anaerobic pond a total number of 180 samples were collected in five different points at each sampling. The pond was considered positive to HEV when at least one of the five sampling points was positive. The rest of the ponds were equipped with a homogenization system which was activated 10 min before the sampling, therefore only one sample of a larger volume was taken. Solid samples were homogenized by mechanical processes included in the periodic aeration system performed in the integrated treatment.

It was impossible to take the same number of samples in type B as in type A plants, due to a premature closing of the same ones because of technical and economical problems.

Liquid samples (50 ml), were collected directly from the ponds and kept in sterile falcon tubes until processing. All the samples were transported refrigerated immediately to the laboratory. Solid samples were diluted at 10% (w/v) in sterile phosphate buffer saline (PBS) pH 7.2.

VIRUS PARTICLE CONCENTRATION, RNA EXTRACTION AND REVERSE TRANSCRIPTION-NESTED PCR (RT-NESTED PCR) WITH INTERNAL CONTROL

All the samples collected were initially centrifuged 1 min at 1000 *g* and 15 ml of the supernatant were centrifuged at 4°C, 1 h at 3000 *g* to concentrate the virus particles. Remaining supernatant (1.5 ml) was transferred to a sterile eppendorf tube and it was centrifuged at 12 100 *g* for 10 min, and 1 ml of the supernatant was stored at −80°C.

RNA was extracted from 140 µl of each concentrated sample according to the method described by Fernández-Barredo and colleagues (2006). Two pairs of degenerated oligonucleotide primers based on human and swine HEV sequences, were used to amplify a 348-bp-long fragment from the HEV open reading frame 2 (ORF-2) using a reverse transcription-nested polymerase chain reaction (RT-nested PCR) (Huang *et al.*, 2002). A negative and a positive control from a naturally infected pig (GenBank Accession No. AY323506) were included in each assay. The different stages of the procedure were performed in different places to avoid the possibility of cross-contamination. The faecal samples may contain RT-PCR inhibition substances like phenolic and methabolic compounds, the concentration and presence of these inhibitors is different and heterogeneous from sample to sample (Rutjes *et al.*, 2007). To detect the presence of these substances an internal control was included within RT-PCR reaction. The internal control used is a modified 77 bp PCR product cloned into a plasmid containing a sequence that can be amplified simultaneously with the target using the same primers set in the first PCR performed according to the protocol described by Huang and colleagues in 2002. The 77 bp amplified fragment was detected in the electrophoresis gel. PCR inhibition was not detected. The sensitivity of the PCR was estimated in 31.6 PID_{50} of infectious swine HEV (Huang *et al.*, 2002).

The PCR products were separated by electrophoresis in a 2% agarose gel and detected by staining with ethidium bromide (0.5 $\mu g\,ml^{-1}$). The samples were considered positive to HEV when a band of 348 bp was seen in the agarose gel. Amplicons from all positive samples were purified and confirmed by sequence analysis. Sequences obtained in this study have been submitted to the GenBank database under Accession No. KC145131–KC145147.

ACKNOWLEDGEMENTS

We are grateful to Ms Beatriz Suay for the assistance in translating the manuscript.

CONFLICT OF INTEREST

None declared.

REFERENCES

1. Ahn, J.M., Kang, S.G., Lee, D.Y., Shin, S.J., and Yoo, H.S. (2005) Identification of novel human hepatitis E virus (HEV) isolates and determination of the seroprevalence of HEV in Korea. *J Clin Microbiol* 43: 3042–3048.
2. Banks, M., Heath, G.S., Grierson, S.S., King, D.P., Gresham, A., Girones, R., *et al.* (2004)Evidence for the presence of hepatitis E virus in pigs in the United Kingdom. *Vet Rec* 154:223–227.
3. Barnaud, E., Rogee, S., Garry, P., Rose, N., and Pavio, N. (2012) Thermal inactivation of infectious hepatitis E virus in experimentally contaminated food. *Appl Environ Microbiol*78: 5153–5159.
4. Bolado-Rodriguez, S., Garcia-Sinovas, D., and Alvarez-Benedi, J. (2010) Application of pig slurry to soils. Effect of air stripping treatment on nitrogen and TOC leaching. *J Environ Manage* 91: 2594–2598.
5. Cole, D., Todd, L., and Wing, S. (2000) Concentrated swine feeding operations and public health: a review of occupational and community health effects. *Environ Health Perspect*108: 685–699.
6. Costantini, V.P., Azevedo, A.C., Li, X., Williams, M.C., Michel, F.C., Jr, and Saif, L.J. (2007)Effects of different animal waste treatment technologies on detection and viability of porcine enteric viruses. *Appl Environ Microbiol* 73: 5284–5291.
7. Emerson, S.U., Arankalle, V.A., and Purcell, R.H. (2005) Thermal stability of hepatitis E virus. *J Infect Dis* 192: 930–933.
8. EUROSTAT DATABASE (2009) Agricultural statistics 2009 edition. Main results — 2007–08. http://epp.eurostat.ec.europa.eu/cache/ITY_OFFPUB/KS-ED-09-001/EN/KS-ED-09-001-EN.PDF.
9. Fernández-Barredo, S., Galiana, C., Garcia, A., Vega, S., Gomez, M.T., and Perez-Gracia, M.T. (2006) Detection of hepatitis E virus shedding in feces of pigs at different stages of production using reverse transcription-polymerase chain reaction. *J Vet Diagn Invest* 18:462–465.
10. Fernández-Barredo, S., Galiana, C., Garcia, A., Gomez-Munoz, M.T., Vega, S., Rodriguez-Iglesias, M.A., and Perez-Gracia, M.T. (2007) Prevalence and genetic characterization of hepatitis E virus in paired samples of feces and serum from naturally infected pigs. *Can J Vet Res* 71: 236–240.
11. Galiana, C., Fernández-Barredo, S., Garcia, A., Gomez, M.T., and Perez-Gracia, M.T. (2008)Occupational exposure to hepatitis E virus (HEV) in swine workers. *Am J Trop Med Hyg* 78:1012–1015.
12. Hazam, R.K., Singla, R., Kishore, J., Singh, S., Gupta, R.K., and Kar, P. (2010) Surveillance of hepatitis E virus in sewage and drinking water in a resettlement colony of Delhi: what has been the experience? *Arch Virol* 155: 1227–

1233.

13. Huang, F.F., Haqshenas, G., Guenette, D.K., Halbur, P.G., Schommer, S.K., Pierson, F.W.,*et al.* (2002) Detection by reverse transcription-PCR and genetic characterization of field isolates of swine hepatitis E virus from pigs in different geographic regions of the United States. *J Clin Microbiol* 40: 1326–1332.

14. Ijaz, S., Arnold, E., Banks, M., Bendall, R.P., Cramp, M.E., Cunningham, R., *et al.* (2005)Non-travel-associated hepatitis E in England and Wales: demographic, clinical, and molecular epidemiological characteristics. *J Infect Dis* 192: 1166–1172.

15. Kasorndorkbua, C., Opriessnig, T., Huang, F.F., Guenette, D.K., Thomas, P.J., Meng, X.J., and Halbur, P.G. (2005) Infectious swine hepatitis E virus is present in pig manure storage facilities on United States farms, but evidence of water contamination is lacking.*Appl Environ Microbiol* 71: 7831–7837.

16. Khuroo, M.S., and Kamili, S. (2003) Aetiology, clinical course and outcome of sporadic acute viral hepatitis in pregnancy. *J Viral Hepat* 10: 61–69.

17. Meng, X.J. (2011) From barnyard to food table: the omnipresence of hepatitis E virus and risk for zoonotic infection and food safety. *Virus Res* 161: 23–30.

18. Pardo, T., Clemente, R., and Bernal, M.P. (2011) Effects of compost, pig slurry and lime on trace element solubility and toxicity in two soils differently affected by mining activities. *Chemosphere* 84: 642–650.

19. Perez-Gracia, M.T., and Rodriguez-Iglesias, M. (2003) [Hepatitis E virus: current status].*Med Clin (Barc)* 121: 787–792.

20. Perez-Gracia, M.T., Garcia-Valdivia, M.S., Galan, F., and Rodriguez-Iglesias, M.A. (2004)Detection of hepatitis E virus in patients sera in southern Spain. *Acta Virol* 48: 197–200.

21. Perez-Gracia, M.T., Mateos, M.L., Galiana, C., Fernández-Barredo, S., Garcia, A., Gomez, M.T., and Moreira, V. (2007) Autochthonous hepatitis E infection in a slaughterhouse worker. *Am J Trop Med Hyg* 77: 893–896.

22. Pesaro, F., Sorg, I., and Metzler, A. (1995) *In situ* inactivation of animal viruses and a coliphage in nonaerated liquid and semiliquid animal wastes. *Appl Environ Microbiol* 61:92–97.

23. Pina, S., Buti, M., Cotrina, M., Piella, J., and Girones, R. (2000) HEV identified in serum from humans with acute hepatitis and in sewage of animal origin in Spain. *J Hepatol* 33:826–833.

24. Pischke, S., and Wedemeyer, H. (2013) Hepatitis E virus infection: multiple faces of an underestimated problem. *J Hepatol* 58: 1045–1046.

25. Purcell, R.H., and Emerson, S.U. (2000) Hepatitis E virus infection. *Lancet* 355: 578.

26. Purdy, M.A., and Khudyakov, Y.E. (2011) The molecular epidemiology of hepatitis E virus infection. *Virus Res* 161: 31–39.

27. Rutjes, S.A., Lodder, W.J., Bouwknegt, M., and Husman, A.M. (2007) Increased hepatitis E virus prevalence on Dutch pig farms from 33 to 55% by using

appropriate internal quality controls for RT-PCR. *J Virol Methods* 143: 112–116.

28. Seminati, C., Mateu, E., Peralta, B., de Deus, N., and MartIn, M. (2008) Distribution of hepatitis E virus infection and its prevalence in pigs on commercial farms in Spain. *Vet J*175: 130–132.

29. Sobsey, M., Gwangpyo, K., Simmons, O., Likirdopulos, C., and Worley-Davis, L. (2003)Evaluation of alternative swine waste treatment and management technologies for control of pathogens. In Proceedings North Carolina Animal Waste Management Workshop. Oct. 16–17, 2003, Research Triangle Park, NC. Havenstein, G.B. (ed.). Raleigh, NC, USA: College of Agriculture and Life Sciences, NCSU, pp. 112–125.

30. Spencer, J.L., and Guan, J. (2004) Public health implications related to spread of pathogens in manure from livestock and poultry operations. *Methods Mol Biol* 268:503–515.

31. Takahashi, K., Kang, J.H., Ohnishi, S., Hino, K., Miyakawa, H., Miyakawa, Y., *et al.* (2003)Full-length sequences of six hepatitis E virus isolates of genotypes III and IV from patients with sporadic acute or fulminant hepatitis in Japan. *Intervirology* 46: 308–318.

32. Tolari, F., Chiaro, L.D., Card, R., Mazzei, M., Bandecchi, P., and Banks, M. (2006) Phylogenetic study of viral isolates of Swine and human hepatitis E virus. *Vet Res Commun* 30 (Suppl. 1): 273–276.

33. Wagner, A.O., Gstraunthaler, G., and Illmer, P. (2008) Survival of bacterial pathogens during the thermophilic anaerobic digestion of biowaste: laboratory experiments and in situ validation. *Anaerobe* 14: 181–183.

34. Ziemer, C.J., Bonner, J.M., Cole, D., Vinje, J., Constantini, V., Goyal, S., *et al.* (2010) Fate and transport of zoonotic, bacterial, viral, and parasitic pathogens during swine manure treatment, storage, and land application. *J Anim Sci* 88: E84–E94.

11
CHAPTER

INCREASING BIOGAS PRODUCTION RATE FROM CATTLE MANURE USING RUMEN FLUID AS INOCULUMS

Budiyono, I.N. Widiasa, Seno Johari, and Sunarso

Keywords: Rumen fluid, inoculums, anaerobic digestion, biogas production

INTRODUCTION

Energy is one of the most important factors to global prosperity. The dependence on fossil fuels as primary energy source has lead to global climate change, environmental degradation, and human health problems. In the year 2040, the world predicted will have 9–10 billion people and must be provided with energy and materials [1]. Moreover, the recent rise in oil and natural gas prices may drive the current economy toward alternative energy sources such as biogas. Anaerobic digestion (AD) is a technology widely used for treatment of organic waste for biogas production. AD that utilizes manure for biogas production is one of the most promising uses of biomass wastes because it provides a source of energy while simultaneously resolving ecological and agrochemical issues. The anaerobic fermentation of manure for biogas production does not reduce its value as a fertilizer supplement, as available nitrogen and other substances remain in the treated sludge [2].

Numerous studies had been conducted by several researchers in order to increase biogas yield in AD. An effort to improve biomass conversion efficiency and biogas yield conducted by several researchers i.e. by using two continuously stirred tank reactors (CSTR) in series [3]-[4]; selectively retaining the solids within the reactor by holding

Citation: Budiyono, Widiasa, I.N., Johari, S. and Sunarso. (2014). Increasing Biogas Production Rate from Cattle Manure Using Rumen Fluid as Inoculums. International Journal of Science and Engineering, 6(1),31-38. Doi: 10.12777/ijse.6.1.31-38.

mixing prior to effluent removal [5]; pretreatment of manure by separating solids from digested material in order to improve biodegradability and accessibility [5]-[7]; and improving bacterial nutritional requirement [8]-[9]. In addition, an effort to increase biogas yield also has been done by improving contact between bacteria and substrate using stirring [10]-[12]; immobilizing microbe using fixed film reactor [13]-[14] as well as Anaerobic Sequencing Batch Reactor (ASBR) [15]; improving substrate composition by codigesting with others substrate [11], [16]-[17]; and controlling ammonia inhibition [18]. Different with other researchers mentioned before, an effort to improve methane yield was carried out by increasing the inoculums content in biodigester [19]-[23]. Several results from these study i.e inoculums are substantially relevant in process kinetics of biogas production [19]; amount of methane produced seemed proportional to the initial cattle manure as inoculums [20]; a strong influence of the bovine rumen fluid inoculums on anaerobic biostabilization of fermentable organic fraction of municipal solid waste [22]; and the higher percentage of inoculums gave the higher production of biogas [23]. However, almost all of AD studied before, inoculums used were dominated by digested sludge from anaerobic digester. In addition, until right now, data concerning the study of the effect of inoculums content to biogas production rate are very limited. Due to the highly anaerobic bacteria content in the rumen of the ruminant animals [24] and the abundance of rumen waste disposal from slaughterhouse, this study focuses on the use of rumen fluid as inoculums in anaerobic digestion of cattle manure. Biogas production with cattle manure as substrate on slaughterhouse has special condition that rumen as inoculums is supplied continuously from rumen waste disposal. To our best knowledge, so far there is very limited academic literature available on using rumen fluid as inoculums in anaerobic digestion of cattle manure. The aim of the current work was to obtain more data on the digestion characteristics of cattle manure under different total solid (TS) and rumen fluid content to biogas production.

MATERIALS AND METHODS

Sample preparation. The cattle manures and rumen fluids used in this research were taken randomly from slaughterhouse located on Semarang city. The fresh raw manure was collected from animal holding pen unit while rumen was collected from evisceration unit. Rumen fluid was prepared as follows: rumen content is poured to 100 L tank and added 25 liter tap water. Solid content then be separated from slurry by filter cloth. To assure that solid content in solution are dominated by bacteria, solution obtained then be filtered by 10 micron cartridge filter. Before using, all of raw manure collected is homogenized by mixing with propeller mixer. Raw manure and rumen fluid sample was analyzed its dry matter (DM) and volatile solid (VS) content by mean heating at 105 and 600°C, respectively. DM and VS content of fresh cattle manure and rumen fluid are presented in Table 1. B. Experimental apparatus set up. A series laboratory test of 400 ml biodigester was operated in batch system. The main experiment apparatus consists of biodigester and biogas measurement. Biodigester were made from polyethylene bottle plugged with tightly rubber plug and was equipped with valve for biogas measurement. The temperature of biodigester was maintained at certain value thermostatically controlled electrically heated water bath. Biogas formed was measured by 'liquid displacement method' as also has been used by Yetilmezsoy and Sakar [25].

The schematic diagram of experimental laboratory set up as shown in Figure 1.

Table 1: DM and VS characteristics of fresh

Parameter	Unit	Fresh manure	Rumen fluid
DM	%	22.75	1.3
VS	%	19.49	1.04
VS/DM	%	85.57	80

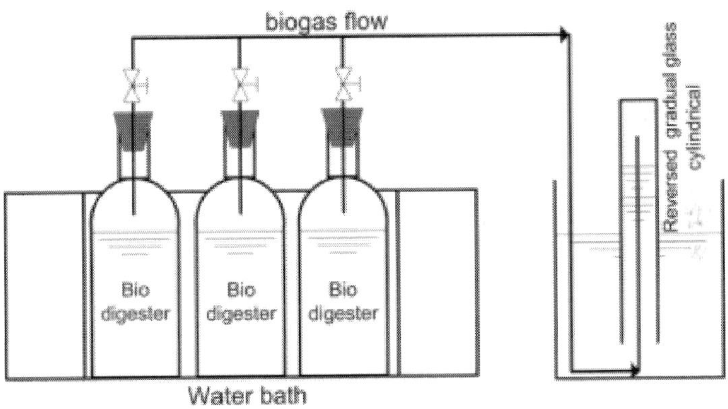

Figure 1: Schematic diagram of series laboratory batch assessment of anaerobic digestion

Experimental design

The influence of rumen fluid inoculums to biogas production rate was studied by varying rumen fluid and TS contents in biodigester. A series of laboratory experiments using 400 ml biodigester were performed in batch operation mode. Manure used fixed on 100 gram. Given 100 grams of fresh cattle manure (M) was fed to each biodigester and mixed with rumen fluid (R) and tap water (W) in several ratio resulting six different M:W:R ratio contents i.e. 1:1:0; 1:0.75:0.25; 1:0.5:0.5; 1:0.25:0.75; and 1:0:1 (correspond to 0; 12.5; 25, 37.5; 50, and 100 % rumen, respectively). Composition of six manure samples used in the study as presented in Table 2. In order to study the influence of total solid (TS) content to biogas production, a series laboratory biodigester in several TS level in feed was investigated. Given 100 grams of fresh cattle manure was fed to each biodigester and mixed with fixed 50 ml of rumen fluid and different volumes of tap water resulting six different TS contents i.e. 2.64, 4.61, 6.15, 7.38, 9.23, 12.30, AND 18.40 % (Equivalent to Volatile Solid VS of 2.31, 4.4, 5.38, 6.46, 8.07, 10.76, and 16.74 %, respectively). Composition of six manure samples used in the study as presented in Table 3. Operating temperature was at room temperature. The biodigester performance was measured with respect to cumulative volume of biogas produced after corrected to standard pressure (760 mm Hg) and temperature 0°C. All of treatment was carried out by triplication.

The experimental procedures

The manure sample with certain TS and rumen fluid content as research variables was fed to biodigester and homogenized with mixer propeller. CO2 gas was bubbled to biodigester to assure that biodigester in anaerobic condition. Biogas formed was measured every two days and stopped after biogas was insignificantly produced. The similar procedure was performed in three replications. Significance difference between treatments was determined statistically by Duncan Multiple Range Test (DMRT).

Table 2: Composition of six manure samples in different rumen fluid content

M:W:R ratio (% R)	Manure, g	Water, ml	Rumen, ml
1:1:0 (0% R)	100	100	0
1:0.75:0.25 (12.5% R)	100	75	25
1:0.5:0.5 (25% R)	100	50	50
1:0.25:0.75 (37.5% R)	100	25	75
1:0:1 (50% R)	100	0	100
0:0:1 (100% R)	0	0	100

Remarks: M=manure;W=water; R=rumen fluid

RESULTS AND DISCUSSIONS

The influence of rumen fluid to cumulative biogas production

This research step was directed to study either the effect of liquid rumen to cumulative biogas production is significant or not. The substrate consists of 100 gram manure and 100 ml rumen (MR 11) was fed to the digester and compared to substrate of manure and water in equal weight ratio (MW 11). The research was carried out in triplication. The cumulative volume of biogas production was observed during 60 days as depicted in Figure 2(a). In other term, the cumulative biogas production per total VS added (specific biogas production) is presented in Figure 2(b). Figure 2 shows that, in general, biogas production rate tend to obey sigmoid function (S curve) as generally occurred in batch growth curve (this is especially more clearly for MW 11 sample). Biogas production is very slow at the beginning and the end period of observation. This is predicted due to the biogas production rate in batch condition is directly corresponds to specific growth rate of methanogenic bacteria in the biodigester [26]. In the around of the first 12 days observation, biogas production is very low or indeed do not formed yet due to the lag phase of microbial growth. In the range of 12 to 50 days observation, biogas production is sigificantly increase due to exponential growth of microorganisms. After 50 days observation, especially for manure without rumen fluid (MW 11), biogas production tend to decrease and this is predicted tend due to stationary phase of microbial growth.

From Figure 2 (a) and (b) also can be seen that after 60 days observation still there is the tendency to increase biogas production and don't stop yet especially for manure mixed with rumen fluid (MR 11). This is predicted that the carbons contained by all waste constituents are not equally degraded or converted to biogas through anaerobic digestion. According to Richard [27] and Wilkie [28], anaerobic bacteria do not or very slow degrade lignin and some other hydrocarbons. In other word, the higher lignin content will lower biodegradability of waste. Animal manure such as waste used in this study include lignocellulosic rich materials, so anaerobically degradation also rather unoptimum [18].

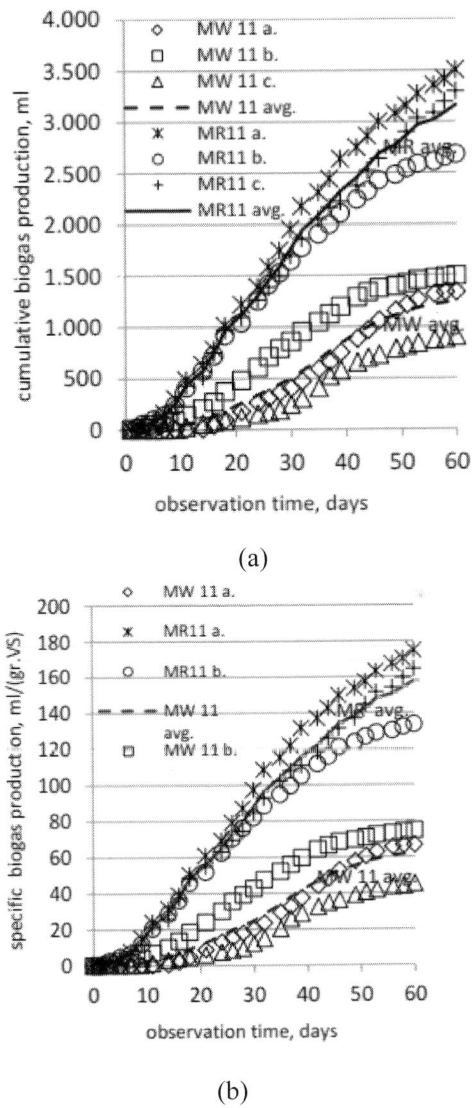

Figure 2: The influence of rumen fluid to biogas production; average from three bloc research with triplication; room temperature.

Figure 2(a) and (b) also shows that, generally, substrates consist of manure and rumen (MR11) exhibit higher biogas production than substrates contain manure and water (MW11). In other terms, specific biogas production per gram VS added (Figure 2.b) of MR11 is higher than MW11. The same behaviour is also shown in average biogas production curve. In the 60 days observation, average biogas production observed from MW11 and MR11 substrates were around 60 and 160 ml/(grVS). This result shows that the presence of liquid rumen in feed cause cumulative biogas production more than twice fold in compare to feed without liquid rumen. In other term, the substrates contain manure are statistically gave the significant effect to biogas production ($P< 0.05$) This is suggest that high concentration of anaerobic bacteria content in liquid rumen works effectively to degrade organic substrate from manure. According to Aurora [24], rumen of the ruminant animals contains the highly anaerobic bacteria dominated by cellulolytic bacteria able to biodegrade cellulose material from manure. This is agree with other results of researcher before that amount of biogas produced seemed proportional to the initial inoculums [20] and the bovine rumen fluid inoculums had a strong effect on anaerobic biostabilization of fermentable organic fraction of municipal solid waste [22]; as well as the higher percentage of inoculums gave the higher production of biogas [23]. From Figure 2 also can be seen that the line slope of MR11 curve is sharper than MW11 line. The implication is that, biogas production rate (ml/grVS.day) of MR11 is higher than MW11. This indicated that the addition of liquid rumen to feed will increase biogas production rate in compare to feed without liquid rumen. Similar with this results, inoculums are substantially relevant in process kinetics of biogas production [19]. Finally, the most important finding from this research can be drawn the conclusion that the liquid rumen seeded to biodigester has significant effect to cummulative biogas production and biogas production rate. Mathematically, the discussion concerning the effect of inoculums to kinetics constant of biogas production rate will be presented in the further section.

From Figure 2 also can be seen that the line slope of MR11 curve is sharper than MW11 line. The implication is biogas production rate (ml/grVS.day) of MR11 is higher than MW11. This is indicated that the addition of rumen fluid to feed will increase biogas production rate in compare to feed without rumen fluid. Similar with this results, inoculums are substantially relevant in process kinetics of biogas production [19]. Finally, the most important finding from this research can be drawn the conclusion that the rumen fluid seeded to biodigester has significant effect to cummulative biogas production and biogas production rate. Hence, the research to study the effec of rumen fluid concentration to biogas production will be carried out in the further step research.

The effect of rumen fluid content to biogas production

The effect of rumen fluid content to biogas production was studied by varying MWR ratio giving percent rumen fluid in mixed samples from 0 to 100 % rumen with fixed 100 gram manure. The TS content was presented in term of dry matter (DM). The research was carried out in triplication. The data obtained from the study then is averaged and the cumulative volume of biogas production was observed during 90 days as depicted in Figure 3(a). In other term, the cumulative biogas production per total VS added (specific biogas production) is presented in Figure 3(b). Numerical values of biogas yield in several days observation time is presented in Table 2. Figure 3 shows that, in general, substrates consist of manure and rumen (12.5 to 50 %R) exhibit higher cumulative biogas production than substrates contain manure and water only (0 %R). In other terms, specific biogas production per gram VS added (Figure 2.b) of sampel contain rumen fluid is higher than sample no contain rumen fluid. The same behaviour is also shown in average biogas production curve. In the 80 days observation, cumulative biogas production of 12.5; 25; 37.5 and 50 %R are 112.5; 144,48; 162.18; and 191.38 ml/gVS, respectively. While sample with 0 %R give cumulative biogas production of 68.61 ml/gVS. In the fisrt 50 days observation, there is no significant differences between 25, 37.5 and 50 %R ($P>0.05$). While sample of 12.5 %R give the significant differences in biogas production with samples of 25, 37.5 and 50 %R as well as 0 %R ($P< 0.05$).

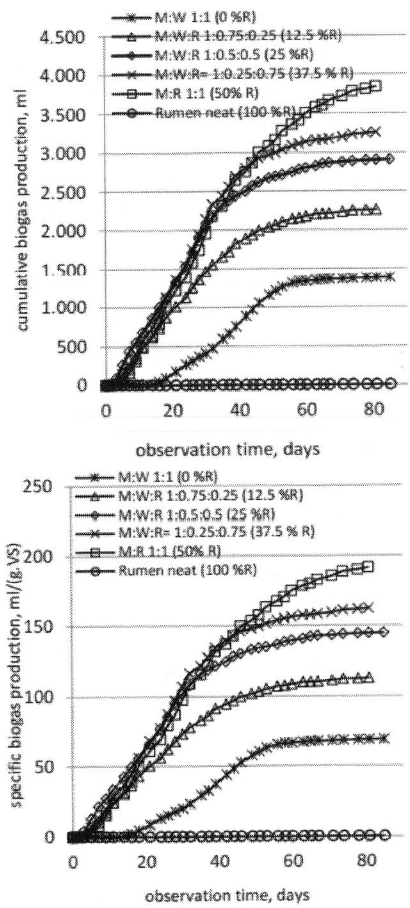

(a)

(b)

Figure 3: The effect of rumen fluid content to biogas production (room temperature).

These above result suggest that the optimum rumen fluid content for giving the best performance of biogas production is in the range of 25-50 %. Similar to this results, Lopes et al. (2004) reported that (a). no substantial difference was in evidence when 5% and 10% of the inoculum were used in preparation of the substrate; (b). in the range of 0 to 15 % rumen tested, the sample with the highest rumen content (15 %) gave the highest biogas production. Unfortunately, Lopes el at. [22] is not extensively investigate yet in using inoculums content more than 15 %. Hence, of course this study doesn't give data concerning optimum content of inoculums for biogas production. On the other hand, according to Foster- Carneiro et al. [23], when treated food waste restaurant with 20 – 30 % inoculums, the best performance for food waste biodegradation and methane generation was the reactor with 30% of inoculums. However, we cannot call this 30 % inoculums is the optimum condition because the research is not extensively investigate yet in using inoculums content more than 30%.

Relatively different with other samples, samples with 50 %R exhibit still there is the tendency to increase biogas production after 90 days observation. This is suggest that, in case of very abundance of rumen fluid such as occur in slaughterhouse, the rumen fluid content of 50 % (Manure : Rumen fluid ratio 1:1) will give the best performance for biogas production.

Table 2: Biogas yield in several days observation time

Observation time, days	M:W:R ratio (%R)					
	1:1:0 (0 %R)	1:0.75:0.25 (12.5 %R)	1:0.5:0.5 (25 %R)	1:0.25:0.75 (37.5 % R)	1:0:1 (50% R)	Rumen neat (100 %R)
0	0.00	0.00	0.00	0.00	0.00	0.00
10	0.07	24.19	33.62	28.33	24.15	0.00
20	8.66	50.38	65.92	68.01	60.79	0.00
30	20.02	73.00	101.58	106.91	97.38	0.00
40	37.29	91.21	121.39	134.31	131.58	0.00
50	60.29	103.11	134.06	149.35	157.33	0.00
60	66.42	108.36	139.46	156.23	174.65	0.00
70	67.85	111.05	143.14	159.32	185.39	0.00
80	68.61	112.50	144.48	162.18	191.38	0.00

From Figure 3 (a) and (b) also can be seen that after 90 days observation still there is the tendency to increase biogas production and don't stop yet. This is predicted that the carbons contained by all of waste constituents are not equally degraded or converted to biogas through anaerobic digestion. According to Richard [27] and Wilkie [28], anaerobic bacteria do not or very slow degrade lignin and some other hydrocarbons. In other word, the higher lignin content will lower biodegradability of waste. Animal manure such as waste used in this study include lignocellulosic rich materials, so anaerobically degradation also rather unoptimum [18]. Even, in other case, AD of organic matter such as municipal solid waste will not stop completely after 360 days observation [22]. From Figure 3 also can be seen that rumen neat (100% R) do not contribute the biogas production. Hence, all of biogases produced during all of treatment are originated only from substrate contained by manure. The substrate content by rumen fluid estimated has been degraded to biogas durung storage. This is because rumen fluid used in this research has been stored in several months. However, although rumen fluid has been stored in several months, is predicted there is no deterioration in activities of microorganism contained. This is suitable with the information of Rajeswari [29] and Speece [30] that decay rate of anaerobic bacteria is very low below 45°C. Even, anaerobic biomass can be preserved for months or years without serious deterioration in activity. Finally, the conclusion can be drawn from this research that the best performance of biogas production will be obtained if rumen fluid is in the range of 25-50%. Increasing rumen content will also increase biogas production. Due to the optimum TS content for biogas production between 7-9% (or correspond to more and less manure and total liquid 1:1) [31]-[33], the rumen fluid content until 50% will give the best performance for biogas production. However, intensively research need to be carried in further research to study interaction effect of TS and rumen content to biogas production. The further research is directed to verify the effect of rumen fluid content to biogas production at higher temperature.

The effect of total solid (TS) content to biogas production

The effect of TS content to biogas production was studied by varying TS from 2.64 to 18.40%. The TS content was presented in term of dry matter (DM). The research was carried out in triplication. The data obtained from the study then is averaged and the cumulative volume of biogas production was observed during 90 days as depicted in Figure 4(a). In other term, the cumulative biogas production per total VS added (specific biogas production) is presented in Figure 4(b). Numerical values of biogas yield in several days observation time is presented in Table 3. Figure 4 shows that, in general, biogas production rate tend to obey sigmoid function (S curve) as generally occurred in batch growth curve and as also has be resulted by Budiyono et al. [34]. Biogas production is very slow at the beginning and the end period of observation. This is predicted due to the biogas production rate in batch condition is directly corresponds to specific growth rate of methanogenic bacteria in the biodigester [26]. In the around of the first 10 days observation, biogas production is very low or indeed do not formed yet due to the lag phase of microbial growth. In the range of 10 to 50 days observation, biogas production is significantly increase due to exponential growth of microorganisms. After 50 days observation, biogas production tend to decrease and this is predicted tend due to stationary phase of microbial growth [34].

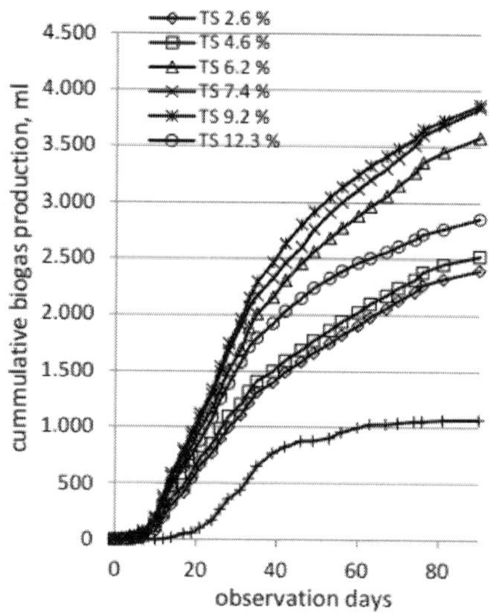

(a)

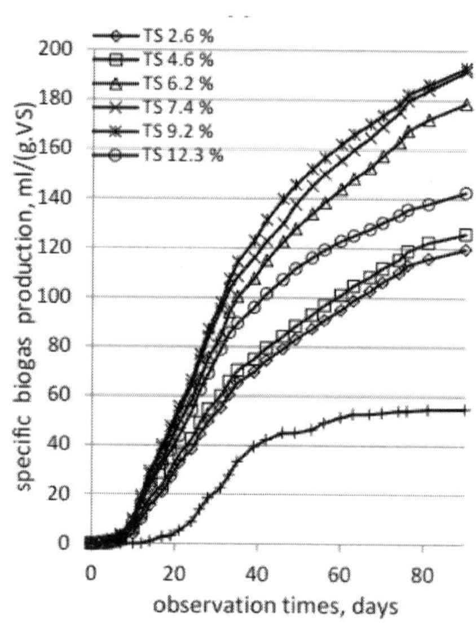

(b)

Figure 4: The effect of TS content to biogas production.

From Fig. 4 (a) and (b) also can be seen that after 90 days observation still there is the tendency to increase biogas production and don't stop yet. This is predicted that the carbons contained by all of waste constituents are not equally degraded or converted to biogas through anaerobic digestion. According to Richard [27] and Wilkie [28], anaerobic bacteria do not or very slow degrade lignin and some other hydrocarbons. In other word, the higher lignin content will lower biodegradability of waste. Animal manure such as waste used in this study include lignocellulosic rich materials, so anaerobically degradation also rather unoptimum [18]. Even, AD of cattle manure will cease completely after 360 days observation.

Table 3: Composition of six manure samples in different TS content

TS, %	VS, %	Cattle manure, g	Water, ml	Rumen Fluid, ml
2.64	2.31	100	550	50
4.61	4.04	100	250	50
6.15	5.38	100	150	50
7.38	6.46	100	100	50
9.23	8.07	100	50	50
12.30	10.76	100	0	50
18.40	16.74	100	0	0

Furthermore, as shown in Fig. 4, the best performance for biogas production was the digester with 7.4 and 9.2 % of TS i.e. give biogas yield 184.09 and 186.28 ml/gVS, respectively after 90 days observation. While the other TSs content of 2.6, 4.6, 6.2, 12.3, and 18.4 % give the biogas yield 115.78, 122.33, 172.34, 137.99, 54.87 ml/gVS, respectively. In addition, in the range of all of the observation time, TS contents of 7.4 and 9.2 % are also exhibit the best performance in biogas yield as presented in Table 3. These results suggest that, based on TS content effects to biogas yield, rumen fluid inoculum exhibit the similar behaviour with other inoculums, respectively. This is similar with the information from Balsam [31] and Zennaki et al. [33] that the optimum solid content obtained for biogas production is in the range 7-9 %. Furthermore, Baserja [32] reported that the process was unstable below a total solids level of 7% (of manure) while a level of 10% caused an overloading of the fermenter. These results suggest that, based on TS content effects to biogas yield, rumen fluid inoculums exhibit the similar effect with other inoculums. These results is predicted due the function of water in biodigester since the TS content will be directly correspond to water content. According to Sadaka and Engler [35], water content is one of very important parameter affecting AD of solid wastes. There are two main reason i.e (a). Water make possible the movement and growth of bacteria facilitating the dissolution and transport of nutrient; and (b) water reduces the limitation of mass transfer of non-homogenous or particulate substrate. Mathematically, the function of water in AD processes organic wastes consists of elements of carbon (C), Hydrogen (H), and Oxygen (O) reflected by reaction as follows [30]:

$$C_cH_hO_o + \left[\frac{4c-h-2o}{4}\right] H_2O ====> \left[\frac{4c+h-2o}{8}\right] CH_4 + \left[\frac{4c-h+2o}{8}\right] CO_2 \quad .... (1)$$

Furthermore, the water needed for biogas production from organic wastes consists of elements of carbon (C), Hydrogen (H), and Oxyigen (O) and Nitrogen (N) is reflected by reaction as follows [16].

$$C_cH_hO_oN_n + \left[\frac{4c - h - 2o + 3n}{4}\right] H_2O \text{ ======>}$$
$$\left[\frac{4c+h-2o-3n}{8}\right] CH_4 + \left[\frac{4c-h+2o+3n}{8}\right] CO_2 + nNH_3 \quad \text{......................... (2)}$$

The both two above equation depicted how important the water need in AD process for biogas production. Finally, the most important finding of this research that that the best performance for biogas production was the digester with 7- 9 % of TS similar with conventional processes used other inoculums. However, although the same optimum concentration of TS, rumen fluid inoculums caused biogas production rate and efficiency increase two to three times in compare to manure substrate without rumen fluid, as has been stated by Budiyono et al. [34].

CONCLUSIONS

Biogas production rate was studied by performing a series laboratory experiment using rumen fluid of animal ruminant as inoculums. The most important finding from this research is that the rumen fluid seeded to biodigester has significant effect to cummulative biogas production and biogas production rate. Rumen fluid inoculums caused biogas production rate and efficiency increase two to three times in compare to manure substrate without rumen fluid. The best performance for biogas generation will be obtained if rumen fluid is in the range of 25-50 %. Increasing rumen content will also increase biogas production. In addition, the best performance for biogas generation was the digester with 7.4 and 9.2 % of total solid i.e. give biogas yield 184.09 and 186.28 ml/gVS, respectively after 80 days observation. These results suggest that, based on TS content effects to biogas yield, rumen fluid inoculum exhibit the similar effect with other inoculums. Due to the optimum TS content for biogas production between 7-9 % (or correspond to more and less manure and total liquid 1:1), the rumen fluid content until 50 % will give the best performance for biogas production. However, intensively research need to be carried in further research to study interaction effect of TS and rumen content to biogas production. The further research is directed to verify the effect of rumen fluid content to biogas production at higher temperature.

REFERENCES

1. Okkerse, C. and H. V. Bekkum. From fossil to green. Green Chem., 1: 107–114, 1999
2. Alvarez, R. and G. Lide′n. 2008. The effect of temperature variation on biomethanation at high altitude, Biores. Technol. 99: 7278–7284, 2008
3. Boe, K. and Angelidaki, I. 2009. Serial CSTR digester configuration for improving biogas production from manure, Water Research, 43(1): 166-172, 2009
4. Kaparaju, P. and Angelidaki, I., 2008. Effect of temperature and microbial activity on passive separation of digested cattle manure. Bioresour. Technol. 99: 1345–

1352, 2008

5. Kaparaju, P., Buendía, I., Ellegaard, L., Angelidaki, I., Effects of mixing on methane production during thermophilic anaerobic digestion of manure: labscale and pilot-scale studies. Bioresour. Technol. 99: 4919–4928, 2008

6. Liao, P.H., K.V. Lo, S.T. Chieng. 1984. Effect of liquid—solids separation on biogas production from dairy manure. Energy in Agriculture, 3: 61-69, 1984

7. Møller, H.B., A.M. Nielsen, R. Nakakubo, H.J. Olsen. Process performance of biogas digesters incorporating pre-separated manure, Livestock Science, 112: 217–223, 2008

8. Kayhanian, M. and D. Rich. Pilot-scale high solids thermophilic anaerobic digestion of municipal solid waste with an emphasis on nutrient requirements, Biomass and Bioenergv, 8(6): 433-444, 1995

9. Demirci, G. G. and G. N. Demirer. Effect of initial COD concentration, nutrient addition, temperature and microbial acclimation on anaerobic treatability of broiler and cattle manure, Bioresource Technology 93: 109–117, 2004

10. Krylova, N.I., R.E. Khabiboulline, R.P. Naumova, and M.A. Nagel, M.A. The influence of ammonium and methods for removal during the anaerobic treatment of poultry manure. J. Chem. Technol. Biotechnol. 70 (1): 99-105

11. Callaghan, F.J., D.A.J. Wase, K. Thayanithy, and C.F. Forster. Codigestion of waste organic solids - batch studies. Bioresource Technol. 67 (2), 117-122, 1999

12. Karim, K., R. Hoffmann, K. T. Klassonb, M.H. Al-Dahhan. Anaerobic digestion of animal waste: Effect of mode of mixing, Water Research 39: 3597–3606, 2005

13. Lo, K.V., A. J. Whitehead, P. H. Liao, N. R. Bulley. Methane production from screened dairy manure using a fixed-film reactor. Agricultural Wastes, 9(3): 175-188, 1984

14. Vartak, D.R., C. R. Engler, M. J. McFarland, and S. C. Ricke. Attachedfilm media performance in psychrophilic anaerobic treatment of dairy cattle wastewater, Bioresource Technology 62: 79-84, 1997

15. Ndegwa, P.M., D.W. Hamilton, J.A. Lalman, and H.J. Cumba. Effects of cycle-frequency and temperature on the performance of anaerobic sequencing batch reactors (ASBRs) treating swine waste, Bioresource Technology, 99: 1972–1980, 2008

16. Gelegenis, J., D. Georgakakis, I. Angelidaki, V. Mavris. Optimization of biogas production by co-digesting whey with diluted poultry manure. Renewable Energy, 32(13): 2147-2160, 2007

17. Lehtomaki, A., S. Huttunen, J.A. Rintala. Laboratory investigations on co-digestion of energy crops and crop residues with cattle manure for methane production: Effect of crop to manure ratio, Resources, Conservation and Recycling, 51: 591–609, 2007

18. Nielsen, H.B. and I. Angelidaki. Strategies for optimizing recovery of the biogas process following ammonia inhibition. Bioresource Technology. 99(17):7995-8001, 2008

19. Luengo, P.L. and J. M. Alvarez. Influence of temperature, buffer, composition and straw particle length on the anaerobic digestion of wheat straw-pig manure mixtures. Resources, Conservation and Recycling, 1(1): 27-37, 1988

20. Castillo, R.T., P.L. Luengo, and J.M. Alvarez. Temperature effect on anaerobic of bedding manure in a one phase system at different inoculums concentration, Agriculture, Ecosystems and Environment, 54:55-66, 1995

21. Sans, C., J. Mata-Alvarez, F. Cecchi, P. Pavan and A. Bassetti. Acidogenic fermentation of organic urban wastes in a plug-flow reactor under thermophilic conditions. Bioresource Technology 54(2): 105-110, 1995

22. Lopes, W.S., V.D. Leite and S. Prasad, Influence of inoculum on performance of anaerobic reactors for treating municipal solid waste. Bioresour. Technol., 94: 261-266, 2004

23. Forster-Carneiro, T., M. Pérez and L. I. Romero. Influence of total solid and inoculum contents on performance of anaerobic reactors treating food waste. Bioresource Technology 99(15): 6994-7002, 2008

24. Aurora, S.P. Microbial Digestion in Ruminants. Indian Council of Agricultural Research, New Delhi, 1983

25. Yetilmezsoy, K and S. Sakar. Development of empirical models for performance evaluation of UASB reactors treating poultry manure wastewater under different operational conditions, J. Hazardous Materials, 153: 532–543, 2008

26. Nopharatana, A., P. C. Pullammanappallil, W. P. Clarke. Kinetics and dynamic modeling of batch anaerobic digestion of municipal solid waste in a stirred reactor, Waste Management, 27: 595–603, 2007

27. Richard, T. The effect of lignin on biodegradability. Cornell Composting, Cornel Waste Management Institute, In http://www.cfe.cornell.edu/compost/calc/lignin.html, 1996

28. Wilkie, A.C. Anaerobic digestion of dairy manure: design and process consideration. in Dairy Manure Management : Tretament, Handling, and Community Relations, Natural Resource, Agriculture, and Engineering Service, Cornell University, Itaca. pp. 301-3, 2005

29. Rajeshwari, K.V., M. Balakrishnan, A. Kansal, K. Lata and V.V.N. Kishore. State-of-the-art of anaerobic digestion technology for industrial wastewater treatment. Renew. Sustain. Energy Rev., 4: 135-156, 2000

30. Speece, R.E. Anaerobic Technology for Industrial Wastewaters. Archae Press, USA, p.43, ISBN:0-9650226-0-9, 1996

31. Balsam, J. Anaerobic digestion of animal wastes: factors to consider. ATTRA-national sustainable agriculture information service. United States Department of Agriculture's, USA, 2002

32. Baserja, U., Biogas production from cowdung: Influence of time and fresh liquid manure. Swiss-Bio Tech., 2: 19-24, 1984

33. Zennaki, B.Z., A. Zadi, H. Lamini, M. Aubinear and M. Boulif, Methane Fermentation of cattle manure: Effects of HRT, temperature and substrate

concentration. Tropicul. Tural., 14: 134-140, 1996

34. Budiyono, I N. Widiasa, S. Johari, and Sunarso. The Kinetic of Biogas Production Rate from Cattle Manure in Batch Mode, International Journal of Chemical and Biological Engineering, 3(1): 39-45, 2010

35. Sadaka, S.S. and Engler, C.R. Effect of initial total solids om composting of raw manure with biogas recovery. Compost Sci. and Utilization. 11(4): 361-369, 2003

12
CHAPTER

GREENING THE GLOBAL PHOSPHORUS CYCLE: HOW GREEN CHEMISTRY CAN HELP ACHIEVE PLANETARY P SUSTAINABILITY

Paul J. A. Withers, James J. Elser, Julian Hilton, Hisao Ohtake, Willem J. Schipper and Kimo C. van Dijk

INTRODUCTION

"This is no way to run a biogeochemical cycle."

(J. Elser quoted in Lougheed (2011)).[1]

Phosphorus (P) is a critical element for our food production systems, manufacturing industries, and general economic growth whose long-term security of supply is of major concern for regional and national economies. Phosphate rock (PR) is mined for processing into P derivates, such as phosphoric acid (PA, H_3PO_4) and white phosphorus (P_4), for final use in various products that society uses ranging from fertilizers to toothpaste to car batteries. The mineable reserves of PR are essentially finite for the human era because of the geological timescales over which the natural cycling and therefore renewability of P occurs. Current high rates of P consumption (over 20 Tg P $year^{-1}$)[2] driven largely by fertilizer use are putting increasing pressure on the global supplies of this vital resource, leading to rising and volatile prices.[3] Estimates of how long global PR reserves will last are uncertain and currently vary from <100 up to 400 years, but their accessibility and cost are a major concern for many countries with no PR reserves, such as Europe.[4,5] Future demand for P in emerging economies, notable in Africa and Asia, is anticipated to be very high, while rising demands are also imposed by expanding biofuel production.[6] Ironically, a more immediate environmental problem is the widespread leakage of P to waterbodies where it causes nuisance algal blooms, loss of aquatic biodiversity, and increased risk to human health.[7] These pressing problems

Citation: Paul J. A. Withers, James J. Elser, Julian Hilton, Hisao Ohtake, Willem J. Schipper and Kimo C. van Dijk. Greening the global phosphorus cycle: how green chemistry can help achieve planetary P sustainability. DOI: 10.1039/C4GC02445A (Perspective) Green Chem., 2015, 17, 2087-2099.

have been evidenced quite tangibly in the bloom-induced drinking water crises that have emerged in recent years in Lake Taihu (China)[8] and Lake Erie (USA/Canada).[9] Phosphorus is therefore both a critical element and a pollutant and must be used more efficiently and sustainably in the future to help safeguard future food, energy and water security.[5,10–12]

Phosphorus has no substitute, but can be continually re-used, and is thus a prime example of a critical resource that could be utilized more efficiently in a circular economy to support sustainable growth with less pollution.[13,14] The current P cycle is inherently inefficient because the vast majority of the P that is mined each year becomes dissipated in low-grade PR ores, manures, by-products and residues that are not fully re-used or re-used uniformly, and in surface waters and seas. Actual rates of P re-use will clearly vary considerably between sectors and countries, but a recent review of P budgets across different geographical scales suggests it is <20% of total P inflows.[15,16] This is a low percentage and there continues to be large scale disposal of P to landfill in developed countries. A large proportion of dissipated P (wrongly termed waste) could potentially be re-used as secondary P resources subject to technological and financial constraints.[17–19] A distinction has been made between strategies that enhance P re-use through more uniform recycling directly back to land, and those that first require recovery through innovation in bio- and chemical engineering.[5] True wastage occurs when P is dissipated to the oceans. These permanent P losses have quadrupled due to anthropogenic activity in the 20th century and must also be reduced.[20]

Achieving long-term sustainability of P use in society will require insights from a variety of emerging approaches, including "green chemistry" and "green engineering".[21,22] In the context of P, these two concepts adopt the same core principles of (a) the development of benign products and processes, (b) the elimination of waste, (c) the use of renewable (secondary) resources, and (d) the design of output-driven production systems with minimum requirements and maximum efficiency.[23] These green principles will be critical in developing new strategies for how we use P in the technosphere and how products can be designed in the future with a focus on recycling. For example, an immediate green chemistry challenge is to remove and re-use the potentially harmful elements naturally present in PR, such Cadmium (Cd), Uranium (U) and Lead (Pb), that might persist in the environment or through the food chain.[24] The cradle-to-cradle philosophy implicit in green chemistry has yet to be fully adopted by sectors in the P cycle, although the conceptual basis for achieving this transition exists through P accounting and recent drivers towards industry sustainability and a circular economy.[13,25–27] This probably relates to the relatively recent recognition of P as a critical resource by government and industry, and P sustainability as a serious societal problem. For example, PR has only just been recognized in Europe as a critical raw material.[28] Considering the environmental impact of the whole life cycle, and afterlife, of P products and not just the impact of their initial manufacture will also require a large paradigm shift in attitude by all stakeholders and actors in defining societal P needs and how systems can be best designed to meet them.[29–31]

This paper outlines a variety of avenues for the application of green chemistry principles and practice to transform P use in society. We describe a variety of existing and emerging technologies and strategies for elemental re-use in the P cycle. If not constrained by cost, their adoption will help deliver a sustainable P system that can sustain future generations with abundant food, feed, fibre, fuel, clean water and other essential P products society relies on. After a synopsis of global P use, we consider sustainability and green chemistry challenges in recycling and recovery technologies within different sectors and then system design with some regional examples of progress. Our ultimate objective is to inspire green chemists and engineers to take a major role in developing and implementing a more efficient and sustainable P cycle.

Phosphorus distribution in the technosphere

Information on how mined PR is dispersed in the technosphere is needed to identify where opportunities for green chemistry and engineering are most likely to occur in the P cycle (Fig. 1). Numerous attempts at global P budgets have been made, but they are all simplified to some degree or another, and contain large errors because accurate data for all the stores and flows are very difficult to obtain or estimate. However some important general trends emerge. Anthropogenic flows of P are much greater than in natural systems, but are still relatively small (<50 Tg year^{-1}) compared to the very large global stores of P present in the pedosphere, (40 000–200 000 Tg), biosphere (550–3050 Tg) and hydrosphere (8300–123 000 Tg).[20,33] They are also very widely dispersed because PR derivates have a very wide range of uses in industrialized society and are therefore traded globally: for example in fertilizers, feeds, foods, pesticides, plastics, computer chips, lubricants, chemical extractants, detergents, toothpaste, car batteries and flame retardants. This P becomes further dispersed in industrial and municipal wastewater, solid waste, livestock manures and crop residues after its first use. Variation in demography, agricultural intensity and industrial activity in different regions, countries and catchments will therefore lead to different priorities for improving P resource use efficiency, choosing the most economic recycling and recovery options and reducing environmental impacts.[15,34] This regional variation is an important aspect regarding achieving progress in global P sustainability.

Two major issues emerge from global P budgeting. The first is that food consumption patterns and global trade in food, feed and livestock products have a major impact on P flows and their global distribution.[35] In particular, increasing demand for meat is a major driver of fertilizer and feed P consumption and a major cause of global inefficiency of P use.[36] This is because large areas of agricultural land are needed to grow the fodder and grain to feed animals, and 70–80% of the P consumed by animals is excreted and must be handled. The second issue is that the amount of new PR mined each year is currently about 30 Tg of P, whilst the actual incremental demand for new P by a growing world population is estimated at only 1 Tg P year^{-1}.[2,37]This highlights the gross inefficiency of the global P cycle and the necessary infrastructure and societal costs of handling dissipated P. Sheldon introduced the *E* Factor (or Environmental Factor) to describe the ratio of the amounts of waste generated per unit of product processed.[23] Although not strictly comparable, PR production has an *E* factor of *at least* 30 making it similar to the fine chemical industry in terms of waste generation.

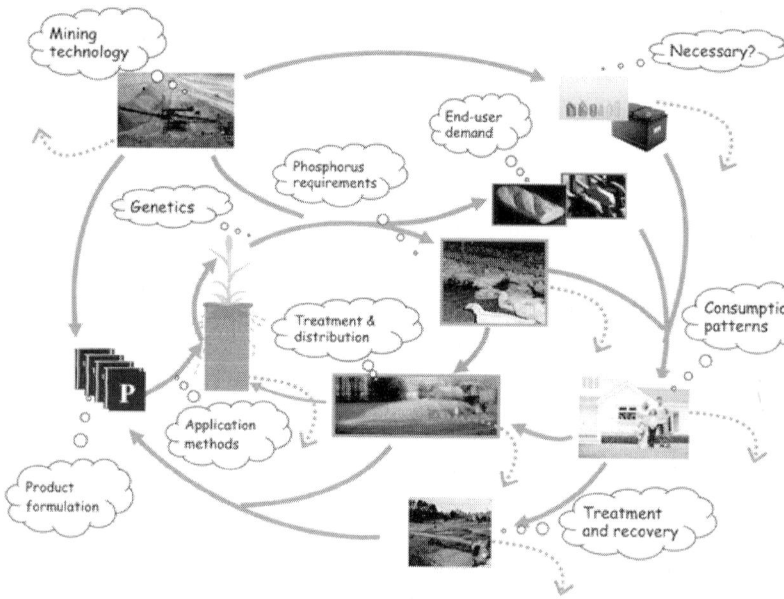

Figure 1: Phosphorus cycling in the technosphere showing thought bubbles suggesting where green technologies and innovations could improve P sustainability by reducing P fluxes, losses and wastage. Points of P loss are shown by dotted lines. Adapted from Sylvester-Bradley and Withers.[32]

The first stage of waste generation is in the mining operation and subsequent processing to PA, with global estimates of unrecovered P amounting to 7–11 Tg P year^{-1}.[2] Over 80% of mined PR is used to manufacture fertilizers (chiefly through PA), with the remainder used in feed and food additives mainly as calcium and sodium phosphates (through PA), and in the chemical industry as white phosphorus (from PR).[38] The majority of fertilizer P accumulates in the soil from where crops take up their P requirements to be harvested or grazed into food for animals and humans. Excreted P is either recycled back to the land, incinerated, taken to landfill after treatment or discharged to water (Fig. 1). Fluxes of P through animals are similar in size to those through crops (arable crops and managed grass, *ca.* 20–25 Tg year^{-1}, and dominate the technosphere P budget.[33,39] Globally there is more P in animal excreta than in manufactured fertilizers, but both grazing returns and collected manure are not recycled uniformly. Animal slaughter also leads to significant amounts of slaughter waste which is often co-incinerated for its fuel value in cement works and power stations. Global estimates of P in slaughterhouse waste suggest it could be 10–15% of total P imports and therefore significant.[40]For example, for Europe it represents about 70% of the P imported into the livestock sector.[15]

High animal stocking rates in some areas lead to an oversupply of collected manure P for the available land area. This misplacement of P, together with its intrinsically high P content, is both a source of localized P surplus, P accumulation in soil and P loss in land runoff causing a high eutrophication risk.[41,42] Surpluses of P in soil also arise due to excessive fertilizer use.[43,44] Global surpluses of P accumulating in agricultural soils are estimated at 11–16 Tg P year^{-1}.[45,46] The legacy soil P associated with cumulative past P surpluses therefore represents a considerable P resource base to reduce P fertilizer use.[5,47] For example, Ringeval *et al.* recently estimated that over 80% of the total P in French soils (0–35 cm) was of anthropogenic origin.[48] Estimates of global P losses in land runoff to water associated with erosion and direct losses from the fertilizers and manures applied are typically quoted at 13–24 Tg year^{-1}.[2,33,46] Although there is large uncertainty around these runoff losses, there is clearly potential to reduce them.

The flux of P associated with consumed food, with smaller contributions from detergents and other materials is *ca.* 5 Tg year^{-1}.[45,49] The majority of this P input ends up in wastewater, or in solid (largely food) waste. Wastewater P is either discharged directly to watercourses, or treated in a septic tank or collectively at a treatment centre. Wastewater treatment plants (WWTPs) are estimated to receive about 4 Tg P year^{-1}, of which up to 90% is removed in the form of sewage sludge depending on local effluent regulations. Of this P rich sludge, *ca.* 50% is returned to the land as biosolids, but this figure can vary considerably; for example in Europe between 0 and 80%.[15] The remainder of sewage P, *i.e.* that is not captured in the sludge, is discharged to water as treated sewage effluent amounting to *ca.* 1.5 Tg P year^{-1}.[50] Globally consumer waste is increasingly being recycled, but the majority is still disposed of in landfill or incinerated.[51] Total amounts of household and industrial P that are disposed of to landfill, or incinerated are estimated at 2–3 Tg P year^{-1}.[2] The total amounts of P potentially recoverable from wastewater effluent, landfill and incineration are therefore only *ca.* 5 Tg P year^{-1}, and much lower than the larger global fluxes of P retrievable from agricultural systems.

Global usage of P in the chemical industry is estimated at only 1–3 Tg year^{-1}, but there are still opportunities to recycle industrial residues containing P.[18,49,52] The use of P in applications that might be considered non-essential are also relatively minor (1 Tg P year^{-1}), but still contribute to the P cycle.[2] This evidently raises debate on which applications of P are considered unnecessary and what the consequences of replacement would be. Detergent P is one example of unnecessary P that is now being phased out in the USA and Europe.[2,5]

GREENING THE PHOSPHORUS CYCLE

The wide dispersal of P in the technosphere and the large variability in P flows across sectors in the P cycle suggests that P sustainability targets are best set and met at the sector and regional level.[15,34,53] A large proportion of dissipated P is contained in mined ores that do not pass the bone phosphate of lime standard for processing (BPL <68), and in sector wastes, by-products, residues and wastewaters. These secondary materials provide a range of accessible resources from which P could be recovered by green chemical and bio-technologies into safe, renewable P sources for re-use, and for substitution of primary PR, with minimum waste (Fig. 2). Many technologies are still in their infancy (*i.e.* pilot stage), whilst others have been already implemented.[18,55,56] These secondary sources are also increasingly being considered as sources for other essential elements and compounds that society uses. For example, Westerhoff *et al.*[57]

recently identified a range of valuable metals in municipal sludge and suggested that the 13 most lucrative (Ag, Cu, Au, P, Fe, Pd, Mn, Zn, Ir, Al, Cd, Ti, Ga, and Cr) had a value of US $280 per tonne of sludge. Wastes and wastewaters generated in the food chain are also renewable sources of energy (electricity and fuel), and a number of essential 'functionalized' compounds that society needs including flavonoids, waxes, fatty acids and biopolymers (Fig. 3).[58–60] This raises the question over whether first generation valorization of secondary bioresources from the food chain through recycling P to land is the most sustainable route.

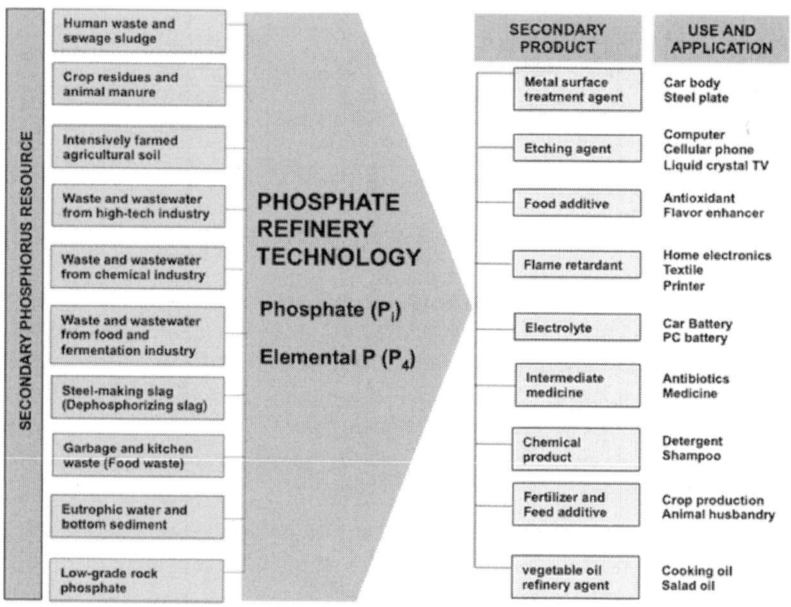

Figure 2: The Phosphate Refinery for production of renewable secondary P sources in agriculture and industry. After Ohtake.[18,54]

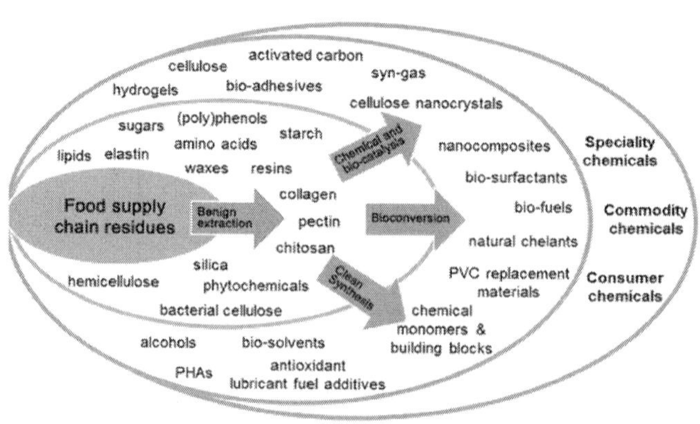

Figure 3: Components present in food chain waste and their uses in common consumer appli-

cations, highlighting sectors of the chemical industry that could benefit from the use of such a renewable resource. After Clark et al.[58]

The mining sector

The first stage in the P cycle for progress in sustainability is in the PR mining operation. A recent green chemistry development by the fertilizer industry has been a renewed interest in comprehensive extraction (CX), Table 1. CX sees PR not just as a source of P, but as an ore "which can contain the entire periodic table". Since PR contains U and a number of rare earths essential to society, the co-extraction U and rare earths from PR and PA is a prime example of how the mining industry can gain added value from PR processing.[62] Technologies for co-extraction of P and U from lower grade ores by adding solvent extraction circuits to existing wet process technology for PA manufacture, or though ion exchange, are now being revisited with commercial-scale projects starting in Brazil and India in 2017.[63] Simpler, gravity separation techniques (shaking table) have been shown to work for extracting both heavy and light rare earths from flotation tailings (heavy) and from phosphogypsum (PG) (light). Thorium (Th) is another metal in PR ores and PA that could also be potentially recovered to provide a long-term supply of safe nuclear energy, with pioneering plants in China, India and Brazil.[63] Co-extraction technologies also help to remove the environmental hazards associated with PR-derived products making them more benign for future P re-use; a clear green chemistry goal. A future challenge for these technologies is to remain economic.

Table 1: The operating principles of comprehensive extraction developed by the PR industry have a number of similarities to those of green chemistry. Adapted from Hilton *et al.*[61]

• address all available resources from a given site/deposit in an integrated resource management strategy
• disturb the ground once
• construct regulation of naturally occurring radioactive materials (NORM) industries based on shared values between operators and regulators
• obtain and keep a social licence to operate, focused on equitable distribution of benefits between stakeholders and stockholders
• sequence extraction procedures and select extraction technologies to optimise deposit returns, *e.g.* by classifying and progressing resources on a whole "energetic basin" project management basis
• extract and store resources that would otherwise be wasted or dispersed for future use
• manage resources across the whole life-cycle seeking to conserve primary resources and substitute secondary resources for primary where feasible
• align to the waste hierarchy seeking all opportunities for re-use and recycling of by-products, residues and "wastes", resulting in a zero waste outcome

• promote new product development as strategic alternative to waste disposal (*e.g.* from recycling tailings or residues)
• ensure a net positive contribution to food, energy and water security as part of a wider commitment to sustainable development

The challenge of recovery of P from low-grade PR ores (BPL values of 50 or above) is stimulating innovation in an industry that has been very change averse. Ecophos and the Improved Hard process (IHP) are typical examples of such innovation. The Ecophos process expands on the well-known hydrochloric acid route to dissolve phosphate from rock (http://www.ecophos.com). This serves as an alternative to the standard sulphuric acid method (wet process), which is applied worldwide to produce PA. Unlike the latter, the hydrochloric acid route does not create an insoluble stream of PG to separate the calcium oxide from the PA, but instead produces soluble calcium chloride which remains mixed with the PA product. This poses a separation challenge which can be surmounted by solvent extraction of the PA, or precipitation with lime. The latter chemistry yields feed grade phosphates in the form of dicalcium phosphate. This process has a marked operational cost advantage if hydrochloric acid waste is locally available and is more sustainable than the market procurement of sulphur (S), which is needed to produce sulphuric acid for the conventional wet process of PA manufacture. The IHP uses the local formation of P_4 in a heated mixture of low-grade rock and a low-cost reduction agent such as petcoke. The off gases are oxidized immediately, giving off sufficient heat to keep the process going without significant external energy input. The now-oxidized P is hydrolysed to PA which is used in fertilizer and industrial phosphates manufacture. The process is eminently suited to use low-grade rocks, being competitive in variable cost for these, and might use other locally available P-rich waste also as input. This process has seen a very long development phase and is now at pilot stage near Fort Meade, Florida. The operator anticipates this will become commercially viable sometime in the coming years based on a 200 000 t PA yr^{-1} production plant.[64]

Another sustainable development within the mining industry has been the focus on registering mine and PA processing wastes, reducing their volume and re-using them. The efficiency of P extraction from PR ore is very variable, with estimates of P losses and wastage during mining, beneficiation, chemical processing and handling of between 15 and 50% (average 30%) depending on PR quality and the methods deployed.[38,49,63] This wastage could be substantially reduced by optimising the PR extraction process and recovering P from mine tailings and the by-product PG produced during PA manufacture.[19] About 20 Tg (dry weight) of mine tailings equivalent to 1 Tg P $year^{-1}$, and about 160–170 Tg of PG (dry weight) equivalent to 0.5 Tg P $year^{-1}$ are produced each year.[63] Reducing ore waste at the mine face is being explored through laser technology to improve ore quality detection, and mine tailings can be flocculated and thickened using polymers and re-used to strengthen concrete.[63] As a result of an evidence-based review by the International Atomic Energy Agency,[65] PG is also no longer classed as a hazardous waste (it contains the radionuclide Radium, Ra), but as a co-product, encouraging its re-use rather than indefinite disposal to stacks. By 2015, annualised re-use of PG as a soil amendment and fertilizer in agriculture, as a building and road base material in the construction industry and as a source of ammonium sulphate (through the Merseburg process) will be at least 30 Tg $year^{-1}$ from a near zero base in 2008. However, although their re-use has increased, the recovery of P from mine tailings and PG remains a challenge and one potentially for green chemistry to tackle.

The agricultural sector

The agricultural sector generates the largest amounts of potentially recoverable and recyclable P. Most of the dissipated P from past fertilizer and recycled manure inputs has accumulated in farmed topsoil (50 000 Tg),[53] and in some es subsoil.[66] Progress in the recovery of this legacy soil P *in situ* is largely dependent on improving its accessibility to plants through bio-engineering based on the inherent traits present in plants and microorganisms for soil P acquisition.[67,68] A more accessible regional source of secondary P in agriculture is livestock manure. Although livestock manure is a multifunctional fertilizer, containing organic matter and essential major nutrients and trace elements, its nutrient composition is not ideally balanced for many crops. In areas with intensive livestock farming and limited land area (*e.g.* large pig and poultry units), uniform spreading of manure on fields is limited by its P content and costly transport over longer distances, or treatment, is needed.[69] Due to the necessity for local disposal, spreading allowances quickly exceed crop P requirement due to their low N : P ratio. In such es, manure application becomes diffuse landfilling (push model) rather than a demand-driven application. A selective technology for partial chemical removal of P from manure through acidulation and solid/liquid separation may remedy this and enable more demand-driven application of P-poor solid fraction of manure as organic fertilizer, and the separate re-use of a precipitated P-rich concentrate (*e.g.* struvite or calcium phosphates) elsewhere as an inorganic P fertilizer (Fig. 4). The recoveries of P from manure and liquid manure fractions by struvite and calcium phosphate precipitation can be high (50–90%).[55] For example, Suzuki recovered 171 g of 95% pure struvite from 1 m^3 of pig slurry in a crystallization plant.[70] Potential alternative technologies for partial removal of P from manures to decrease their N : P ratio for land application and reduce the need for costly manure export pose a challenge for green chemistry. One green engineering option might be to use simple water extraction and micro-filtration of animal slurries to manipulate their N : P ratios since a significant proportion P in manures and slurries is water-extractable.[71,72]

Drying, or prior separation before drying and pelletizing may pose an environmentally acceptable solution, allowing the dried manure (nutrient concentrate) to be transported over longer distances to crop producing areas with P demand. Separation of the P-poor liquid fraction to allow more efficient drying of the P-rich solid fraction requires separate treatment and/or disposal of the liquid which increases cost. Green chemistry techniques to combust whole (wet) manures without separation by supercritical and subcritical oxidation or and water gasification for energy production could provide a solution, but are still in their infancy.[55]

Alternatively, bulky manures or separated solids can be anaerobically digested, incinerated or pyrolysed to provide energy (Fig. 4). Pyrolysis can promote the conversion of animal manure into charcoal (*i.e.* biochar) by heating to 300–550 °C in the absence of oxygen, thereby reducing the solid volume. Biochar is proving a useful soil amendment and fertilizer with low greenhouse gas emissions.[73] Pyrolysis has also been used as a pre-treatment for raw biomass which is not suitable for direct combustion because of a low energy density and high moisture content.[74]

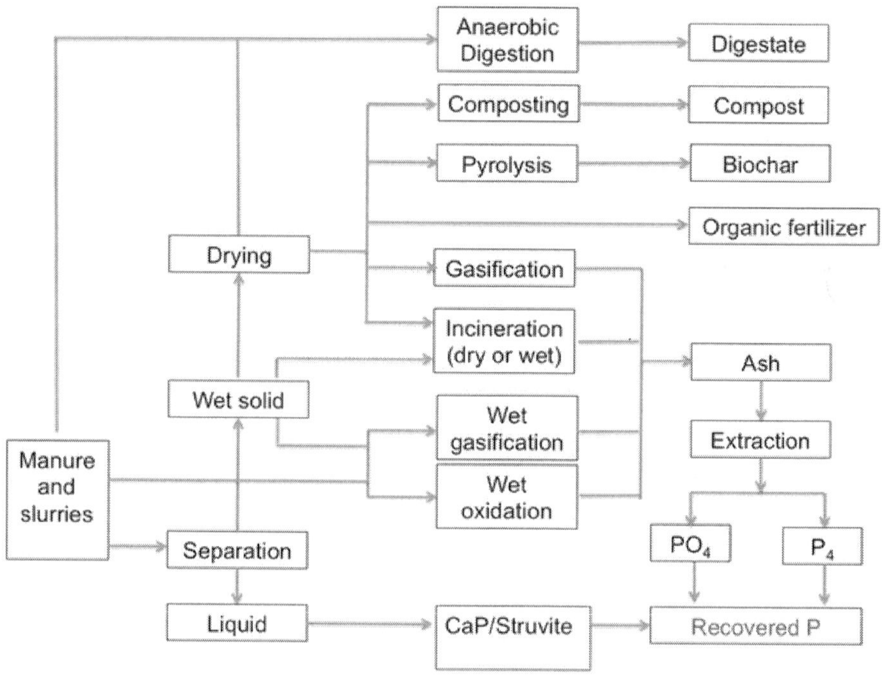

Figure 4: Schematic overview of the main options to recover P from manures and biosolids. Adapted from Oenema et al.53 and Schoumans et al.[55]

Gasification heats manure at higher temperatures (800–1000 °C) but loses more of the carbon as CO_2.[55] Incineration of animal manure is an attractive option in regions with a high animal density, particularly for handling chicken manure, and the incineration ash, contains a considerable amount of P and can be processed into fertilizers (see wastewater sector below).[75]

The food processing sector

After leaving the farm significant amounts of P are dissipated in slaughter wastes, food processing waste and wastewater. After various steps aiming at recovery of valuable by-products such as animal fats, gelatine and proteins, livestock slaughter waste becomes Meat and Bone Meal (MBM). This is a mixture of calcium phosphate (bones) and proteins (meat) potentially usable as a fertilizer or feed. Various streams of MBM occur worldwide with different protein/bone phosphate ratios, and generally have the highest P concentrations of waste flows suitable to substitute PR. After the Bovine Spongiform Encephalopathy (BSE) crisis, Europe required a BSE-sensitive fraction MBM, Category I, to be thermally destroyed instead of being used as feed ingredients for proteins and minerals.[40,76] Currently, these materials with high calorific value are often used as bio-co-fuel in power plants, to give carbon dioxide (CO_2) credits. This unfortunately dilutes their considerable P content beyond recovery. MBM mono-incineration and smart development to recover P from the ashes (as for other incinerated wastes) needs to be implemented to avoid this P wastage. In one process, MBM is incinerated with alkaline earth compounds at 1000 °C in a rotary kiln to improve P availability in the ash and marketed for use in agriculture (ULOPHOS®).[77]

Wastewater from the vegetable industry has been conventionally treated with iron salts to remove P to allow discharge to rivers, but alternative green chemistry approaches are now being explored. In the potato industry after anaerobic digestion of the organic matter to produce biogas, P is recovered from the P-rich wastewater as struvite by adjusting pH and adding magnesium chloride in the NuReSyS-P process (http://www.nuresys.com).[56] The process is suitable for any anaerobic digestate or P-rich wastewater (>55 mg PO_4-P L^{-1}) with up to 80% P recovery and produces a crystalline struvite product suitable for re-use as fertilizer. A similar process used in the potato industry uses a fluid-bed crystallizer partially filled with sand or mineral to seed the crystallization process to struvite or calcium phosphate (Crystalactor®).[78] The process is accurately controlled to limit the growth of the crystals to approximately 1 mm which then move to the bottom of the bed for removal and drying. The advantage of this process is that there is no residual waste because the nearly dry phosphate pellets are fully recovered and useable. However, the pH adjustments to the inflow necessary before entry to the bed are quite demanding in terms of energy (CO_2 stripping) and chemicals (acids/bases).[56] A comparable approach currently under investigation is recovering P from vegetable wastewater as calcium phosphate for subsequent conversion to PA for use in the chemical industry (http://www.biorefine.eu).

Food waste contains non-negligible amounts of P, not least because society wastes at least a third of its food, whilst in rich countries it can be up to 40%.[79,80] As food waste has by definition been safe for consumption before its expiry, it is suited for composting (provided packaging can be separated) and the small amount of P in the ensuing compost contributes to its wider agricultural value. As such, it may not be necessary to selectively extract P from food waste, but rather see it as an essential if minor part of its wider agricultural value as a soil amendment, or as a source of other second generation critical compounds.[81]

The wastewater sector

Phosphorus recovery from human sewage (*e.g.* excreta, detergents and food washings) has generated most P recovery research not least because accessibility is already guaranteed. WWTP need to remove P to reduce effluent P concentrations discharged to water, most commonly by conventional anaerobic digestion and Fe/Al dosing. The remaining P-rich sludge (biosolid) can be directly recycled to land as another multifunctional fertilizer, but the presence of pathogens, pharmaceuticals and hormone residues and heavy metals are a continual cause for concern.[82]The plant availability of recycled Fe-bound P to agricultural crops can also be low.[83] Urine diversion from human excreta *via* a specially designed toilet is an option for improving the direct recycling of human waste to agriculture, especially in developing countries where the majority of households are not already directly connected to WWTP.[84,85] Mihelcic *et al.*[86] suggested there is 1.7 Tg P in global urine, and recycling urine as a fertilizer seems to be socially acceptable.[87]

To avoid potential environmental and human health risks, there is an increasing global trend towards sludge incineration with the ashes offering an entry point for P recovery and metal removal to produce fertilizers that are safer to store, handle and apply.[88,89] The green chemistry advantage of this route is the near-complete collection of sewage P provided WWTPs are well-designed and meet operational criteria. There is also potential to recover other non-renewable elements of value to society (*e.g.* K, Zn, Cu, Se).[57] While P recovery from liquid phases at WWTP is only 40–50% at most, recovery of P from ashes is up to 90%.[56] Thermo-chemical and wet chemistry technologies using acid or caustic digestion of these ashes can produce P products with high P-availability suitable for animal feed or fertilizers, or as elemental P, whilst also removing any heavy metal contamination that might otherwise reduce the recycling value.[52,55] For example an innovative refinement of the hydrochloric acid technology used in the Ecophos process to produce feed phosphate from sewage sludge ashes is currently being investigated and may achieve full scale in 2017 (http://www.ecophos.com). Two full-scale plants have been implemented in Japan to recover P from incinerated sludge ash using alkaline (NaOH) leaching technology to minimize the leaching of heavy metals from the ash that would otherwise contaminate the recovered product.[18] The relatively low level of Ca (typically less than 10% CaO by weight) in the ash makes alkaline leaching sufficiently efficient. Such routes are economical wherever a local source of attractively priced NaOH (*e.g.* as by-product) is available.

An alternative green engineering approach to P removal in WWTPs is by biological treatment, a setup where little or no further chemical P removal is needed.[54] In enhanced biological P removal (EBPR), alternate anaerobic and aerobic cycles facilitate polyphosphate accumulation in microorganisms.[90] Inorganic P (P_i) can be released from EBPR sludge in a more concentrated form by various technologies, including heat treatment,[91] anaerobiosis,[92] anaerobic digestion,[93] and incineration followed by chemical leaching.[94,95] P can also be recovered from P_i-rich solution using precipitation technologies with inorganic cations such as Ca^{2+} or Mg^{2+}, producing either calcium hydroxyapatite ($Ca_{10}(PO_4)_6(OH)_2$), or struvite ($MgNH_4PO_4{\cdot}6H_2O$).[96,97] Struvite poses a scaling issue in EBPR plants, and its targeted precipitation therefore offers considerable operational cost savings. The recovered product has so far proved a useful slow-release fertilizer, but recovery of struvite from liquid phases is only applicable in WWTP with EBPR.[55] Recovery of struvite directly from the digested sludge is also in operation (*e.g.* Airprex, and Seaborne processes), but not on a large scale due to economic feasibility and national legislation.[56] In the Seaborne process, P and metals are recovered separately. Digested sludge is first acidified with sulphuric acid to mobilise P and heavy metals, the metals are removed with sulphur-rich digester gas and the P is precipitated as struvite by addition of sodium hydroxide. Both struvite and ammonium sulphate produced from the process can be re-used in agriculture. A more green chemistry approach at the pilot stage is the Budenheim process which uses carbon dioxide rather than acid to dissolve the P in the sludge before precipitating the P as calcium phosphate.[77] The CO_2

extractions are very efficient at mobilising the P with 60–70% recovery and the used CO_2 is recycled in the process (http://www.budenheim.com).

A number of potential green and bio-engineering solutions to recover P from dilute wastewaters are at the pilot scale. Amorphous calcium silicate hydrates (A-CSHs) synthesized using natural and low cost materials, such as siliceous shale and calcium hydroxide ($Ca(OH)_2$) have proved an effective option to recover P from aqueous solutions by simple adsorption.[98] Similar suitable materials can even be obtained from construction material production waste. The lack of any need for pH adjustment and the high settling rates, filterability, and dewaterability of recovered P are the advantages of A-CSHs over conventional $CaCl_2$ and $Ca(OH)_2$ chemistry.[99] No chemical coagulants are required for P recovery by A-CSHs, and, unlike $Ca(OH)_2$, no significant carbonate inhibition occurs with P recovery with A-CSHs. Other novel phosphate-binding materials for treating wastewater have included polymeric hydrogels synthesized by chemically crosslinking linear poly(allylamine) PAA···HCl chains with epichlorohydrin, and ion-exchangeable ceramic beads (0.55 mm diameter and 85% porosity) which exhibit a high specificity for fast phosphate adsorption over a wide pH range (pH 2 to 14), and can be used more than 100 times.[100,101] With an increasing need to further lower WWTP effluent P concentrations for eutrophication control, sorption and ion exchange may become more attractive green engineering approaches for enhanced wastewater P recovery in the future.[102,103]Microfiltration (0.2 μm) and nanofiltration technology together with various pre-treatment steps also have the potential to produce recovered products with specific N : P ratios from wastewaters.[60,71] Microalgal recovery of wastewater P for re-use in biodiesel production is another bio-based technology, but the economics of this route still remain unfavourable.[104,105]

The industrial sector

The waste-based production of white P, a key intermediate for many industrial products (flame retardants, plasticizers, battery ingredients, catalyst ligands, pharmaceuticals, lubricant additives, specialty fertilizers, herbicides and metals and electronics etchants) has been pioneered by Thermphos International and is subsequently being developed by the Recophos consortium (http://www.recophos.org). The raw material is sewage sludge ash (or other P-rich waste), which is converted to white phosphorus in a novel, lean manufacturing setup which shows a clear departure from the classical large scale furnace design practised for over a century. The process will be entering pilot stage in 2015. The processing route involves reduction of phosphates by means of coke to the element (P_4), with carbon monoxide and a calcium silicate slag as sellable/usable by-products.

Elemental P from this process, or the classical production route, serves as the base for a large number of derivatives. Processing routes include the intermediate product phosphorus trichloride (PCl_3), the workhorse for organic P chemistry. The chlorine in this e merely serves as a means to activate the P atom and does not turn up in most – if not all – final products. Therefore, more direct and green routes to such derivatives are needed. Chlorine free routes have been pioneered by CNR-ICCOM, Thermphos International and are now being pursued by the SusPhos project, aimed at green nonphosphate (*i.e.* white P) based chemistry especially in fine chemistry applications (http://www.susphos.eu). For example, triphenylphosphine (TPP) is commonly used in the Wittig reaction for the industrial production of Vitamin A, carotenoids and many other alkenes. Yet, the P-containing by-product, triphenylphosphine oxide (TPPO), does not currently have a large-scale application, which results in thousands of tonnes of solid P-containing waste.[106] SusPhos aims to achieve an economic recycling protocol to create a P waste-free Wittig reaction, which can be applied throughout a wide range of industrial processes. More generally, using white P in an atom efficient way remains a challenge for the non-fertilizer P industry. Designing products to allow re-use over and over again is also a goal for such industries.

High-grade PA is widely used in the chemical and manufacturing industries and a considerable portion could be recovered and recycled from different industrial wastes.[11] In terms of quantity, P emitted into steel-making slag is a particularly important secondary P resource and is an input to the P cycle outside the usual PR-based inputs. Coal and iron ore, which are essential raw materials for the manufacture of iron and steel, contain small amounts of P (typically less than 0.03% P_2O_5 by weight). Since P has detrimental effects on the mechanical properties of steel, it is removed into dephosphorizing slag at concentrations as high as 2–10% P_2O_5 by weight.[107] The global production of iron ore is approximately 2000 Tg year^{-1}. Steel slags therefore contain approximately 0.6 Tg P year^{-1}, which is equivalent to one-fifth of the annual world industrial P demand (2.8 Tg P year^{-1}). Removing P from these slags would allow their iron content to be recovered in the process, adding economic benefits to this P recovery route. As elemental P is reported in some of these es as a nuisance, its targeted recovery poses a challenge for green chemistry. The separation of iron and other metals from the product (P_4 or PA) may be difficult to achieve.

Losses in the P cycle

Dissolved and particulate P transfer in land (urban and agricultural) runoff represents a considerable loss of P from the cycle. Whilst improvement in land management practices can help reduce these losses, by-product chemical amendments are increasingly viewed as a potential green chemistry strategy for prevention and/or control.[108] Applications of moderate-charge-density anionic polyacrylamide (co-polymerised with sodium acrylate) to the soil surface, or in irrigation water, have been highly effective (up to 90%) at reducing sediment and nutrient losses in land runoff from agricultural land and from construction sites.[109] The water-soluble polymer works by flocculating soil particles through co-bridging with divalent cations already present in the soil or co-applied. Stabilizing the soil in this way prevents surface crusting, increases water infiltration rates and reduces the risk of soil erosion, which is the main process of P loss in many situations. High P solubility in over-fertilized soils is also a significant eutrophication risk, which can be reduced by application of by-product chemical amendments that bind P. For example, coal combustion by-products from technologies to improve air quality, such as fluidized bed combustion fly ash and flue gas desulfurization gypsum, have been shown to be effective (20–40%) at reducing soluble P concentrations in soils without affecting plant P availability, or increasing soil or runoff contamination from heavy metals and arsenic.[110] When targeted at the small critical source areas in catchments that generate the majority of the P loss, by-product re-use provides a cost-effective green chemistry approach to preventing P release to runoff. Similarly a range of natural materials (*e.g.* Fe-rich sand), synthetic filtration materials (*e.g.* calcinated clay) and industrial by-products (*e.g.* steel slag, gypsum, red mud) with high P binding capacity can be used to remove P in land runoff, providing the runoff can be channelled, the flow velocity is not too high and is sufficiently P-rich.[102,108] For example from livestock hard standing areas.[111] The recovered P is either directly suitable for re-use as a fertilizer, or the P is stripped from the binding agent using acid–base technology and the binding agent re-used. However, the amounts of recoverable P at each site are usually very small. Direct recovery of P from eutrophic waters using algae or aquatic plants, has also been considered with the advantage of also lowering eutrophication risk.[112] Whether this is a feasible and worthwhile P recovery route in itself still needs to be established. The synergy between P removal and potentially using the harvested material as P source is the main advantage here.

Designing a new food system

Progress towards greening the global P cycle also requires consideration of sector or process design, and utilization efficiency.[22] Food and industry production systems could be better designed for maximum efficiency and to optimize recovery of secondary P as a substitute for primary PR; a key green chemistry goal. Agriculture dominates regional and global P flows, but it is an input-driven rather than a demand-led sector that overuses P unnecessarily because the P demands of the food chain have no governance. This leads to unsustainable P surpluses, continued P accumulation in soils and increased eutrophication risk, as is currently being evidenced most dramatically in China.[113] Animal product consumption accounts for 72% of global dietary P demand and P demands could be reduced, and P efficiency substantially increased, by reducing meat consumption.[36,114] For example, in Europe and the USA average dietary P intake is double actual P requirements due to a high proportion of meat in the diet.[15,115] It may also be possible to reduce crop P requirements by 20–30% through breeding more P efficient plants.[116,117] Reducing the intake and improving the utilization of P in animal feeds, for example through precision (phase) feeding, phytase addition and feeding low phytate crops, could reduce both P inputs and excretion rates by up to a third.[118,119] The addition of the phytase enzyme allows the phytate in crops to be more effectively utilized by monogastrics without the need for inputs of highly-soluble feed P supplements. Refining livestock (and human) feeds to improve digestibility through other green biotechnologies could also help to reduce the volumes of livestock (and municipal) manure that requires recovery and recycling. For example the use of biorefinery to separate out the main constituents of feeds (proteins, enzymes and phosphates) to increase their digestibility and absorption.[55] The microfiltration of liquid feeds to manipulate their P content might be another green chemistry option.

Similarly fertilizer use efficiency could be increased at the field scale through improved product formulation design and by more precise application methods targeting the crop rather than the soil.[120,121] The synthesis of nano-particle P fertilizers that are able to supply P to crops efficiently and with lower leaching risk represents a significant advance. Stable hydroxyapatite nanoparticles (15.5 nm in diameter) have been successfully synthesized using sodium carboxymethyl cellulose solution and used to fertilize soybean.[122] Fertilizer P nanoparticles (28.2 nm in diameter) have also been recently biosynthesized from tricalcium phosphate using mycelium from the Aspergillus tubingensis fungus.[123] The green engineering advantage of these nanoparticles is that they can be potentially be (bio)-synthesized from a range of secondary P products, and engineered so that the particle size (and hence P uptake rate) can be matched to the P uptake patterns of different crops, thereby improving P efficiency.

It has been suggested that society could potentially substitute at least 50% of its PR processing requirement based on recovered secondary P at the regional scale.[15,18,45,77] Such predictions belie the huge financial and social challenges in developing and marketing recovered products that have suitable physical and chemical consistency, good P availability, are safe to use and are economically viable. In view of these constraints, progress in greening the global P cycle will be more realistically achieved in the short term through sector and regional level initiatives.[21,29,53] For example, green chemistry principles have already been adopted in regional government policies towards circular and bio-based economies,[13] and in sustainable food production initiatives such as Origin Green in Ireland.[124] New smarter, more diversified and customer-focused business models are rapidly emerging in the mining industries that take account of the their wider social responsibilities, the need to raise safety and environmental standards and reduce wastage.[27] For example, under a government-industry covenant, Amsterdam-based ICL has committed to substituting its entire PR feedstock, amounting to 0.5 Tg year^{-1} with secondary P, initially from human wastewater. Yet more ambitious are the aspirations of leading edge Chinese producer Wengfu, who have taken a top-down policy-led decision to reach zero waste by 2015, which includes 100% re-use of PG as ammonium sulphate and calcium carbonate.[63] This has entailed restructuring the company from being a fertilizer only company to having three divisions, fertilizers, chemical products and construction materials. It has also started to recover a wide range of materials from phosphate ores, including 100 t year^{-1} of iodine.

CONCLUSIONS

Although it has taken at least one generation to realize the unintended consequences of PR processing, there is compelling economic, environmental and ethical justification for more efficient and sustainable use of P to safeguard PR resources and the environment for future generations. We conclude that the potential opportunities for green chemistry in achieving planetary P sustainability goals are large and fully consistent with societal migration to bio-based and circular economies based on smart science, practical policies and innovative technologies. We define three major areas for progress: (a) maximizing the economic and resource value of PR in fertilizer and other uses, (b) recover and recycle P from the vast array of secondary P resources though innovative green technologies so that they can increasingly be re-used as renewable materials in the future, and with minimum waste, and (c) provide P governance in the food chain and define more precisely what societal (end user) P requirements are to provide a firm foundation for designing and investing in new smarter demand-driven production systems that use only what is needed and with maximum efficiency and minimum P losses to the oceans.

ACKNOWLEDGEMENTS

The authors are grateful to Prof. Dr E Meers, Ghent University, Belgium (Biorefine project), and Dr C. Slootweg and M. de Boer of Vrije Universiteit, Amsterdam, The Netherlands (SusPhos project) for helpful information in the preparation of this manuscript.

NOTES AND REFERENCES

1. T. Lougheed, *Environ. Health Perspect.*, 2011, 119(5), A208–A213.
2. L. Reijnders, *Resour., Conserv. Recycl.*, 2014, 93, 32–49.
3. J. J. Elser, T. J. Elser, S. R. Carpenter and W. A. Brock, *PLoS One*, 2014, 9, e93998.
4. J. Cooper, R. Lombardi, D. Boardman and C. Carliell-Marquet, *Resour. Conserv. Recycl.*, 2011, 57, 78–86.
5. P. J. A. Withers, K. C. van Dijk, T.-S. S. Neset, T. Nesme, O. Oenema, G. H. Rubæk, O. F. Schoumans, A. L. Smit and S. Pellerin, *AMBIO*, 2015, 44(Suppl. 2), S163–S179.
6. L. Hein and R. Leemans, *AMBIO*, 2012, 41, 341–349.
7. V. H. Smith and D. W. Schindler, *Trends Ecol. Evolut.*, 2009, 24, 201–207.
8. B. Qin, G. Zhu, G. Gao, Y. Zhang, W. Li, H. W. Paerl and W. W. Carmichael, *Environ. Manage.*, 2010, 45(1), 105–112.
9. D. F. Millie, G. L. Fahnenstiel, J. D. Bressie, J. Dyble, R. J. Pigg, R. R. Rediske, D. M. Klarer, P. A. Tester and R. W. Litaker, *Aquat. Ecol.*, 2009, 43(4), 915–934.
10. J. Elser and E. Bennett, *Nature*, 2011, 478, 29–31.
11. W. Schipper, *Eur. J. Inorg. Chem.*, 2014, 1567–1571.
12. R. W. Scholz, A. H. Roy, F. S. Brand, D. T. Hellums and A. E. Ulrich, *Sustainable Phosphorus Management: A Global Transdisciplinary Roadmap*, Springer, Dordrecht, 2014.
13. E. C., Memo 14/377, http://europa.eu/rapid/press-release_IP-14-599_en.htm, 2014.
14. M. A. Sutton, A. Bleeker, C. M. Howard, M. Bekunda, B. Grizzetti and W. de Vries, *et al.*, *Our Nutrient World: The challenge to produce more food and energy with less pollution*, The Centre for Ecology and Hydrology, Edinburgh, U. K., 2013.
15. K. C. Van Dijk, J. P. Lesschen and O. Oenema, *Sci. Tot. Environ*, In press.
16. R. B. Chowdhury, G. A. Moore, A. J. Weatherley and M. Arora, *Resour., Conserv. Recycl.*, 2014, 83, 213–228.

17. D. Cordell, A. Rosemarin, J. J. Schroder and A. L. Smit, *Chemosphere*, 2011, 84, 747–758.
18. H. Ohtake and K. Okano, *Global Environ. Res.*, 2015, 19(1), 1–30.
19. J. Hilton and C. Dawson, *Waste Resour. Manage.*, 2012, 165(4), 179–189.
20. V. Smil, *Ann. Rev. Energy Environ.*, 2000, 25, 25–53 .
21. P. T. Anastas and J. C. Warner, *Green Chemistry: Theory and Practice*, Oxford University Press, Oxford, UK, 1998.
22. P. T. Anastas and J. B. Zimmerman, *Environ. Sci. Technol.*, 2003, 37, 94A.
23. R. A. Sheldon, *Chem. Commun.*, 2008, 3352–3365.
24. W. Jiao, W. Chen, A. C. Chang and A. L. Page, *Environ. Pollut.*, 2012, 168, 44–53.
25. K. Syers, A. E. Johnston and D. Curtin, *FAO Fertilizer and Plant Nutrition Bulletin 18*, Food and Agriculture Organization of the United Nations, Rome, Italy, 2008.
26. D. Seyhan, *Resour., Conserv.. Recycl.*, 2009, 53(12), 698–709.
27. R. Batterham, *Procedia Eng.*, 2014, 83, 8–15.
28. E. C., COM/2014/0398, http://eur-lex.europa.eu/legalcontent/EN/TXT/?uri=CELEX:52014DC0398, 2014.
29. J. Voorthuis and C. Gijbels, *Sustainability*, 2010, 2, 371–382.
30. J. Hilton, A. E. Johnston and C. J. Dawson, *The Phosphate Life-Cycle: Rethinking the options for a finite resource*, Proceedings 668, International Fertiliser Society, York, U. K., 2010.
31. J. S. Guest, S. J. Skerlos, J. L. Barnard, M. B. Beck, G. T. Daigger and H. Hilger, *et al.* , *Environ. Sci. Technol.*, 2009, 43(16), 6126–6130.
32. R. Sylvester-Bradley and P. J. A. Withers, *Innovation in Crop Nutrition*, Proceedings 700, International Fertiliser Society No. 700, Leek, UK, 2013.
33. J. Peñuelas, B. Poulter, J. Sardans, P. Ciais, M. van der Velde and L. Bopp, *et al.* , *Nat. Commun.*, 2013, 4, 2934.
34. K. Senthilkumar, T. Nesme, A. Mollier and S. Pellerin, *Nutr. Cycling Agroecosyst.*, 2012, 92, 145–159.
35. M. E. Schipanski and E. M. Bennett, *Ecosystems*, 2012, 15, 256–268 .
36. G. S. Metson, E. M. Bennett and J. J. Elser, *Environ. Res. Lett.*, 2012, 7, 044024.
37. C. J. Dawson and J. Hilton, *Food Policy*, 2011, 36, S14–S22 .
38. M. Prud'Homme, World Phosphate Rock Flows, Losses and Uses, in *Proceedings of Phosphate 2010 International Conference*, Brussels, 2010 .
39. L. Bouwman, K. K. Goldewijka, K. W. Van Der Hoekc, A. H. W. Beusena, D. P. Van Vuurena, J. Willemsa, M. C. Rufinoe and E. Stehfesta, *Proc. Natl. Acad. Sci. U. S. A.*, 2011,110(52), 20882–20887 .
40. H. Lamprecht, D. J. Lang, C. Binder, R. Scholz and W. Roland, *GAIA*, 2011, 20(2), 112–121 .

41. A. C. Edwards and P. J. A. Withers, *Soil Use Manage.*, 1998, 14(Suppl. 1), 124–130 .

42. A. Sharpley, H. P. Jarvie, A. Buda, L. May, B. Spears and P. Kleinman, *J. Environ. Qual.*, 2013, 42, 1308–1326 .

43. G. K. MacDonald, E. M. Bennett, P. A. Potterc and N. Ramankuttyd, *Proc. Natl. Acad. Sci. U. S. A.*, 2011, 108(7), 3086–3091 .

44. H. Li, G. Huang, Q. Meng, L. Ma, L. Yuan, F. Wang, W. Zhang, Z. Cui, J. Shen, X. Chen, R. Jiang and F. Zhang, *Plant Soil*, 2011, 349, 157–167 .

45. R. H. E. M. Koppelaar and H. P. Weikard, *Global Environ. Change*, 2013, 23, 1454–1466 .

46. E. M. Bennett, S. R. Carpenter and N. F. Caraco, *Bioscience*, 2001, 51(3), 227–234 .

47. S. Sattari, A. Bouwman, K. E. Giller and M. van Ittersum, *Proc. Natl. Acad. Sci. U. S. A.*, 2012, 109, 6348–6353 .

48. B. Ringeval, B. Nowak, T. Nesme, M. Delmas and S. Pellerin, *Global Biogeochem. Cycles*, 2014, 28, 743–756 .

49. G. Villalba, Y. Liu, H. Schroder and R. U. Ayres, *J. Ind. Ecol.*, 2008, 12(4), 557–569 .

50. G. van Drecht, A. F. Bouwman, J. Harrison and J. M. Knoop, *Global Biogeochem. Cycles*, 2009, 23, Gb0a03.

51. T. Karak, R. M. Bhagat and P. Bhattacharyya, *Crit. Rev. Environ. Sci. Technol.*, 2012, 42(15), 1509–1630 .

52. W. J. Schipper, A. Klapwijk, B. Potjer, W. H. Rulkens, B. G. Temmink, F. D. G. Kiestra and A. C. M. Lijmbach, *Environ. Technol.*, 2001, 22, 1337–1345 .

53. O. Oenema, W. Chardon, P. Ehlert, K. van Dijk and O. Schoumans, *Phosphorus Fertilizers from By-Products and Wastes*, Proceedings 717, International Fertiliser Society, Leek, UK, 2012 .

54. H. Ohtake, in *Biochemical Engineering and Bioprocessing*, Wiley, 2015, vol. 3, pp. 1–15 .

55. O. F. Schoumans, F. Bouraoui, C. Kabbe, O. Oenema and K. van Dijk, *AMBIO*, 2015, 44(Suppl. 2), 180–192 .

56. E. Desmidt, K. Ghyselbrecht, Y. Zhang, L. Pinoy, B. van der Bruggen, W. Verstraete, K. Rabaey and B. Meesschaert, *Crit. Rev. Environ. Sci. Technol.*, 2015, 45, 336–384.

57. P. Westerhoff, S. Lee, Y. Yang, G. W. Gordon, K. Hristovski, R. U. Halden and P. Herckes, *Environ. Sci. Technol.*, 2015.

58. J. H. Clark, L. A. Pfaltzgraff, V. L. Budarin, A. J. Hunt, M. Gronnow, A. S. Matharu, D. J. Macquarrie and J. R. Sherwood, *Pure Appl. Chem.*, 2013, 85(8), 1625–1631.

59. C. O. Tuck, E. Pérez, I. T. Horváth, R. A. Sheldon and M. Poliakoff, *Science*, 2012, 337, 695–699 .

60. M.-P. Zacharof and R. W. Lovitt, *Waste Biomass Valorization*, 2013, 4, 557–581.
61. J. Hilton, B. K. Birky and M. Moussaid, *Comprehensive Extraction, a Key Requirement for Social Licensing of NORM Industries?* Proceedings, NORM 7, Beijing, China, 2013.
62. P. Christmann, *Procedia Eng.*, 2014, 83, 19–26 .
63. P. Zhang, *Procedia Eng.*, 2014, 83, 37–51 .
64. Anon, JDC Phosphate starts hot operations, *Fert. Int.*, 2014, 458, 40–41 .
65. IAEA, *Radiation Protection and Management of NORM Residues in the Phosphate Industry*, Safety Reports Series No. 78, International Atomic Energy Agency, Vienna, Austria, 2013 .
66. G. H. Rubaek, K. Kristensen, S. E. Olesen, H. S. Østergaard and G. Heckrath, *Geoderma*, 2013, 209–210, 241–250 .
67. A. R. Richardson and R. J. Simpson, *Plant Physiol.*, 2011, 156, 989–996 .
68. L. Ramaekers, R. Remans, I. M. Rao, M. W. Blaier and J. Vanderleyden, *Field Crop Res.*, 2010, 117, 169–176 .
69. A. Bateman, H. van der Horst, D. Boardman, A. Kansal and C. Carlliel-Marquet, *Resour., Conserv. Recycl.*, 2011, 55, 1146–1153 .
70. K. Suzuki, Y. Tanaka, K. Kuroda, D. Hanajima, Y. Fukumoto, T. Yasuda and M. Waki, *Bioresour. Technol.*, 2007, 98, 1573–1578 .
71. M. L. Gerardo, M. P. Zacharof and R. W. Lovitt, *Water Res.*, 2013, 47, 4833–4842 .
72. P. J. A. Kleinman, A. N. Sharpley, A. M. Wolf, D. B. Beegle and P. A. Moore Jr., *Soil Sci. Soc. Am. J.*, 2002, 66, 2009–2015 .
73. T. Wang, M. Camps-Arbestain and M. Hedley, *Plant Soil*, 2014, 375, 61–74 .
74. Z. Liu and R. Balasubramanian, *Appl. Energy*, 2014, 114, 857–864 .
75. K. S. Ro, K. B. Cantrell, P. G. Hunt, T. F. Ducey, M. B. Vanotti and A. A. Szogi, *Bioresour. Technol.*, 2009, 100, 5466–5471 .
76. C. J. De Vos and L. Heres, Risk Anal., 2009, 29(4), 541–557.
77. C. Kabbe, in Factor X. Re-source – Designing the Recycling Society, Eco-Efficiency in Industry and Science, ed. M. Angrich, A. Burger and H. Lehmann, Springer, Dordrecht, The Netherlands, 2013, vol. 30, pp. 261–273 Search .
78. A. Giesen, Environ. Technol., 1999, 20(7), 769–775.
79. D. Gunders, Wasted: How America is Losing Up to 40 Percent of Its Food from Farm to Fork to Landfill, NRDC, Washington, USA, 2012 Search .
80. J. Parfitt, M. Barthel and S. Macnaughton, Philos. Trans. R. Soc. London, Ser. B, 2010, 365, 3065–3081.
81. L. A. Pfaltzgraff, M. de Bruyn, E. C. Cooper, V. Budarin and J. H. Clark, Green Chem., 2013, 15, 307–314 RSC.
82. S. R. Smith, Philos. Trans. R. Soc. London, Ser. A, 2009, 367, 4005–4041.
83. H. A. Elliott and G. A. O'Connor, Soil Biol. Biochem., 2007, 39, 1318–1327.

84. W. Pronk and D. Kone, Desalination, 2009, 248(1–3), 360–368.
85. B. Etter, et al. , Water Res., 2011, 45(2), 852–862.
86. J. R. Mihelcic, L. M. Fry and R. Shaw, Chemosphere, 2011, 84(6), 832–839.
87. J. Lienert and T. A. Larsen, Environ. Sci. Technol., 2009, 44(2), 556–566.
88. C. Adam, B. Peplinski, M. Michaelis, G. Kley and F. G. Simon, Waste Manage., 2009, 29, 1122–1128
89. Z. Tan and A. Lagerkvist, Renewable Sustainable Energy Rev., 2011, 15, 3588–3602.
90. R. Hirota, A. Kuroda, J. Kato and H. Ohtake, J. Biosci. Bioeng., 2010, 109, 423–432.
91. A. Kuroda, N. Takiguchi, T. Gotanda, K. Nomura, J. Kato, T. Ikeda and H. Ohtake, Biotechnol. Bioeng., 2002, 78, 333–338.
92. P. L. Bond, L. Keller and L. L. Blackall, Biotechnol. Bioeng., 1999, 63, 507–515.
93. Y. Cao and A. Pawlowski, Renewable Sustainable Energy Rev., 2012, 16, 1657–1665.
94. A. Petterssona, L. E. Amand and B. M. Steenari, Biomass Bioenergy, 2008, 32, 224–235.
95. S. Donatello and C. R. Cheeseman, Waste Manage., 2013, 33, 2328–2340.
96. Y. Song, P. G. Weidler, U. Berg, R. Nüesch and D. Donnert, Chemosphere, 2006, 63, 236–243.
97. N. Marti, L. Pastor, A. Bouzas, J. Ferrer and A. Seco, Water Res., 2010, 44, 2371–2379.
98. K. Okano, M. Uemoto, J. Kagami, K. Miura, T. Aketo, M. Toda, K. Honda and H. Ohtake, Water Res., 2013, 47, 2251–2259.
99. K. Okano, S. Miyamaru, A. Kitao, H. Takano, T. Aketo, M. Toda, K. Honda and H. Ohtake, Sep. Prif. Technol., 2015, 144, 63–69.
100. D. R. Kioussis, F. W. Wheaton and P. Kofinas, Agric. Eng., 2000, 23, 315–332 Search .
101. I. Midorikawa, H. Aoki, A. Ohmori, T. Shimizu, Y. Kawaguchi, K. Kassai and T. Murakami, Proc. 5th IWA Leading-Edge Conference on Water and Wastewater Technologies, 2008.
102. P. Loganathan, S. Vigneswaran, J. Kandasamy and N. S. Bolan, Crit. Rev. Environ. Sci. Technol., 2014, 44, 847–907.
103. B. E. Rittmann, B. Mayer, P. Westerhoff and M. Edwards, Chemosphere, 2011, 84, 846–853.
104. J. K. Pittman, A. P. Dean and O. Osundeko, Bioresour. Technol., 2011, 102, 17–25.
105. W. Mulbry, S. Kondrad, C. Pizarro and E. Kebede-Westhead, Bioresour. Technol., 2008, 99, 8137–8142.
106. S. P. Marsden, Nat. Chem., 2009, 1, 685–687.

107. E. Yamasue, K. Matsubae, K. Nakajima, S. Hashimoto and T. Nagasaka, J. Ind. Ecol., 2013, 17, 722–730.

108. A. R. Buda, G. F. Koopmans, R. B. Bryant and W. J. Chardon, J. Environ. Qual., 2012, 41, 621–627.

109. J. A. Entry and R. E. Sojka, Trans. Am. Soc. Agric. Eng., 2003, 46, 75–83 .

110. W. L. Stout, A. N. Sharpley and J. Landa, J. Environ. Qual., 2000, 29, 1239–1244.

111. C. Penn, J. McGrath, J. Bowen and S. Wilson, J. Soil Water Conserv., 2014, 69(2), 51A–56A .

112. R. S. Quilliam, M. A. van Niekerk, D. R. Chadwick, P. Cross, N. Hanley, D. L. Jones, A. J. A. Vinten, N. Willby and D. M. Oliver, J. Environ. Manage. DOI:10.1016/j.jenvman.2015.01.046.

113. L. Ma, G. L. Velthof, F. H. Wang, W. Qin, W. F. Zhang, Z. Liu, Y. Zhang, J. Wei, J. P. Lesschen, W. Q. Ma, O. Oenema and F. S. Zhang, Sci. Total Environ., 2012, 434, 51–61.

114. S. D. Donner, Global Environ. Change, 2007, 17, 105–113.

115. A. Flynn, T. Hirvonen, G. B. M. Mensink, M. C. Ocke, L. Serra-Majem, K. Stos, L. Szponar and I. Tetens, et al. , Food Nutr. Res., 2009, 53, 1–51 Search .

116. T. J. Rose, J. Pariasca-Tanakaa, M. T. Rosea, Y. Fukutab and M. Wissuwaa, Field Crops Res., 2010, 119, 154–160.

117. V. Raboy, Plant Sci., 2009, 177, 281–296.

118. R. O. Maguire, D. A. Crouse and S. C. Hodges, J. Environ. Qual., 2007, 36, 1235–1240.

119. E. Kebreab, A. V. Hansen and A. B. Strathe, Curr. Opin. Biotechnol., 2012, 23, 872–877.

120. P. J. A. Withers, R. Sylvester-Bradley, D. L. Jones, J. R. Healey and P. J. Talboys, Environ. Sci. Technol., 2014, 48, 6523–6530.

121. M. J. McLaughlin, T. M. McBeath, R. Smernik, S. P. Stacey, B. Ajiboye and C. Guppy, Plant Soil, 2012, 349, 69–87 .

122. R. Q. Liu and R. Lal, Sci. Rep., 2014, 4, 5686,DOI:10.1038/srep05686.

123. J. C. Tarafdar, R. Raliya and I. Rathore, J. Bionanosci., 2012, 6, 1–6.

124. Origin Green, http://www.origingreen.ie, 2014.

13
CHAPTER

EFFECT OF COMPOSTING ON DISSOLVED ORGANIC MATTER IN ANIMAL MANURE AND ITS BINDING WITH CU

Fengsong Zhang, Yanxia Li, Xiong Xiong, Ming Yang, and Wei Li

INTRODUCTION

Animal manure was usually applied to arable soils in order to improve soil fertility and increase the organic matter content. However, in recent years, high concentration of heavy metal such as Cu in animal manure has been frequently reported in China due to abuse of mineral additives [1, 2]. Because part of the organic substances in animal manure are water soluble, a direct impact of the application of animal manure to agricultural land is the release of dissolved organic matter (DOM) into soil solution [3]. DOM could complex with heavy metals and then improve their transport to surface water [4, 5].

In natural water, humic acids and fulvic acids are major components and represent up to 70% of DOM, which contributed the most organic ligands to Cu complexing [6]. However, a recent study indicates that nonhumic substances such as amino acids are likely engaged in the Cu complexation [7]; as one type of protein-like materials, amino acids are important components in DOM of organic wastes [8–10]. In addition, de Zarruk et al. [9] verified that the fraction in vinasse with the highest proteinaceous fluorescence has the greatest ability to bind with Cu. According to previous research, raw pig manure and cattle manure DOM also have a large number of proteinaceous materials [8, 10]. Whether the proteinaceous materials in manure DOM play a key role in complexing with Cu was not clear.

Composting is a useful method for organic wastes stabilization [11–13]. A decline in DOM has been reported by Inbar et al. [14] and Huang et al. [12] for cattle manure

Citation: Fengsong Zhang, Yanxia Li, Xiong Xiong, Ming Yang, and Wei Li, "Effect of Composting on Dissolved Organic Matter in Animal Manure and Its Binding with Cu," The Scientific World Journal, vol. 2012, Article ID 289896, 9 pages, 2012. doi:10.1100/2012/289896.

and pig manure composting, respectively. The common characteristic is that DOM composition may undergo significant transformation after the composting process [10, 15]. For example, domestic organic wastes (coffee residues and garden trimmings) had a reduction of carbohydrates and increase of aromatic, phenolic, carboxylic, and carbonylic C in DOM after composting [16]. Accordingly, a decrease in the tyrosine-like and tryptophan-like materials and an increase in the humic, and fulvic-like materials were observed by excitation-emission matrix (EEM) fluorescence spectroscopy during composting of winery residues and municipal solid wastes [10, 13]. However, little is reported about the transformation of animal manure DOM during composting process at present. Furthermore, variation of binding ability of manure DOM with heavy metals after composing was also unknown.

As a selective, sensitive, and nondestructive analytical technique, EEM fluorescence spectroscopy has been always used to characterize the DOM composition using contour plots, number of fluorescence peaks, position of wavelength-independent fluorescence maxima (Ex_{max}/Em_{max}), and fluorescence intensity at Ex_{max}/Em_{max} [8, 17–19]. However, it is limited to quantifying the properties based on one, two, or three data points from the fluorescence spectra. Chen et al. [20] developed the fluorescence regional integration (FRI) technique for quantitative analysis, which has been successfully used to study the evolution of organic waste DOM during the composting process [10, 13]. In addition, fluorescence spectroscopy has been revealed as a very promising technique for the study of metal ion binding to DOM [7, 21, 22]. Metal ions, especially paramagnetic metals, that is, Cu and Hg, are able to quench the intrinsic fluorescence of DOM [23, 24]. Therefore, together with the fluorescence quenching titration, the metal ion complexing capacities of DOM and stability constants of metal-DOM complexes can be examined.

The objectives of this study were (1) to explore the composition evolution of animal manure (pig manure and cattle manure) DOM during composting by EEM fluorescence spectroscopy; (2) to investigate the effect of composting process on manure DOM complexation with Cu.

MATERIALS AND METHODS

Composting Procedure and Sample Preparation

The composting experiment was conducted in the same way with our previous study [22]. Manures and bulking agents were collected at local farms. Sawdust and corn stalks were chopped into 2-3 cm pieces and air dried before composting. The treatments of the composting piles on a dry volume basis were as follows. Treatment A: 50% pig manure + 50% sawdust; treatment B: 50% pig manure + 50% corn stalks; treatment C: 50% cattle manure + 50% sawdust; treatment D: 50% cattle manure + 50% corn stalks.

The composting experiments were performed in cylindrical vessels (diameter: 500 mm; height: 600 mm). The uniform forced ventilation was equipped at a rate of 0.1 m^3/min for 10 minutes at 60-minute intervals through perforated plates fixed at the bottom of the vessels to provide oxygen. The moisture content of each pile was kept at 50–60% (weight/weight) during composting. In the first 30 days of composting, the piles were turned periodically to keep the temperature under 60°C. Afterwards, the forced ventilation was stopped, and the piles were stirred daily for further humification.

The composting process was stopped when the compost temperature equaled the ambient temperature with no measurable changes for approximately 20 days. The pig manure and cattle manure were composted for 71 days and 46 days, respectively. Samples were collected from treatment A and treatment B on days 1, 8, 11, 17, 32, and 71 (A1, B1, A8, B8, A11, B11, A17, B17, A32, B32, A71, and B71), whereas sampling was done on days 1, 6, 10, 13, 29, and 46 (C1, D1, C6, D6, C10, D10, C13, D13, C29, D29, C46, and D46) for treatment C and treatment D, respectively. The subsamples were taken at different positions within the vessel and then thoroughly mixed as a composite sample. Prior to extracting the DOM, the samples were air dried.

Extraction of DOM and Fluorescence Analysis

Two grams of subsamples were extracted with 40 mL of deionized water and shaken for 24 hours. The solution was then centrifuged at 10,000 rpm for 10 minutes. The supernatant was then filtered using Whatman GF/F glass microfiber filter papers that had previously been heated at 450°C to remove any possible organic matter. The extracts were immediately analyzed for dissolved organic carbon using the TOC analyzer (Liqui TOC, Elementar, Germany).

The fluorescence of the filtered DOM samples was determined with a model F-4500 fluorescence spectrophotometer (Hitachi, Japan) with a 150-W Xe arc lamp. Prior to fluorescence analysis, all sub-samples for fluorescence analysis were diluted to the uniform concentration of 10 mg C/L to reduce inner filter effects [9]. To generate an EEM, excitation wavelengths were scanned from 200 to 400 nm in 2 nm steps, and the emitted fluorescence was detected between 300 and 550 nm in 5 nm steps. The band-pass width was 5 nm for excitation and 10 nm for emission, and the scan speed was 2400 nm/min [24].

Fluorescence regional integration method was applied for spectral comparison to thoroughly explore the transformation of DOM composition during manure composting [20]. To avoid the scattering effects of fluorescence data, the treatment method of the first-order Rayleigh, Raman and second-order Rayleigh scatters was applied, which was proposed by Bahram et al. [25]. The EEM plots were generated from the fluorescence spectral data using Sigmaplot 10.0 software (Systat Software, Inc.).

Fluorescence Quenching Titration and Complexation Modeling

Fluorescence quenching titration was carried out to characterize the complexation of manure DOM with Cu according to the research of Plaza et al. [21]. Experiments were carried out by adding 0.01 M $Cu(NO_3)_2$ solutions to a series of glass bottles that contained 50 mL of DOM solution. The pH value was then adjusted to 7.0. All samples were shaken in the dark for 24 hours under a nitrogen atmosphere at constant temperature (°C) to ensure complexation equilibrium.

The selection of the wavelengths for the fluorescence titration was based on the highest fluorescence intensity observed from the EEM of the samples [21]. The complexation model of Ryan and Weber was used to determine the binding parameters between DOM and Cu ions [26]. The model assumes a simple 1 : 1 equilibrium between a metal ion and an organic ligand

$$I = I_0 + (I_{CuL} - I_0)\left(\frac{1}{2K_{Cu}C_L}\right) \times \left(1 + K_{Cu}C_L + K_{Cu}C_{Cu} - \sqrt{(1 + K_{Cu}C_L + K_{Cu}C_{Cu})^2 - 4K_{Cu}^2 C_{Cu}C_L}\right), \quad (1)$$

where I and I_0 are the fluorescence intensity (arbitrary units) at the Cu concentration of C_{Cu} and at the start of the titration, respectively, I_{CuL} is the limiting value below which the fluorescence cannot decrease with the addition of Cu^{2+}, C_L is the total ligand concentration, CC_{Cu} is the total Cu ion concentration, and K_{Cu} is the conditional stability constant.

The complexation capacity (CC_{Cu}), that is, the amount of active binding sites per unit mass of DOM, was calculated as

$$CC_{Cu} = \frac{C_L}{(DOM)_{total}}, \quad (2)$$

where $(DOM)_{total}$ is the total concentration of DOM. K_{Cu} and C_L were solved by a nonlinear regression analysis with the software 1st Opt 1.5 (7D-soft High Technology Inc., China). The optimum set of fitting parameters for each DOM sample was obtained by iteratively varying the adjustable parameter values until the sum of the squares of the differences between the observed and fitted values of I was minimized. Full, unconstrained optimization was achieved using the quasi-Newton algorithm.

Statistical Analysis

Correlations were analyzed between percentages of fluorescence response ($P_{i,n}$) to elucidate the transformation of DOM during the composting. Meanwhile, correlation analysis was also used to determine the correlations between $P_{i,n}$ and parameters of DOM binding with Cu (i.e., CC_{Cu} and Log K_{Cu}). Statistical analyses were performed with the software SPSS 11.5 (SPSS Inc., Chicago, IL, USA) for Windows.

RESULTS AND DISCUSSION

Change of DOM Concentrations during Composting

The total concentration of DOM in pig and cattle manure was reduced after composting (Figure 1), which was similar to results obtained by Inbar et al. [14] and Huang et al. [12]. Cattle manure DOM concentrations continuously declined during the composting process, with a sharp reduction occurring in the initial stage of composting. During co-composting with corn stalk and sawdust, DOM in cattle manure was reduced by 27.4%

and 31.4%, respectively. However, during cocomposting with exhausted grape marc, cattle manure DOM only reduced by 18.3% [27].

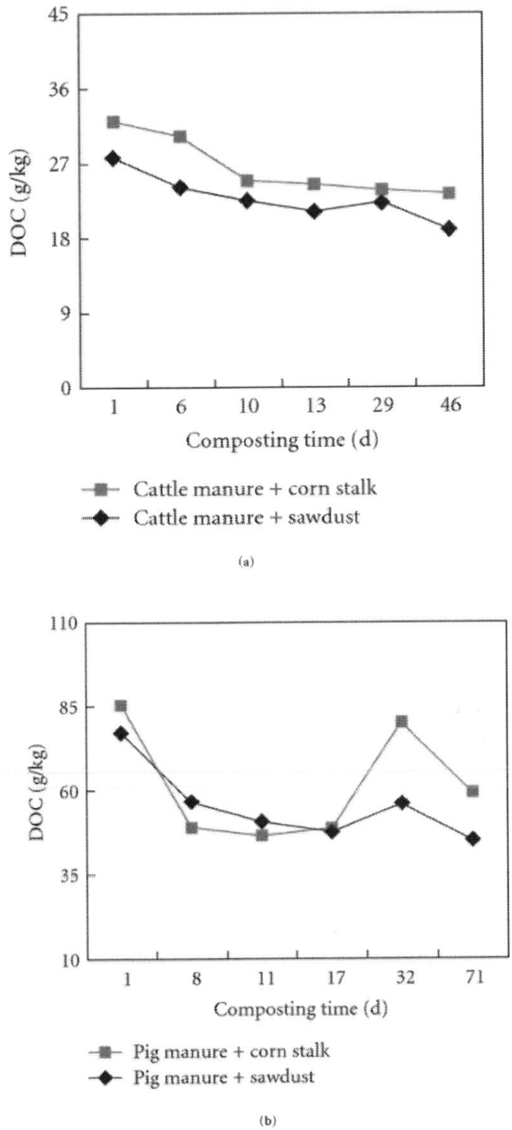

Figure 1: Changes of dissolved organic matter (DOM) concentrations during manure composting.

During composting of pig manure, DOM increased after a sharp initial decrease. At the end of the composting process, DOM concentration in pig manure with addition of sawdust and corn stalk decreased to 58.6% and 69.5%, respectively, of the raw materials after composting. In comparison to the cattle manure, there is more DOM degraded during pig manure composting than cattle manure, using the same composting method. This observation may reflect that pig manure contains far more DOM than cattle manure, but that the DOM in pig manure is also more easily degradable. In contrast, it was reported that about 95.8% of DOM in municipal solid waste had been degraded at the end of composting [13]. Our results and the cited study show that the rate of decrease in DOM concentration depends not only on the composting technique utilized, but also on the composition of the source material.

DOM Fluorescence Characteristics

As the EEM spectra evolution of DOM from treatment B and treatment D were similar to that from treatment A and treatment C, the DOM spectra of treatment A and treatment C are displayed as the representative in Figure 2. According to the research of Chen et al. [20], the fluorescence of regions I, II, and IV in manure DOM are related to tyrosine-like, tryptophan-like, and soluble microbial byproduct-like materials while the fluorescence of regions III and V are related to fulvic-like and humic-like materials. Soluble microbial byproduct-like materials also contained another kind of tyrosine-like and tryptophan-like compounds, which were different from materials associated with region I and region II [28]. In the raw pig manure DOM (A1), the most intense fluorescence peak of Ex/Em = 280 nm/342 nm centered at region IV. Protein-like fluorescence peaks such as tyrosine-like and tryptophan-like peaks were associated with growth of living organisms in marine water [17]. However, these peaks were also detected in organic wastes, such as animal slurry and landfill leachates [8, 29]. In addition, we observed a peak of Ex/Em = 245 nm/399 nm centered at region III in raw pig manure DOM (A1), which may be attributed to aromatic and aliphatic groups in the DOM and widely labeled as fulvic-like substances [19]. However, there was no obvious humic-like fluorescence peak in pig manure DOM. Similar to raw pig manure, cattle manure DOM (C1) had intense tryptophan-like and tyrosine-like fluorescence peaks. However, cattle manure also had two intense fulvic acid-like and humic acid-like peaks centered at Ex/Em = 245 nm/400 nm and 305 nm/412 nm.

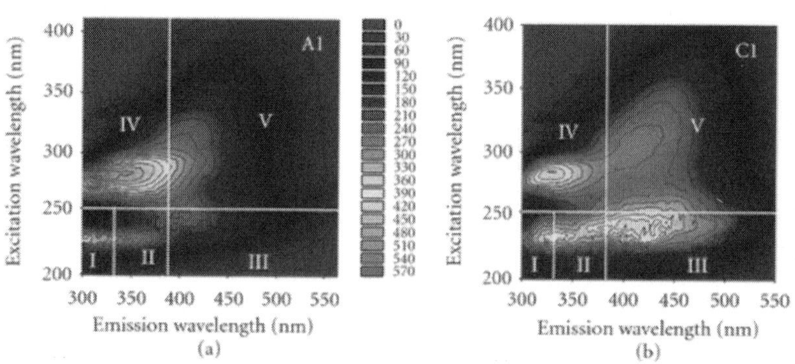

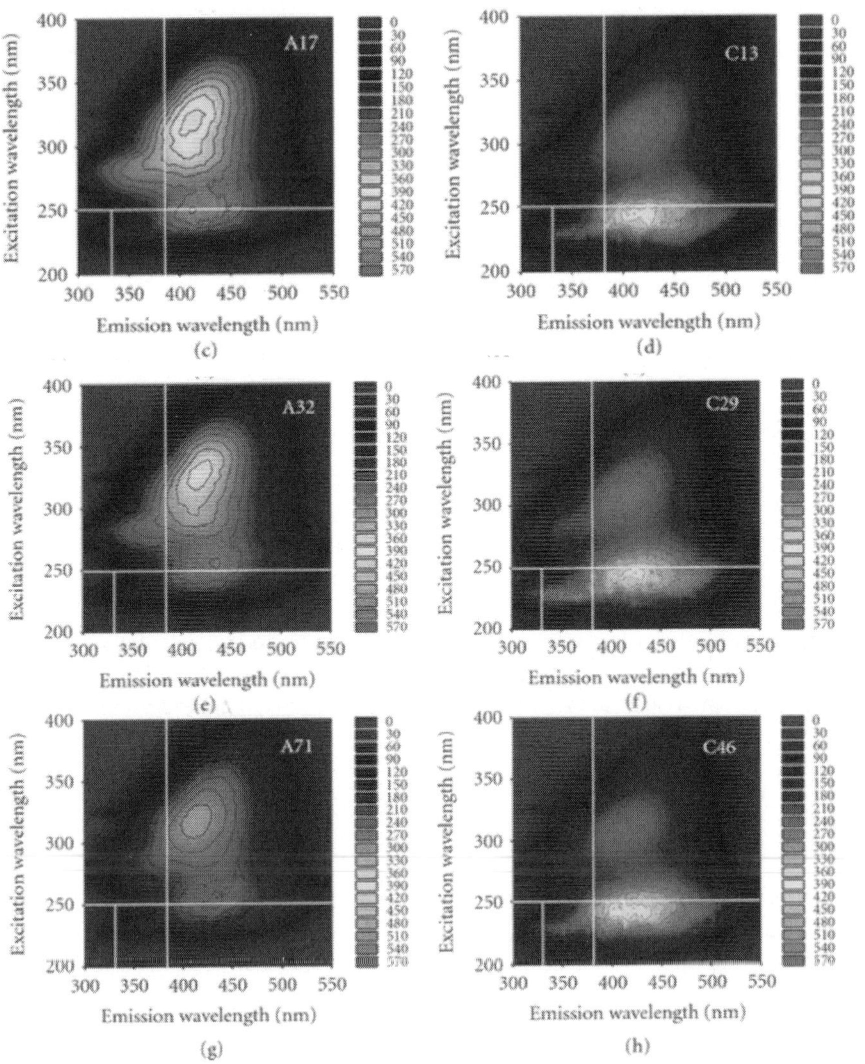

Figure 2: Excitation-emission matrix (EEM) spectra of DOM from treatment A (pig manure + sawdust) on days 1, 17, 32, and 71 (A1, A17, A32, and A71) and treatment C (cattle manure + sawdust) on day 1, 13, 29, and 46 (C1, C13, C29, and C46).

The percentage fluorescence response ($P_{i,n}$) of the five regions in the different composting samples is displayed in Figure 3. In the raw pig manure DOM (A1), the quantities of $P_{1,n}$, $P_{2,n}$, and $P_{4,n}$ in raw pig manure were approximately 16.6%, 19.9%, and 40.0% of total DOM, respectively, while the fulvic-like, and humic-like materials only contribute 10.9%, and 12.5%, respectively. In contrast, $P_{i,n}$ of tyrosine-like, tryptophan-like and soluble microbial byproduct-like materials in cattle manure DOM were 17.1%, 23.2%, and 26.3%, while the fulvic acid-like and humic acid-like substances were 18.1% and 15.3%. Thus, there are more humic substances in cattle manure DOM than in pig manure DOM in the manures obtained for this study.

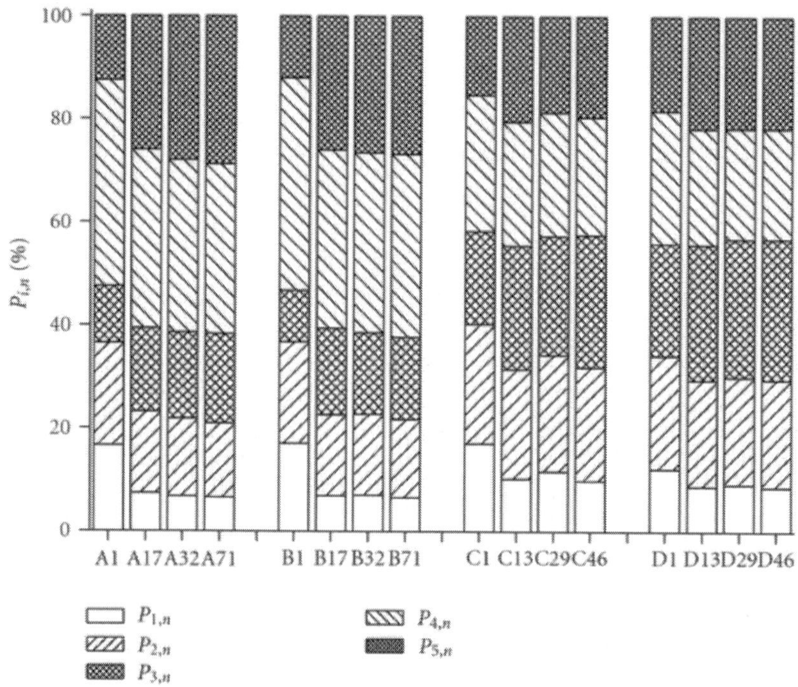

Figure 3: Evolution of the percentage fluorescence response ($p_{i,n}$) during composting.

DOM from treatment A (pig manure + sawdust) and treatment B (pig manure + corn stalk) was on days 1, 17, 32, and 71 (A1, B1, A17, B17, A32, B32, A71, and B71), treatment C (cattle manure + sawdust) and treatment D (cattle manure + corn stalk) on days 1, 13, 29, and 46 (C1, D1, C13, D13, C29, D29, C46, and D46).

During the composting process, a sharp decrease of tyrosine-like, tryptophan-like, and soluble microbial byproduct-like fluorescence occurred in the initial stage of composting for pig manure and cattle manure, while fulvic-like and humic-like fluorescence displayed reverse trends. At the end of composting, there were only fulvic-like and humic-like fluorescence peaks left both in cattle manure and pig manure DOM. The total $P_{5,n}$ values of pig manure or cattle manure cocomposting with sawdust increased by 16.3% and 4.2%, respectively, compared with the raw materials, while the values of $P_{3,n}$ increased by 6.4% and 7.6%, respectively. Similar results were obtained for DOM evolution of exhausted grape marc cocomposted with cattle manure and poultry manure in the study of Marhuenda-Egea et al. [10]. Shao et al. [13] also reported that $P_{5,n}$ increased from 18.8% to 54.8%, and humic acid-like compounds became the main component in municipal solid wastes DOM at the end of the composting process.

Implications of the DOM Change on Cu-DOM Complexation

The high determinative coefficients (DCs) in Table 1 indicated that the experimental data fit well with the Ryan-Weber model [26]. The complexing capacity (CC_{Cu}) is an indicator presenting the amount of active binding sites for complexation with metal ions in a unit mass of DOM. The CC_{Cu} of raw pig manure ranged from 50.6 to 55.0 mmol/g, which was far higher than that of raw cattle manure. After the composting process, the manure DOM featured much lower CC_{Cu} values in the range of 0.41 to 1.99 mmol/g. The increases in $P_{3,n}$ and $P_{5,n}$ confirmed that humification of manure DOM occurred during composting. Previous study has shown that carboxylic- and phenolic-type groups are two kinds of binding sites in humic acids and fulvic acids [30]. Plaza et al. [31] investigated that total acidic functional group contents including phenolic OH group and carboxyl group increased in cattle manure after composing. Caricasole et al. [16] also reported the increasing of phenolic, carboxylic, and carbonylic C occurred in DOM from domestic organic wastes after composting. However, there were negative linear relationships between CC_{Cu} and $P_{5,n}$ ($R^2 = 0.58, P < 0.01$) or $P_{3,n}$ ($R^2 ==0.62, P < 0.01$) shown in Figure 4. In contrast, significantly positive linear correlations were found between CC_{Cu} and P_{4n} ($R^2 = 0.59, P < 0.01$) or $P_{1,n}$ ($R^{2=0.54}, p < 0.01$). The results may suggest that raw manure DOM has higher CC_{Cu} than that of composted manure DOM that can be attributed to its more tyrosine-like organic compounds and soluble microbial byproduct-like materials. The remarkable diversity of Cu complex capacity may be attributed to the difference of protein-like materials in manure DOM. In surface water, fulvic acid and humic acid in DOM were accepted as main components which complexed with Cu [6]. However, Yamashita and Jaffé [7] reported that nonhumic substances such as amino acids are likely engaged in the Cu complexation in surface water. de Zarruk et al. [9] have also verified that the fraction with high proteinaceous fluorescence in vinasse formed the most DOM-Cu complexes. Thus, the binding of Cu to DOM was inhibited properly due to the degradation of protein-like materials during composting, particularly in DOM from composted pig manure.

Table 1: Fitting parameters of Ryan-Weber model

DOM samples[a]	Ex/Em (nm)[b]	I_{CuL} (AU)[b]	CC_{Cu} (mmol/g)[b]	log K_{Cu}[b]	DC[b]
A1	276/306	260.98	50.60	5.24	0.992**
A71	315/412	79.04	1.99	5.14	0.998**
B1	276/306	296.51	55.00	5.17	0.987**
B71	315/412	91.00	0.41	5.02	0.996**
C1	305/412	39.90	6.52	4.90	0.997**
C46	310/408	46.48	0.53	5.08	0.998**
D1	310/414	81.69	7.78	4.93	0.996**
D46	310/420	53.71	1.62	5.00	0.998**

[a]DOM from treatment A (pig manure + sawdust) on days 1 and 71 (A1, A71), treatment B (pig manure + corn stalk) on days 1 and 71 (B1, B71), treatment C (cattle manure + sawdust) on days 1 and 46 (C1. C46), and treatment D (cattle manure + corn stalk) on days 1 and 46 (D1, D46).

[b]Ex/Em: selected wavelengths for the fluorescence titration; I_{CuL}: fluorescence intensity of Cu ion-saturated complex (arbitrary units: AU); CC_{Cu}: complexing capacity (mmol/g); log K_{Cu}: the conditional stability constant; DC: determinative coefficiency.

**Statistical significance value $P < 0.001$.

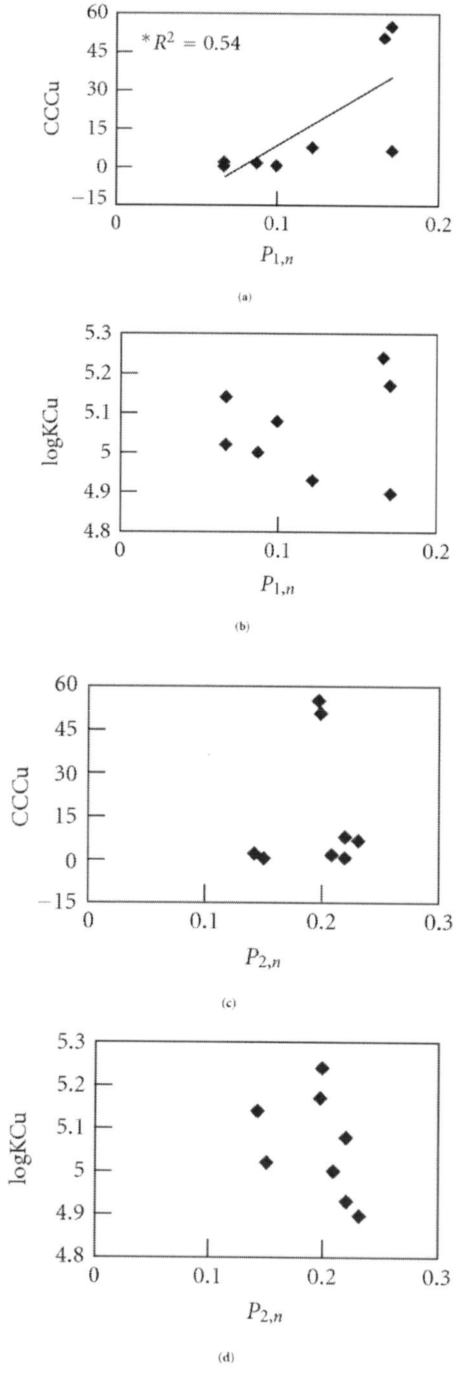
*R^2 = 0.54
CCCu
60
45
30
15
0
−15
0
0.1
0.2
$P_{1,n}$
(a)
logKCu
5.3
5.2
5.1
5
4.9
4.8
0
0.1
0.2
$P_{1,n}$
(b)
CCCu
60
45
30
15
0
−15
0
0.1
0.2
0.3
$P_{2,n}$
(c)
logKCu
5.3
5.2
5.1
5
4.9
4.8
0
0.1
0.2
0.3
$P_{2,n}$
(d)

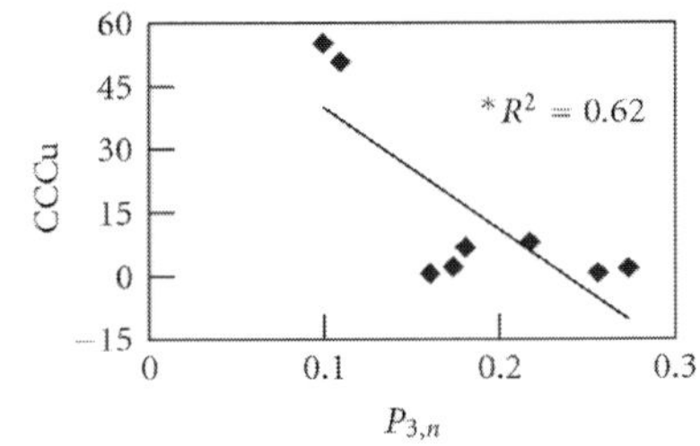
60
45
30
15
0
−15
CCCu
*R^2 = 0.62
0
0.1
0.2
0.3
$P_{3,n}$

(e)

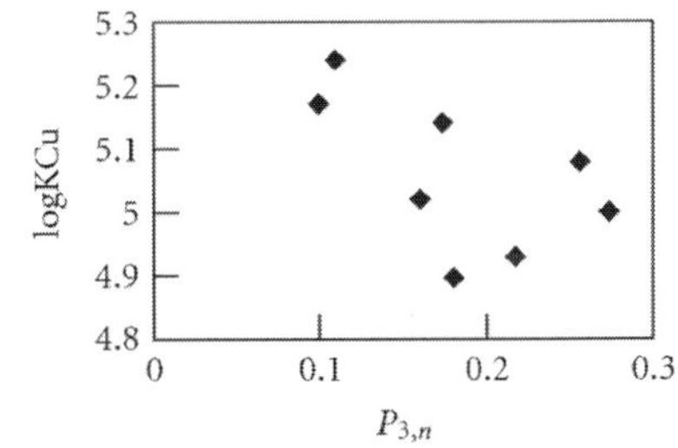
5.3
5.2
5.1
5
4.9
4.8
logKCu
0
0.1
0.2
0.3
$P_{3,n}$

(f)

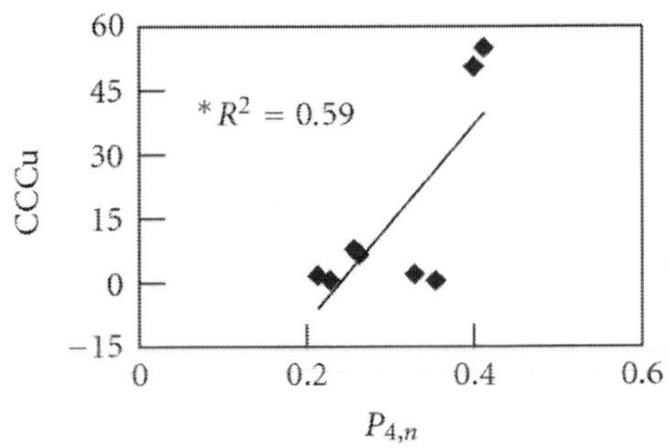
60
45
30
15
0
−15
CCCu
*R^2 = 0.59
0
0.2
0.4
0.6
$P_{4,n}$

(g)

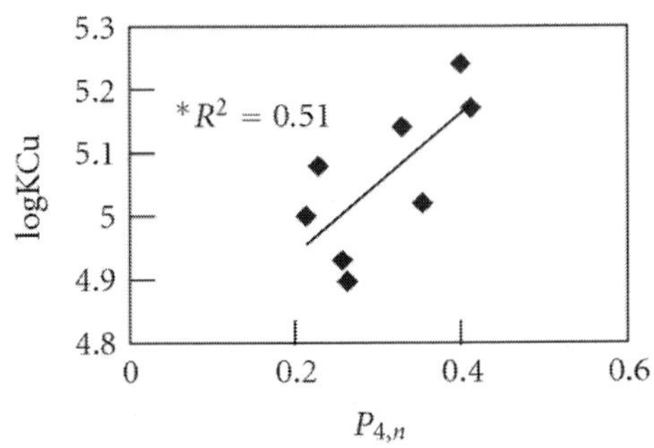
5.3
5.2
5.1
5
4.9
4.8
logKCu
*R^2 = 0.51
0
0.2
0.4
0.6
$P_{4,n}$

(h)

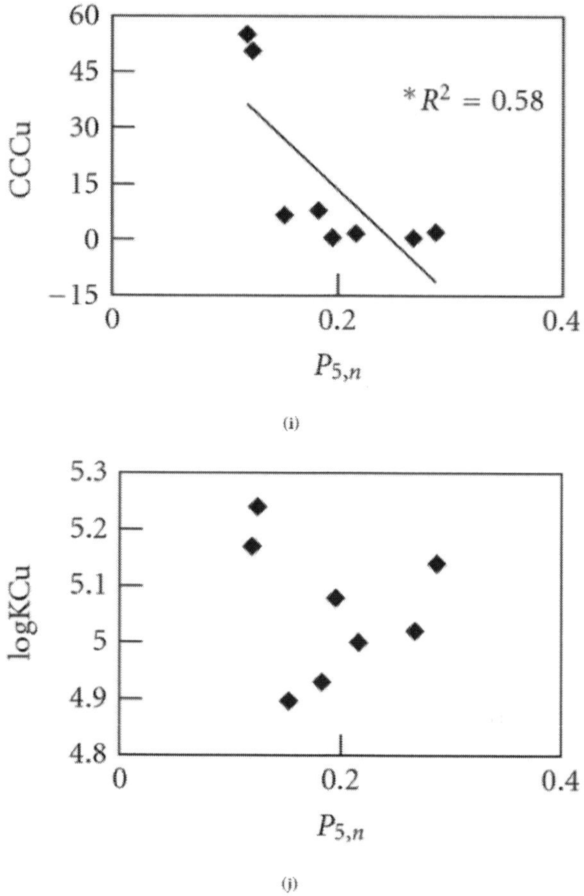

Figure 4: Relationships between percentages of fluorescence response of the five regions ($p_{i,n}$) and complexing capacity (CC_{Cu}) and the conditional stability constant ($\log K_{Cu}$) *Statistical significance value P< 0.05.

The conditional stability constants ($\log K_{Cu}$) of Cu complexes with manure DOM were between 4.90 and 5.24 (Table 1), and the values are similar to those studies on surface water DOM [7]. The DOM of composted cattle manure with sawdust and corn stalk featured higher $\log K_{Cu}$ than that of raw cattle manure. On the contrary, pig manure DOM showed the decreasing trend after composting. The significantly positive linear correlation between $\log K_{Cu}$ and $P_{4,n}$ can be observed in Figure 4 ($R^2 = 0.51, p < 0.01$). This may indicate, in part, that some functional groups in soluble microbial byproduct-like materials play an important role in determining the conditional stability constant $\log K_{Cu}$. Soluble microbial byproduct-like materials contain a large number of proteins and amino acids [20]. It has been reported that protein-like fraction of natural water DOM has the highest $\log K_{Cu}$ [23]. If it was also true for manure DOM, the decrease in the complex stability of pig manure DOM may be attributed to the degradation of large amounts of aromatic amino acid materials.

CONCLUSION

The composting process reduced the amounts of DOM in pig and cattle manure. A majority of the protein-like materials were decomposed, and new humic-like and fulvic-like components were repolymerized. Humic-like materials in composted DOM were mainly transformed from tryptophan-like organic compounds, whereas fulvic-like components were largely transformed from soluble microbial byproduct-like substances.

The complexing capacities of pig and cattle manure DOM decreased after composting, which can be attributed to the degradation of protein-like components. Furthermore, the degradation of protein-like components in pig manure reduced the stability constants of log K_{Cu}. Our study suggests that the composting process might be a way to decrease the bioavailability, mobilization, and transport of manure DOM-Cu complexes, and lower the potential pollution risk to soil and ground water.

ACKNOWLEDGMENTS

The research was funded through the Environmental Protection Public Welfare Program (200909042), National Natural Science Foundation of China (no. 20977010), and the Special Water Pollution Controlling Program of China (2008ZX07209-007).

REFERENCES

1. L. Cang, Y. J. Wang, D. M. Zhou, and Y. H. Dong, "Heavy metals pollution in poultry and livestock feeds and manures under intensive farming in Jiangsu Province, China," Journal of Environmental Sciences, vol. 16, no. 3, pp. 371–374, 2004.
2. Y. X. Li, W. Li, J. Wu, L. C. Xu, Q. H. Su, and X. Xiong, "Contribution of additives Cu to its accumulation in pig feces: study in Beijing and Fuxin of China," Journal of Environmental Sciences, vol. 19, no. 5, pp. 610–615, 2007.
3. G. J. Clark, N. Dodgshun, P. W. G. Sale, and C. Tang, "Changes in chemical and biological properties of a sodic clay subsoil with addition of organic amendments," Soil Biology and Biochemistry, vol. 39, no. 11, pp. 2806–2817, 2007.
4. A. P. Aldrich, D. Kistler, and L. Sigg, "Speciation of Cu and Zn in drainage water from agricultural soils," Environmental Science & Technology, vol. 36, no. 22, pp. 4824–4830, 2002.
5. H. Xue, P. H. Nhat, R. Gächter, and P. S. Hooda, "The transport of Cu and Zn from agricultural soils to surface water in a small catchment," Advances in Environmental Research, vol. 8, no. 1, pp. 69–76, 2003.
6. J. P. Croue, M. F. Benedetti, D. Violleau, and J. A. Leenheer, "Characterization and copper binding of humic and nonhumic organic matter isolated from the South Platte River: evidence for the presence of nitrogenous binding site,"Environmental Science & Technology, vol. 37, no. 2, pp. 328–336, 2003.
7. Y. Yamashita and R. Jaffé, "Characterizing the interactions between trace metals

and dissolved organic matter using excitation-emission matrix and parallel factor analysis," Environmental Science & Technology, vol. 42, no. 19, pp. 7374–7379, 2008.

8. A. Baker, "Fluorescence properties of some farm wastes: implications for water quality monitoring," Water Research, vol. 36, no. 1, pp. 189–195, 2002.·

9. K. K. de Zarruk, G. Scholer, and Y. Dudal, "Fluorescence fingerprints and Cu^{2+}-complexing ability of individual molecular size fractions in soil- and waste-borne DOM," Chemosphere, vol. 69, no. 4, pp. 540–548, 2007.

10. F. C. Marhuenda-Egea, E. Martínez-Sabater, J. Jordá et al., "Dissolved organic matter fractions formed during composting of winery and distillery residues: evaluation of the process by fluorescence excitation-emission matrix,"Chemosphere, vol. 68, no. 2, pp. 301–309, 2007.

11. J. H. Hsu and S. L. Lo, "Chemical and spectroscopic analysis of organic matter transformations during composting of pig manure," Environmental Pollution, vol. 104, no. 2, pp. 189–196, 1999.

12. G. F. Huang, Q. T. Wu, J. W. C. Wong, and B. B. Nagar, "Transformation of organic matter during co-composting of pig manure with sawdust,"Bioresource Technology, vol. 97, no. 15, pp. 1834–1842, 2006.

13. Z. H. Shao, P. J. He, D. Q. Zhang, and L. M. Shao, "Characterization of water-extractable organic matter during the biostabilization of municipal solid waste," Journal of Hazardous Materials, vol. 164, no. 2-3, pp. 1191–1197, 2009.· ·

14. Y. Inbar, Y. Hadar, and Y. Chen, "Recycling of cattle manure: the composting process and characterization of maturity," Journal of Environmental Quality, vol. 22, no. 4, pp. 857–863, 1993.

15. D. Said-Pullicino, F. G. Erriquens, and G. Gigliotti, "Changes in the chemical characteristics of water-extractable organic matter during composting and their influence on compost stability and maturity," Bioresource Technology, vol. 98, no. 9, pp. 1822–1831, 2007.

16. P. Caricasole, M. R. Provenzano, P. G. Hatcher, and N. Senesi, "Chemical characteristics of dissolved organic matter during composting of different organic wastes assessed by ^{13}C CPMAS NMR spectroscopy," Bioresource Technology, vol. 101, no. 21, pp. 8232–8236, 2010.

17. P. G. Coble, S. A. Green, N. V. Blough, and R. B. Gagosian, "Characterization of dissolved organic matter in the Black Sea by fluorescence spectroscopy,"Nature, vol. 348, no. 6300, pp. 432–435, 1990.

18. N. Senesi, T. M. Miano, M. R. Provenzano, and G. Brunetti, "Characterization, differentiation, and classification of humic substances by fluorescence spectroscopy," Soil Science, vol. 152, no. 4, pp. 259–271, 1991.

19. P. G. Coble, "Characterization of marine and terrestrial DOM in seawater using excitation-emission matrix spectroscopy," Marine Chemistry, vol. 51, no. 4, pp. 325–346, 1996.

20. W. Chen, P. Westerhoff, J. A. Leenheer, and K. Booksh, "Fluorescence excitation-

emission matrix regional Integration to quantify spectra for dissolved organic matter," Environmental Science & Technology, vol. 37, no. 24, pp. 5701–5710, 2003.

21. C. Plaza, G. Brunetti, N. Senesi, and A. Polo, "Molecular and quantitative analysis of metal ion binding to humic acids from sewage sludge and sludge-amended soils by fluorescence spectroscopy," Environmental Science & Technology, vol. 40, no. 3, pp. 917–923, 2006.

22. X. Xiong, Y. X. Li, M. Yang, F. S. Zhang, and W. Li, "Increase in complexation ability of humic acids with the addition of ligneous bulking agents during sewage sludge composting," Bioresource Technology, vol. 101, no. 24, pp. 9650–9653, 2010.

23. F. Wu and E. Tanoue, "Isolation and partial characterization of dissolved copper-complexing ligands in streamwaters," Environmental Science & Technology, vol. 35, no. 18, pp. 3646–3652, 2001.

24. P. Q. Fu, F. C. Wu, C. Q. Liu et al., "Fluorescence characterization of dissolved organic matter in an urban river and its complexation with Hg(II)," Applied Geochemistry, vol. 22, no. 8, pp. 1668–1679, 2007.

25. M. Bahram, R. Bro, C. Stedmon, and A. Afkhami, "Handling of Rayleigh and Raman scatter for PARAFAC modeling of fluorescence data using interpolation," Journal of Chemometrics, vol. 20, no. 3-4, pp. 99–105, 2006.

26. D. K. Ryan and J. H. Weber, "Copper(II) complexing capacities of natural waters by fluorescence quenching," Environmental Science & Technology, vol. 16, no. 12, pp. 866–872, 1982.

27. M. A. Bustamante, C. Paredes, F. C. Marhuenda-Egea, A. Pérez-Espinosa, M. P. Bernal, and R. Moral, "Co-composting of distillery wastes with animal manures: carbon and nitrogen transformations in the evaluation of compost stability," Chemosphere, vol. 72, no. 4, pp. 551–557, 2008.

28. A. Baker and R. Inverarity, "Protein-like fluorescence intensity as a possible tool for determining river water quality," Hydrological Processes, vol. 18, no. 15, pp. 2927–2945, 2004.

29. A. Baker and M. Curry, "Fluorescence of leachates from three contrasting landfills," Water Research, vol. 38, no. 10, pp. 2605–2613, 2004.

30. E. Tipping, "Humic ion-binding model VI: an improved description of the interactions of protons and metal ions with humic substances," Aquatic Geochemistry, vol. 4, no. 1, pp. 3–48, 1998.

31. C. Plaza, R. Nogales, N. Senesi, E. Benitez, and A. Polo, "Organic matter humification by vermicomposting of cattle manure alone and mixed with two-phase olive pomace," Bioresource Technology, vol. 99, no. 11, pp. 5085–5089, 2008.

14
CHAPTER

ASSESSING THE INFLUENCE OF SUMMER ORGANIC FERTILIZATION COMBINED WITH NITROGEN INHIBITOR ON A SHORT ROTATION WOODY CROP IN MEDITERRANEAN ENVIRONMENT

Anita Maienza, Giovanni Mughini, Luca Salvati, Anna Benedetti, and Maria Teresa Dell'Abate

INTRODUCTION

Short rotation woody crops (SRWC) include fast growing species (e.g., poplar, willow, eucalyptus, and black locust) capable of producing large amounts of biomass for energy purpose in a relatively short time. Lake Albacutya provenance of Eucalyptus camaldulensis Dehnh is one of the best choices for SRWC in Mediterranean environments [1, 2]; however in low quality soils with dry periods the species is able to cover the soil but with low yields [3]. Some eucalypt clones have been selected as more productive than E. camaldulensis, and for two of these clones (Viglio and Velino) the European Union plant patent has been requested [3, 4].

Two cultivation models are suggested for eucalyptus SRWC in Italy: (i) 5,000–5,500 plants/ha^{-1} for 3-4 rotation cycles *f* 2-3 years each and (ii) 1,100–1,600 plants/ha^{-1} for 2 rotation cycles instead of 5-6 age.

Eucalyptus SRWC with rotation age of 2-3 years each requires higher amount of water and nutrients than conventional Eucalyptus crop. In Brazil SRWC Eucalyptus plantation 2.5 years of age (5,000–6,000 plants/ha^{-1}and 2/3 years of rotation age) is estimated to absorb in the vegetal tissues high amounts of N,P, K, and Ca from soil [5] three times higher than a conventional plantation (600 to 1,600 p ha^{-1} and eight years of

Citation: Anita Maienza, Giovanni Mughini, Luca Salvati, Anna Benedetti, and Maria Teresa Dell'Abate, "Assessing the Influence of Summer Organic Fertilization Combined with Nitrogen Inhibitor on a Short Rotation Woody Crop in Mediterranean Environment," International Journal of Forestry Research, vol. 2014, Article ID 371895, 5 pages, 2014. doi:10.1155/2014/371895.

rotation age). Moreover water contribution was estimated to be four times higher [5]. In conventional E. globulus Labill. plantation in Spain and Brazil, the quantity of water to produce 1 kg of dry biomass is evaluated in 306 and 344 liters, respectively [3, 6].

The use of irrigation and fertilization to satisfy SRWC request is not feasible because it is usually uneconomic [7–9]. As a consequence, fertilization with manure could be an attractive solution to recycle sewage waste and supply nutrients, by increasing fertilization potential.

Animal manure recycling could increase soil organic matter and nutrient content and plant production and preserve water used for irrigation [10]. However one of the earliest pieces of EU legislation aimed at controlling pollution and improving water quality, the European Directive (91/676/EEC) enforced in law in 1991, rules basic agronomic practices regarding the use of animal manure source, livestock, and waste water from small food companies. In fact, the agricultural use of nitrates in organic and chemical fertilizers has been the source of 50% of total nitrogen discharge into surface water [11–13] in Europe.

The application of livestock waste to soil needs appropriate treatments in order to preserve soil and ground water from nitrate pollution [12, 13] and greenhouse gas emissions. In the last years attention has been paid to the use of nitrification inhibitors mixed with effluent with the aim of minimizing the environmental problems related to animal manure disposal [11, 14, 15].

The nitrification inhibitor 3,4-dimethylpyrazole phosphate (3,4 DMPP) delays the bacterial oxidation of ammonia to nitrite by depressing the activity of Nitrosomonas bacteria in the soil (first step of nitrification) for a short time (30 days), whereas the second step of nitrification appears to be not influenced [15]. Thus, losses due to nitrate leaching decreased with nitrification steps. A number of studies [16, 17] reported significant reduction in NO_3^- formation in soil and N_2O gaseous loss has also been reduced due to the addition of nitrification inhibitors.

The objective of the present study was to analyze differences in Eucalyptus plant growth in a Mediterranean SRWC after summer treatments with bovine effluent with (or without) addition of 3,4 DMPP as nitrification inhibitor, in order to establish whether 3,4 DMPP in addition to effluents can offer advantage in agroforestry. An additional aim of the study was to test the 3,4 DMPP effect on NH_3 volatilization from soil in order to highlight possible environmental advantages of such a fertilization strategy. To the best of our knowledge the present study is the first investigation on these issues carried out in a Mediterranean SRWC.

Materials and Methods

The field experiment was conducted in an SRWC Eucalyptus plantation located in Central Italy (Rome), in a farm of Agricultural Research Council (CRA-PLF) (41° 54′ 33.55″ N,12° 21′ 37.83″ E) during June 2011. The soil is a Luvisols with a sandy loam texture (sand 63%, clay 21%, and silt 16%) and pH 7.0. The soil is poor in nutrient with concentrations of organic carbon (TOC) 1.0%, total nitrogen (N_{tot}) 0.06%, and available phosphorus (P) 8 mg kg^{-1} and a high capacity of cation exchange (23 mEq 100 g^{-1} soil).

Soil fertility index calculated using soil chemical and biochemical variables (Table 1) falls in class of "stress" [9]. Eucalyptus pure species (E. gomphocephala DC, E. bridgesiana R.T. Baker, and E. camaldulensis) and a number of hybrid clones of E. camaldulensis × E. bicostata Maiden, Blakely & Simmons, occur in the experimental SRWC. Plant density per hectare amounts to 5,000 p ha^{-1} with 20 plants per plot in row. Plantation was harvested when the plants were three years old and the experiment was carried out during the first growing season after the harvesting. A completely randomized block design with three replicates was used for each treatment. Treatment was carried out with bovine effluent with a N_{tot} content by 0.32% ($N–NH_4^+$ 0.17%, $N–NO_3$ 0.0025%, and $N–NO_2$0.0004%) and TOC 5.7%. Three field tests were carried out: (i) control nonfertilized (C); (ii) bovine effluent only (Ef); (iii) bovine effluent with the addition of 3,4 DMPP (Ef + DMPP). Treatments were done once in June with tanker furrowing plowing with 170 kg N ha^{-1}; 3,4 DMPP was added at the rate of 1% of N_{tot}. A pure species (E. camaldulensis) and a hybrid clone Viglio were selected for the experiment. Ef and Ef + DMPP were put into the ground 40 cm depth at 50 cm distance from plants. NH_3 volatilization from soil was monitored after the treatment with a simple chamber method for field determination (Drager-Tube Method, DTM) [11]. Two measurements per day were carried out, once at morning (7 am) and once at midday (1 pm) with 3 replicates for each plot. Values of NH_3 concentrations on the Drager indicator tubes have a coefficient of variation between 10 and 15%, as indicated by the manufacturer. A meteorological station was put downwind of the field measurement pole. Temperature and rainfall measurements were taken according to international standards.

Table 1: Chemical and biochemical parameters of control soil, samples were collected at (0–30 cm). Data were calculated on averge of six field replicates

Soil parameters	Average soil	SE
Sand (%)	63	2.3
Silt (%)	16	0.01
Clay (%)	21	0.5
N_{tot} (%)	0.06	0.01
TOC (%)	1.06	0.02
pH	7	0.2
qCO_2	0.17	0.01
Cumulative $C-CO_2$ evolved over 28 days (mg kg^{-1} soil)	282.11	10.5
Respiration measured at 28 days (mg $C-CO_2$ kg^{-1} soil)	0.4	0.02
C_{mic} (mg-C kg^{-1} soil)	95.53	2.8

Plant growth analysis was carried out at the end of the first growing season (November 2011) using Basal Area (BA) to establish plants growth on the first growing season. Diameter to the breast height (DBH) for each experimental plant was measured with a digital caliber to 130 cm from the soil and each DBH converted in BA $[BA = \pi(DBH/2)^2]$ and BA total value for each plot was calculated considering 20 neighboring plants. DBH and BA are standard characters used to evaluate plant growth for Eucalypts as well as for also in other forest trees [18]. Kruskal-Wallis H and Mann-Whitney U nonparametric inference testing at $P < 0.05$ was applied to BA values with the purpose of verifying differences in the median of the examined groups.

Results and Discussion

Bagnouls-Gaussen climate diagram (Figure 1) highlighted a drought period from April to October, indicating that plants had been under water stress for a seven-month period, due to lack of ground water. The ammonia flux from soil was detected for 92 hours after the treatments and this loss of flux should be also due to the beneficial effect of the disposal procedure, that is, the burial of fertilizers into the soil. The Ef + DMPP treatment moderately reduces NH_3 emissions from soil during 52 hours, while a significant reduction (Mann-Whitney U, $P < 0.05$) compared with Ef treatment was observed at 26 hours (Figure 2).

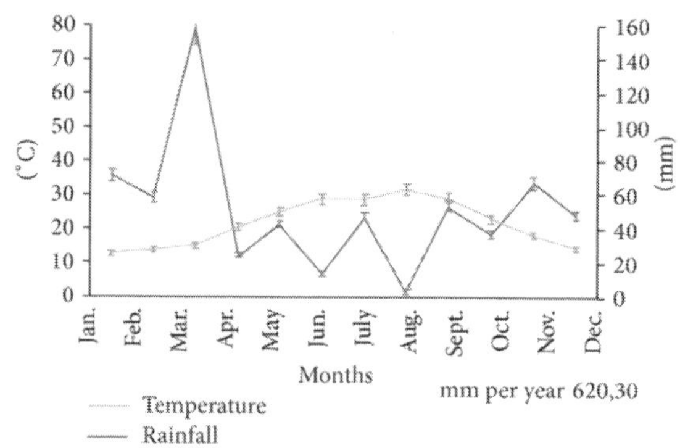

Figure 1: Bagnouls-Gaussen plot. Air temperature and rainfall data were collected by a meteorological station during the summer season 2011 in CRA-PLF experimental site.

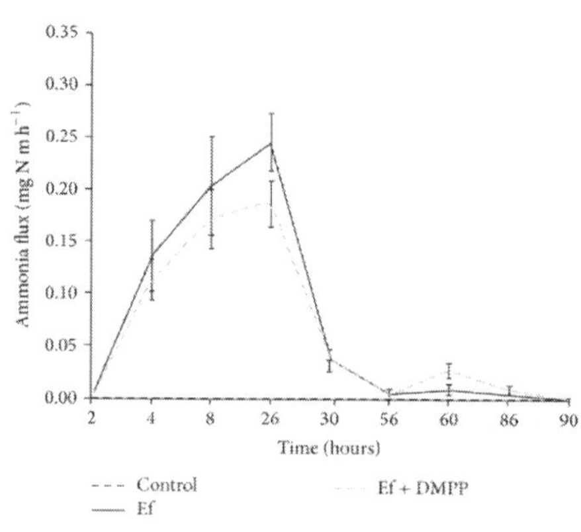

Figure 2: Ammonia volatilization detected by chambers method tube. Manual measurements were done at the beginning of the treatment and for 92 hs after. Control is the soil with any treat-

ment, Ef is bovine manure effluent only, and Ef + DMPP is combination of bovine manure effluent and 3,4 DMPP.

Plant growth estimates showed high variability of the BA values for E. camaldulensis with nonsignificant effects of treatment (Mann-Whitney U, $P < 0.05$). These results are probably due at high genetic variability of E. camaldulensis (in this case each plant is a different genotype) and no sufficient number of replications. The Viglio clone grew significantly (Kruskal-Wallis test, $P < 0.05$) after treatments (BA values, Table 2) and showed the best growth performance in Ef + DMPP treatment which resulted also as being more productive than E. camaldulensis according to Mughini [4] and Facciotto et al. [3].

Table 2: Basal area (BA) in cm^2 of E. camaldulensis and E. camaldulensis × E. bicostata(Viglio clone). Data were collected on 20 trees for each plot replicates, standard errors are also reported

	Treatment	Total BA/plots (mean of 20 trees) (cm^2)	Average	SE
		105.89		
	C	75.69	76.50	16.74
		47.91		
		68.03		
E. camaldulensis	Ef	81.83	61.74	13.78
		35.36		
		55.14		
	Ef + DMPP	51.91	67.32	13.83
		94.91		
		60.65		
	C	79.18	76.82	8.74
		90.64		
		63.20		
Viglio clone	Ef	75.67	72.56	4.77
		78.82		
		95.37		
	Ef + DMPP	129.32	107.30	11.02
		97.22		
Average			85.56	6.95

The Viglio clone BA was compared by fertilization treatment (Figure 3). Ef treatment revealed being ineffective possibly due to water stress and restricted nutrient availability during the experiment; otherwise Ef + 3,4 DMPP treatment showed a clear increase (Mann-Whitney test,). The different responses of Viglio clone to treatments are related to nutrient soil content, as temporarily determined by manure application: lower ammonia losses and longer availability of ammonium to the plants due to delayed oxidation to nitrate acted by 3,4 DMPP should be the main processes affecting the different nutrients uptake. In fact, in Eucalyptus plantations nitrogen in the soil is dominated by ammonia [14, 19] and ammonium is absorbed faster (30–40 times) than the nitric form [19]. In this context, the use of 3,4 DMPP in Ef [11, 15, 20] supports the ammonium availability for the plants during dry season longer and more efficiently than Ef alone.

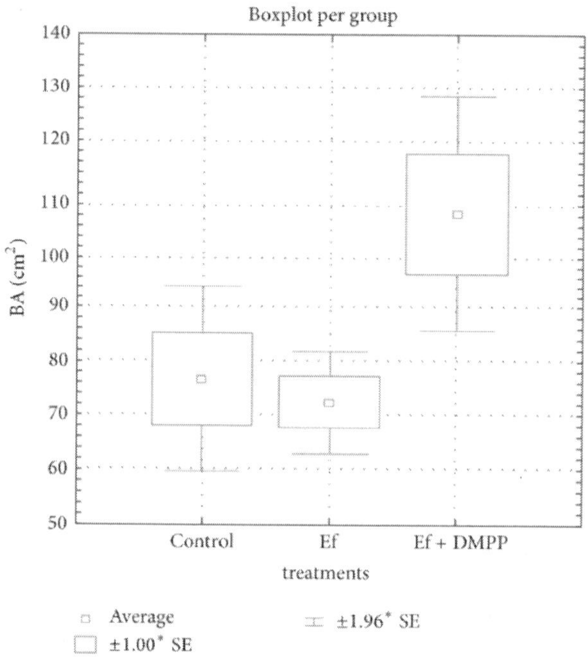

Figure 3: Viglio clone growths measured by Basal Area by groups. Control is Viglio without treatment, Ef is Viglio treated with bovine manure effluent, and Ef + DMPP is Viglio grown with effluent combined with 3,4 DMPP.

CONCLUSIONS

Results of this study support the development of a strategy of organic fertilization in Mediterranean regions based on combination of 3,4 DMPP and animal manure. The main effects were to improve plant growth in Eucalyptus Viglio clone SRWC in a Mediterranean dry environment after the first growing season and to reduce ammonia emissions from soil in the first 26 hours after the fertilization, especially important in summer, when gaseous emissions are enhanced by high temperature. The climate conditions of summer 2011 and low soil fertility of the field could have conditioned the effect on growth of the treatments with Ef alone. Further investigations on biomass productivity, nitrate leaching, and soil ecology can provide more information about the efficiency and the sustainability of this practice.

Conflict of Interests

The authors declare that there is no conflict of interests regarding the publication of this paper.

ACKNOWLEDGMENT

The present research was conducted with the agreement and the financial contribution of K + S Nitrogen (since July 2012 EuroChem Agro).

REFERENCES

1. G. Gemignani, "Observations préliminares sur les provenances australiennes d'Eucalyptus camaldulensis," Sous-commission des Questions Forestieres Méditerranéennes-Comuté del Recherche Forestiere Mediterranéenne, FO:SCM/FR/68-8A/9, FAO, Rome, Italy, 1968.
2. G. Mughini, "Comportamento di alcune specie di eucalitto in tre prove in Italia meridionale," Cellulosa e Carta, vol. 42, pp. 2–7, 1991.
3. G. Facciotto, S. Bergante, G. Mughini, M. Gras, and G. Nervo, "Le principali specie per la produzione di biomassa," L'Informatore Agrario, vol. 40, pp. 36–37, 2007.
4. G. Mughini, "Production of hybrid clones: Italy," Final Report of Improvement of eucalypt management an integrated approach: breeding, silviculture and economics, AIR3-CT93-1678, 1996.
5. F. Poggiani and H. T. Z. Couto, "Biomass and nutrient estimates in short rotation intensively cultured plantation of Eucalyptus grandis," IPEF, vol. 23, pp. 29–36, 1983.
6. E. Jimenez, J. A. Vega, P. Pérez-Gorostiaga, T. Fonturbel, P. Cuinas, and C. Férnandez, "Evaluaciòn de la transpiratiòn de Eucalyptus globulus mediante densidada de flujo de savia y su relaciòn con variables meteorològicas i dendrométricas," Boletin del CIDEU, vol. 3, pp. 119–138, 2007.
7. C. P. Mitchell, E. A. Stevens, and M. P. Watters, "Short-rotation forestry—operations, productivity and costs based on experience gained in the UK," Forest Ecology and Management, vol. 121, no. 1-2, pp. 123–136, 1999.
8. G. Facciotto, T. Zenone, and G. Sperandio, "Dalla colture da biomassa reddito incerto senza aiuti,"L'Informatore Agrario, vol. 59, no. 10, pp. 91–99, 2003.
9. P. Sequi, A. Benedetti, and M. T. Dell'Abate, ATLAS Atlante di Indicatori di qualità del suolo, Ministero delle Politiche Agricole e Forestali, 2011.
10. S. Melero, J. C. R. Porras, J. F. Herencia, and E. Madejon, "Chemical and biochemical properties in a silty loam soil under conventional and organic management," Soil and Tillage Research, vol. 90, no. 1-2, pp. 162–170, 2006.
11. A. Pacholski, G. Cai, R. Nieder et al., "Calibration of a simple method for determining ammonia volatilization in the field—comparative measurements in Henan Province, China," Nutrient Cycling in Agroecosystems, vol. 74, no. 3, pp. 259–273, 2006.
12. C. Plaza, D. Hernández, J. C. García-Gil, and A. Polo, "Microbial activity in pig slurry-amended soils under semiarid conditions," Soil Biology and Biochemistry, vol. 36, no. 10, pp. 1577–1585, 2004.

13. J. C. Neff, A. R. Townsend, G. Gleixner, S. J. Lehman, J. Turnbull, and W. D. Bowman, "Variable effects of nitrogen additions on the stability and turnover of soil carbon," Nature, vol. 419, no. 6910, pp. 915–917, 2002.

14. M. X. Gómez-Rey, M. Madeira, S. J. Gonzalez-Prieto, and J. Coutinho, "Soil C and N dynamics within a precipitation gradient in Mediterranean eucalypt plantations," Plant and Soil, vol. 336, no. 1, pp. 157–171, 2010.

15. W. Zerulla, T. Barth, J. Dressel et al., "3,4-Dimethylpyrazole phosphate (DMPP)—a new nitrification inhibitor for agriculture and horticulture. An introduction," Biology and Fertility of Soils, vol. 34, no. 2, pp. 79–84, 2001.

16. D. Chen, Y. Li, P. Grace, and A. R. Mosier, "N_2O emissions from agricultural lands: a synthesis of simulation approaches," Plant and Soil, vol. 309, no. 1-2, pp. 169–189, 2008.

17. H. J. Di and K. C. Cameron, "Effects of temperature and application rate of a nitrification inhibitor, dicyandiamide (DCD), on nitrification rate and microbial biomass in a grazed pasture soil," Australian Journal of Soil Research, vol. 42, no. 8, pp. 927–932, 2004.

18. N. Borralho, I. M. Almeida, and P. P. Cotterill, "Genetic control of growth of young Eucalyptus globulus clones in Portugal," Silvae Genetica, vol. 41, no. 2, pp. 39–45, 1992.

19. M. Madeira, M. C. Magalhães, A. Azevedo, A. Fabião, M. C. Araújo, and J. P. Pina, "Efeito da gestão dos resíduos de abate nas características do solo e no crescimento de uma plantação de Eucalyptus globulus, em talhandia," Revista de Ciências Agrárias, vol. 27, pp. 414–431, 2004.

20. W. Zerulla, G. Pasda, R. Händle, and A. H. Wissemeier, "The new nitrification inhibitor DMPP (ENTEC) for use in agricultural and horticultural crops—an overview," Plant and Soil Science, vol. 92, pp. 750–751, 2002.

15
CHAPTER

EFFECTS OF BIOSOLIDS AND MANURE APPLICATION ON MICROBIAL WATER QUALITY IN RURAL AREAS

Amira Oun, Arun Kumar, Timothy Harrigan, Andreas Angelakis and Irene Xagoraraki

Keywords: Best management practices, biosolids, manure, contamination of ground and surface waters, microbial source tracking, microorganisms, pathogens, risk assessment, rural systems, waterborne diseases, water quality

WATERBORNE OUTBREAKS IN RURAL AREAS IN THE US

Most of the waterborne disease outbreaks worldwide and in North America are associated with rural drinking water systems. According to Craun *et al.* [1], in the United States during the 12 years period of 1991–2002, 207 waterborne disease outbreaks and 433,947 illnesses were reported; 42% of these outbreaks occurred in non-community water systems, 22% occurred in individual systems such as private wells, and only 36% occurred in community systems. In most cases the drinking water supply in rural areas is provided by groundwater wells that in some cases are shallow. Between 1981 and 1998, 50% (210 of 417) of the reported waterborne disease outbreaks were linked to contaminated groundwater [2,3]. 696 outbreaks have been reported in the US between 1971 and 2000, of which 59% were linked to groundwater [4].

Most of the reported waterborne diseases outbreaks in North America were related to microbial agents (parasites, bacteria and viruses), some to chemical agents, and some were of unknown etiology. For example, a survey by the Centers for Disease Control and Prevention (CDC) reported that one sixth (5 out of 30 cases) of drinking water-associated waterborne disease outbreaks during 2003–2004 were of unknown etiology

Citation: Amira Oun, Arun Kumar, Timothy Harrigan, Andreas Angelakis and Irene Xagoraraki. Effects of Biosolids and Manure Application on Microbial Water Quality in Rural Areas. Water 2014, 6(12), 3701-3723; doi:10.3390/w6123701.

because of a lack of available analytical methods [5]. The US EPA suspects that many of the outbreaks due to unidentified sources were caused by enteric viruses [6]. Zoonotic and human viruses are of particular interest because viruses are the smallest of all pathogens and their small size can facilitate transport through the soil. Viruses have low die-off rates and viral infections may lead to chronic health effects.

Runoff carrying manure-borne microbial pathogens have been implicated in some of the largest water and food borne outbreaks [7,8,9]; in contrast, it is important to state that there are no scientifically documented outbreaks or excess illnesses that have occurred from exposure to pathogens associated with treated biosolids [10]. An outbreak that occurred in a small farming community in Canada demonstrated that the potential of human infections caused by zoonotic pathogens (pathogens that may infect both animals and humans) has serious consequences [7,9]. More than 2300 people in the town of Walkerton, Ontario suffered gastrointestinal illness and seven died when the water supply was contaminated by manure pathogens from a nearby farm after more than five inches of rain fell over a five day period in May 2000 [7,9].

During March and April 1993 water contaminated with *Cryptosporidium* oocysts from Lake Michigan caused a massive waterborne outbreak of cryptosporidiosis among residents of Milwaukee, Wisconsin. The source of contaminated water was most likely runoff from rural areas during high rainfall. It was estimated that 403,000 residents living in a five-county area and numerous visitors to the city of Milwaukee were sick during this outbreak and 58 lost their lives [11].

In recreational waters, fecal contamination from bathers has been an important source of exposure. During 1971–2000, 259 outbreaks associated with recreational activities have been reported, 144 outbreaks (56%) were associated with recreation in untreated natural water such as lakes and streams; the remaining of the recreational outbreaks occurred in treated water such as swimming pools [12]. Fecal contamination from bathers, septic tanks, runoff from agricultural areas, and other sources was identified as the important cause of outbreaks in untreated recreational waters. Zoonotic bacterial agents such as *E. coli* O157:H7 and O121:H19 caused 38% of the outbreaks associated with untreated natural recreational water [1,12].

SOURCES OF MICROBIAL POLLUTION IN RURAL AREAS IN THE US

The nexus between water quality and public health in the rural environment is complex. Non-point sources of contaminants are common and there is little monitoring of the system or the types of pollutants and emerging contaminant risks. Human and zoonotic pathogens and other microbial pollutants are a particular concern. Water quality research tends to focus on urban areas and sophisticated treatment systems. Water and waste management in rural areas is often minimal and water quality monitoring is sporadic. Rural areas typically have high density livestock housing and lack of advanced treatment systems for animal and human waste and wastewater. Animal waste from livestock production facilities is often applied to the land without prior treatment. Domestic wastewater is often treated on-site by septic systems without regular monitoring.

There is a potential risk of human exposure to waterborne contaminants such as human and zoonotic pathogens originating from manure, biosolids, and leaking septic systems in rural areas [13,14,15]. Livestock manure has been spread on the land as fertilizer and also for disposal. Manure and other wastes of various livestock often contain high concentrations of pathogens. Levels of pathogen in manure depend on the source animal, the animal's state of health, and how the manure was stored or treated before use. Non-point sources of contamination by manure include pastured animals, roaming wild animals, and leaching or runoff from agricultural areas. Point sources of manure contamination include animal feedlots, animal housing facilities, and manure storage areas, such as lagoons [16], resulting in soil and water contamination.

Biosolids (treated municipal wastewater sludge that meets standards for use as a fertilizer or soil conditioner); from large wastewater utilities in urban areas are often transported and applied to land in rural areas. Land application of biosolids provides agricultural benefits and presents a cost effective method for disposal of treated sludge following wastewater treatment; however, reuse of this product presents health concerns that must be addressed. Health concerns include pathogen transmission to food, contamination of ground water or surface water from field runoff, and build-up of heavy metals or organic contaminants [17,18]. Because biosolids may contain human pathogens [13,19], exposure to biosolids has raised human health-related concerns. The microbial load of biosolids that are routinely applied on farm land varies depending on the type of sludge treatment and common sludge treatment processes do not completely inactivate pathogens.

CHARACTERIZATION OF BIOSOLIDS AND MANURE

In 2004, a US national biosolids survey indicated that about 6.50 million tons of dry biosolids were produced in the US; and approximately 55% of the total was applied on land as soil amendment [10,20]. In 2006, the US EPA [21] estimated that more than 7.10 million tons of dry biosolids per year are produced in the US; about 50% of which are land applied. Land application of biosolids has increased since restrictions were placed on ocean dumping disposal. However, due to public concern over potential hazards, in some areas of the US land application of Class B biosolids has been banned. This is particularly true in California, where in many areas Class A land application has replaced Class B land application [22].

Biosolids contain organic matter and nutrients, and when applied to farmland can improve productivity and reduce the need for manufactured fertilizer inputs [23,24]. The main benefits of biosolids are through the supply of nitrogen, phosphorus, potassium, calcium, and magnesium, biosolids also supplying the essential plant micro nutrients (e.g., sulfur, manganese, zinc, copper, iron, molybdenum, and boron) [25]. Biosolids are a valuable source of nutrients although it contains lower N, P, and K compared to commercial fertilizers, especially high-grade ones [26]. A study of nutrient levels in biosolids [27], with more than 240 samples collected and analyzed in Pennsylvania (aerobically digested, anaerobically digested, or alkali-treated) between 1993 and 1997 showed average N, P, and K contents of 4.74%, 2.27% and 0.31%, respectively. Nutrient values of biosolids vary with sources of wastewater and wastewater treatment processes, in general compared to cattle manure, biosolids have lower P, the N:P ratio is around (3.1–3.4) [27,28], and have limited amount of K [29].

Even though biosolids are a beneficial soil amendment they may also pose a potential threat to the environment and pose risks to human health. There are potential hazards with land application since several contaminants can be present in biosolids including human pathogens [22,24,30,31]. Pathogens contained in biosolids include viruses, bacteria, and animal and human parasites (protozoa and helminthes), which may cause various human diseases and illnesses [24,32]. The characteristics and properties of biosolids vary depending on the quality and origin of sludge, along with the type of treatment processes [33,34,35]. Incomplete destruction, contamination from external sources, and changes in environmental factors during storage can lead to regrowth or reactivation of pathogens. Typical concentrations of some pathogens and indicators in biosolids are summarized in Table 1.

The rapid growth of concentrated animal feeding operations (CAFOS) has caused an increase in the amount of manure produced annually in US. There are approximately 450,000 CAFOs in the US including dairy, cattle, pig, and poultry farms. An estimated 450 million tons of wet weight manure or 90 million tons of dry solids manure are generated annually from about 92 million swine, 109 million cattle, 292 million turkeys, and 7.5 billion chickens in the US [36,37,38]. US EPA estimates that a single dairy cow produces approximately 55 kg/day of wet manure [39]. Livestock and poultry manure can provide large quantities of valuable nutrients for crop and pasture when applied to land as a fertilizer but most of this manure is applied untreated. Historically, the environmental impacts of animal manure were principally associated with nutrients [40]. Manure application rates are usually based on crop N requirement which greatly increases soil P levels because, the N: P ratios of manure (2:1 to 4:1) are significantly smaller than N: P uptake ratios (6:1 to 8:1) for most crops. The excess P and N (as NO_3) from manure application can be transported in runoff or leached into the groundwater. Recently, however, other constituents such as naturally excreted hormones and pathogens have been of interest [41,42,43].

Although livestock manure is a good source of nutrients for crops and a natural way of recycling waste, raw manure is also a potential source of human pathogens if improperly stored and mishandled, or not managed correctly, fecal contamination from livestock manure handling and storage facilities is one of the most important sources of water microbiological pollution [21]. Animal manure depending on its origin can be a source of zoonotic pathogens such as Hepatitis E virus, Rotavirus A (some strains), Adenovirus (some strains), *Aeromonashydrophila, Yersinia enterocolitica, Vibrio cholerae, Leptospira*, *Campylobacter jejuni*, *Escherichia coli* O157:H7, *Salmonella* spp., *Listeria monocytogenes*, *Cryptosporidium parvum*, and *Giardia lamblia* [44,45,46,47,48,49,50,51,52,53]. The concentration of some pathogens and indicators in manure are summarized in Table 2.

Table 1: Example pathogens and indicators in class B-biosolids

Organism	Detection Method	Concentration	Reference
Pathogens			
Salmonella sp.	culture	1.2–3.2 MPNCU/g	[54]
		0.487–0.954 MPN/4 g	[34]
		40.1 MPN/4 g	[31]
Cryptosporidium	Laser scanning cytometry	4.3/10 g	[35]
Adenoviruses	qPCR	5×10^5 copies/g	[33]
		7.5×10^5 copies/g	[34]
		1.59×10^4 copies/g	[55]
	Cell culture	480 MPN/4g	[34]
Human polyomavirus	qPCR	8.05×10^2 copies/g	[55]
		2.5×10^5 copies/g	[34]
Enteroviruses	qPCR	1.9×10^4 copies/g	[34]
		4.8×10^3 copies/g	[54]
		3.3×10^4 copies/g	[56]
	RT-PCR	1.2×10^4 copies/g	[57]
		$1.0^6 \times 10^4$ copies/g	[58]
	Cell culture	480 MPN/4 g	[34]
		38.2 MPNCU/g	[57]
		9 MPNCU/g	[58]
		15–80 MPNCU/g	[54]
Noroviruses GI	qPCR	5×10^4 copies/g	[34]
	Cell culture	480 MPN/4 g	[34]
Noroviruses GII	qPCR	1.5×10^5 copies/g	[34]
	Cell culture	480 MPN/4 g	[34]
Indicators			
Somatic coliphages	Cell culture	5.5×10^2 PFU/10 g	[35]
		2.09×10^5 PFU/4 g	[31]
Total coliform	Culture	7.64×10^5 MPN/4 g	[31]
Enterococci	Culture	7.2×10^5–2.6×10^6 MPN/g	[54]
		6.4×10^5 MPN/g	[56]
E. coli	Culture	4.4×10^5–1.1×10^6 MPN/g	[54]
		7.2×10^5 MPN/g	[56]
		10^4 MPN/g	[34]

Notes: qPCR-quantitative polymerase chain reaction; MPNCU-most probable number colony forming unit; RT-PCR-reverse transcription polymerase chain reaction; PFU-plaque forming unit.

Table 2: Example pathogens and indicators in animal manure

Organism	Concentration	Reference
	Pathogens	
Salmonella sp.	2.8×10^5 CFU/25g	[59]
Listeria	1.7×10^4 CFU /g	[59]
E. coli O157:H7	2.2×10^6 CFU/g	[60]
	Indicators	
Enterococci	1.5×10 MPN/g	[61]
E. coli	10^5–10^6 CFU /g	[61]
	5.5×10^7 MPN/g	[59]

Notes: CFU-colony forming unit; MPN-most probable number.

In the US, the quantity of manure application as a soil amendment is 30 times higher than the biosolids application [21,23,37,38]. This is mostly due to high manure production and public acceptance of manure application practice for thousands of years [62]. In addition, most of the studies focus on the pathogens in biosolids, and there is lack of comparison studies for pathogens in manure [20,62].

In comparison, the benefits of biosolids as soil amendments are similar to those provided by animal manure, both provide important plant nutrients and organic matter. Animal manure contains more P than crops require if the application rate is based on N needs [40]. Most of the N in biosolids and manure is organic and becomes available to crops as it is mineralized. As a potential source of human pathogens public perception is that, biosolids pose higher risk to the public than animal manure. In fact, the risks associated with their use are not greater than the risks associated with untreated manure use. For example levels of *E.coli* indicator bacteria and *Salmonella sp* pathogens in manure are similar to biosolids (Table 1 and Table 2). Overall quantitative data on total pathogen content and indicators in manure is limited comparing with biosolids.

ASSOCIATION OF BIOSOLIDS AND MANURE APPLICATION WITH MICROBIAL CONTAMINATION OF SURFACE AND GROUNDWATER

Runoff from agricultural areas has been suggested as one of the major sources of nonpoint-source pollution [63]. Rainfall events may carry human pathogens in water runoff from contaminated sites to water bodies serving as recreational, irrigation or drinking water sources. The effects of nonpoint source pollutants on specific waters vary and may not always be fully assessed because they are not easy to identify. However, we know that nonpoint sources for biosolid and manure may have harmful effects on drinking water supplies, and recreation [64]. Nonpoint sources are difficult to control, and they pose a great threat to the integrity of the water bodies [17,64].

Microbial source tracking (MST) methods have recently been used to help identify nonpoint sources, several microbial source tracking studies have pointed to the application of biosolids and manure. Lapen *et al.* [14], and Gottschall *et al.* [15] monitored tile- and ground- water quality for bacteria (*Escherichia coli,* enterococci, *Clostridium perfringens*) after land application of dewatered biosolids (using both spreading and surface injection application methods), and they found that there were no significant differences between using different methods of spreading biosolids, also they measured the concentration of the *E. coli* contamination in ground water.

In May 2000 runoff from a field treated with cattle manure contaminated a groundwater supply with pathogenic bacteria [65,66]. In addition to human pathogenic *E. coli* and *Campylobacter*, other pathogens can also be present and survive in livestock manure. For example, human pathogenic protozoa, *Giardia lamblia* and *Cryptosporidium parvum*, can also be excreted by infected livestock and be transported in runoff events [67,68,69]. Microbial quality of runoff following land application of manure has been reported as an important source of microbial contamination of water bodies [18]. Studies involving fecal bacterial contamination in streams near dairy farms and cattle pastures [70,71], surface runoff from grazed pastures [72,73], and subsurface runoff from manure applied fields [74] demonstrated the ability of rain water runoff to transport bacteria from manure to surface water supplies. Rain events can also flush manure borne bacteria through the soil profile contaminating shallow groundwater [75] and springs and wells within the hydrological catchments of pastures [76].

Land-applied manure has been shown to quickly enter subsurface drains by preferential flow through macro pores [77,78]. Macro pores are large, continuous openings in the soil formed by plant roots, soil fauna, cracks, fissures and other natural phenomena. Shipitalo and Gibbs [34] reported that earthworm holes within 0.5 m of subsurface drains expedited the transmission of injected liquid to these drains. Soils under no-till crop management often have more continuous flow channels (macro pores) than tilled soils [79,80,81], and this may contribute to the rapid movement of injected manure to the subsurface drains.Bacterial contamination of drainage effluent was reported to be most likely in: (a) artificially drained, wet soils; (b) soils receiving high rates of liquid manure; and (c) soils that demonstrated preferential flow [82]. Several studies [77,83,84,85,86] reported that the application of liquid manure to drained fields resulted in elevated levels of nutrients and bacteria in the receiving waters compared to sites where liquid manure was not applied.

MST study on recreational water contamination in southeastern Lake Huron demonstrated that the dominant source of *E. coli* in lake water samples was agriculture, which supplied about 60% of the bacteria to the lake, whereas human sources provided only about three percent [87]. In addition, one study carried out on the South Shore Beach in Milwaukee concluded that the high *E. coli* levels were from local sources of pollution and were rarely affected by regional contamination events such as sewage overflows [88]. Another study by Verhougstraete and Rose [89], demonstrated that two sites in Lake Michigan have been impacted by bovine pollution, (67%) samples were positive for the bacteroides cow marker. The bacteroides results indicate the major source of fecal contamination at both sites was cow manure. Bacteria containing antibiotic resistance gene can also be released to the environment [90]. Although the levels of antibiotics in the environment are usually far below threshold levels to have inhibitory effects on bacterial populations, they still exert selective pressure on the development of antibiotic resistance in bacteria. The presence of antibiotic resistance genes in various environmental settings including river sediments, irrigation ditch water, dairy lagoon water, drinking water treatment plants and wastewater recycling plants has been confirmed [91,92]. Antibiotics and antibiotic-resistant bacteria can be released into streams and groundwater near concentrated animal feeding operation facilities [93,94,95,96]. In a study by Chee-Sanford *et al.* [94]. The presence of tetracycline resistance genes has been confirmed in waste lagoons on two swine farms, and in the groundwater underlying the two farms. These antibiotic-resistant genes may be transferred horizontally between bacterial species, even to animals and humans [97,98].

In addition, Munir and Xagoraraki [99] reported that 24 manure samples from three farms and 18 biosolids samples from seven different wastewater treatment plants across Michigan were analyzed for tetracycline and sulfonamide resistance genes (Tet-W, Tet-O, and Sul-I). They reported high concentrations of antibiotic resistance genes in manure and biosolids samples. The concentrations of antibiotic resistance genes in manure was significantly greater than in biosolids and the background soil samples had significantly less contaminations than the biosolids and manure.

RISK ASSESSMENT

Land application of biosolids and manure may cause human risks of infection for residents and occupational workers. Manure and biosolids, when applied to the land, may contribute to pathogens in surface water, air, soil and groundwater depending on extent of precipitation, aerosolization, and fate and transport of pathogens on surface soil, in subsurface soil media and in air (Figure 1).

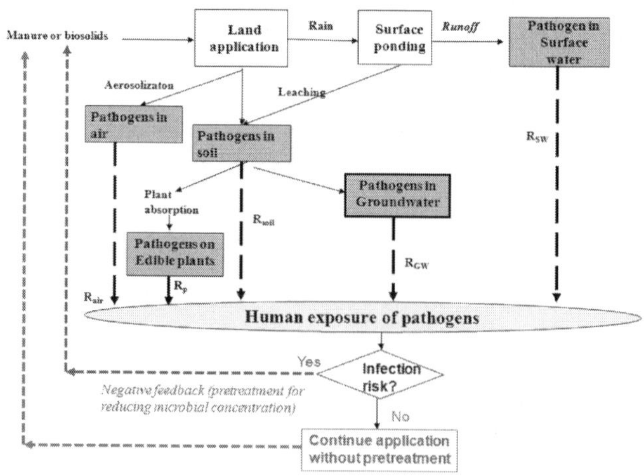

Figure 1: A schematic linking biosolids and manure application to human exposure.

Human exposure to pathogens might occur from one or more of these exposure routes. Once infected, chances of illness and mortality depend on pathogen type and human susceptibility. The risk of human exposure to biosolids-associated pathogens has been assessed in several studies (Table 3). It is important to mention here that the reviewed studies are examples and do not represent a complete list of all published studies.**Table 3:** Example studies focusing on risk assessment during land application activities of biosolids and/or manure.

Ref.	Exposure Scenario	Microorganisms Type	Exposed Population	Findings/Risk Estimate
		Biosolids		
[100]	Ingestion of biosolids-contaminated soil	Rotavirus	Residential population(480 mg/d soil ingestion rate)	Risk of infection: 7.8×10^{-4} (when mixed with soil); 2.11×10^{-1} (without any mixing with soil)
[101]	Inhalation of indicator organisms from air during land application activity	Coliphage MS-2, *E. coli*, coxsackievirus A21	Occupational workers (for 1 to 8 h of exposure; 0.1 virus particles/g biosolids)	Risk of infection at 2 m: 1.64×10^{-7} to 1.31×10^{-6}
[101]	Inhalation of indicator organisms from air during land application activity	Coliphage MS-2, *E. coli* , coxsackievirus A21	Residential population (for 1 to 8 h of exposure; 0.1 virus particles/g biosolids)	Risk of infection at 30.5 m: 1.5×10^{-8} to 1.2×10^{-7}
[102]	Inhalation of indicator organisms from air during land application of class B biosolids	Coliform bacteria, coliphages	Occupational workers	annual risk of infection: 7.8×10^{-4} to 2.1×10^{-2}
[103]	Direct contact of Class B biosolids and subsequent ingestion (without incorporation in soil)	*Ssalmonella*	Residential (soil ingestion = 480 mg/d)	Risk of infection = 5.7×10^{-3}
[103]	Direct contact of Class B biosolids and subsequent ingestion (with 100 times dilution with soil)	*Salmonella*	Residential (soil ingestion = 480 mg/d)	Risk of infection = $5.5\ 10^{-5}$
[103]	Direct contact with class A residuals following regrowth of *Salmonella*	*Salmonella*	Residential(soil ingestion = 480 mg/d)	Risk of salmonella infection = 6×10^{-1}
[103]	Direct contact with class A residuals following regrowth of *Salmonella* (with 100 times dilution with soil)	*Salmonella*	Residential (soil ingestion = 480 mg/d)	Risk of salmonella infection = 2.64×10^{-1}
[104]	Direct contact of Class B biosolids	echovirus-12, enterovirus types 68–71, adenoviruses, rotaviruses, and noroviruses genotype-I	Residential population	Risk of infection: 4.45×10^{-5}
		Manure		
[105]	Exposures from fomite, soil, crop, and aerosol exposures from manure and biosolids	Bacteria and viruses	Occupational and residential population	Greatest risk from direct consumption of contaminated soil ; Greater bacterial risks from manure and greater viral risks from biosolids
[106]	Exposure of dairy wastewater	*Campylobacter jejuni*, *E.coli*O157:H7, non-O157 *E.coli*, *Listeria monocytogenes*, and *Salmonella* spp	Residential	daytime risk is less than 10^{-6} at distance > 1 km
[107]	Inhalation exposure during dairy manure application	*Enterococcus*spp., *E. coli*, *Salmonella* spp.,*Campylobacter* spp., *E. coli* O157:H7	Occupational (8h)	Median risk of infection: 1:500 (at 100 m), 1:100,000 (at 1,000 m)

Risk of infection to residents and occupational workers during land application of biosolids depends on pathogen type, pathogen concentration in biosolids, pathogen concentration in air after aerosolization, pathogen concentration in soil, biosolids application methods and location of receptor from biosolids application activities [18,100,101,102,108]. For example, Gerba *et al.* [100] estimated risk of infection from enteric viruses during mixing of biosolids with soil for an assumed hypothetical exposure scenario. They reported that risk of daily infection from rotaviruses present in biosolids ranged from 7.8×10^{-4} (when mixed with soil) to 2.11×10^{-1} (without any mixing with soil) (enteric virus concentration in biosolids = 5.13 MPN/4g biosolids; soil ingestion rate = 480 mg/d). In addition, some studies conducted sampling of microbial indicators and pathogens to determine concentration of pathogens in air and water media for estimating risk of infection for different hypothetical exposure scenarios [18,101,102,108]. They observed microbial concentration with distances and sampling heights for different weather conditions (wind velocity and direction, relative humidity and temperature) to develop empirical fate and transport models.

These studies have generally focused on estimating risk of microbial infection from one medium only [14,100,101,102,108]. However, different environmental compartments (*i.e.* air, surface water, soil, groundwater, vegetables) may be contaminated during land application of biosolids depending of methods of biosolids application (*i.e.*, surface application, injection method). It is important to consider risk of infection from different environmental compartments to estimate the overall risk of infection during land application of biosolids.

To incorporate risk assessment and fate of biosolids-associated pathogens in different environmental media, Eisenberg [109] and Galada *et al.* [110] developed fate and transport models of pathogens after biosolids application using different application methods and predicted risks of infection for different exposure subpopulations. For example, the model "Spreadsheet Microbial Assessment of Risk: Tool for Biosolids "SMART Biosolids" included estimation of risk of infection due to more than 20 pathogens from surface water, air, groundwater, vegetables and soil during land application of biosolids [109]. In general, the following information is needed to run these comprehensive multi-compartment models: (a) weather-related information; (b) pathogenic concentration in biosolids; (c) biosolids application method; (d) decay of pathogens in different environmental media; (e) exposure- related information; and (f) dose-response information of different pathogens.

During land application of manure, some studies have conducted assessment of risk of pathogenic infection. Findings presented in Table 3. For example, Brooks *et al.*, 2012 [105] compared risks of microbial infection during land application of manure and biosolids and found that risks of bacterial infection were higher during manure application activities and risks of viral infection were higher during biosolids land application activities. They further reported that risks of infection were higher due to land application of biosolids than land application of manure due to high infectivity of viruses. They also mentioned that risk estimates involved uncertainties due to lack of availability of concentration levels, field-specific inactivation rates of pathogens, and pathogens decay and regrowth rates. In another recent study by Jahne *et al.*, 2014 [107] on risk assessment of bioaerosols from a manure application site indicated that peak risks (95th percentile values) were found to be very high (1:250) at 100 m distance from the source. During application of dairy wastewater, Dungan, 2014 [106] estimated inhalation risks of residents due to exposures of pathogens *(Campylobacter jejuni, Escherichia coli* O157:H7, non-O157 *E.coli, Listeria monocytogenes,* and *Salmonella spp.)* in bioaerosols and found that risks were higher near the source and infection, and depended on weather conditions and setback distance. They reported the need for data on: (a) aerosolization efficiency of pathogens during spray irrigation of wastewaters; (b) inactivation and deposition rates of airborne pathogens under various environmental conditions; (c) inhalation transmission and dose–response of enteric pathogens in humans; and (d) exposure frequency and duration of affected populations.

A comparison of findings of risk assessment studies on pathogenic exposure from biosolids and manure indicated that risks of infection varied with setback distance, weather conditions (wind speed and direction, relative humidity, and temperature). Pathogen levels and extent of pathogen release from biosolids and manure during aerosolization and leaching due to rain events [68,111,112,113,114] was found to differ. Pathogen-related information needs to be obtained from fields and for conditions in which biosolids and/or manure are applied. Subsequently, pathogen-related fate and transport and risk models can be used.

BEST MANAGEMENT PRACTICES FOR BIOSOLIDS TO PROTECT WATER QUALITY

In 1993, the US EPA [115] established standards for land-applied biosolids under 40 CFR Part 503. Part 503 describes quantitative standards, management practices, operational standards, monitoring, recordkeeping, and reporting requirements for the use and disposal of biosolids. Individual states may develop additional guidelines so long as the state guidelines meet the minimum requirements of the federal Part 503 rule. Biosolids standards include limitations for metal and other compounds, pathogen reduction, vector requirements, best management practices, and include limitations for the land application of biosolids. Biosolids are physically, chemically and biologically treated to reduce pathogens to levels specified for Class A and B designations.

Class B biosolids (restricted use) are treated but still contain detectible levels of pathogens. Sludge to produce class B biosolids is treated using a "Processes to Significantly Reduce Pathogens" (PSRP), such as aerobic digestion, anaerobic digestion, air drying, and lime stabilization, which reduce but do not eliminate pathogens. Mesophilic anaerobic digestion (MAD) is the most prevalent treatment process for Class B biosolids in the US with a mean reduction in pathogen or indicator cultivability of 1 log [116]. Class B biosolids must meet one of the three alternatives requirements: (1) Monitoring of indicator organisms based on fecal coliform; the geometric mean of the density of fecal coliform must be less than 2 million Colony Forming Units (CFU) or Most Probable Number (MPN) per gram of total dry solid in seven grab samples of sludge within one day, at the time of biosolids use or disposal; (2) Use of PSRP to significantly reduce pathogens; (3) Use of processes equivalent to PSRP, as determined by the permitting authority. Regulations for land application of Class B biosolids limit human exposure to pathogens by delaying harvesting post application, and prohibit the use of Class B biosolids on food crops eaten directly without processing. Class B biosolids can be applied on grain and forage crops, pastures, and grassland.

Class A biosolids (unrestricted use) contain no detectible levels of pathogens and low levels metals. Class A biosolids are treated by one of several "Processes to Further Reduce Pathogens" (PFRP), such as composting, pasteurization, drying or heat treatment, or advanced alkaline treatment, which reduce pathogens to below detectable levels. Class A classification is required for application onto public-use sites and certain food crops. This includes residential areas or home gardens, road banks, parks, golf courses, schools and other similar areas. Class A biosolids must meet one of the following bacteria limitations; fecal coliform is less than 1000 MPN per gram of total dry solid or *Salmonella sp.* bacteria density is less than 3 MPN per 4 grams of total dry solid. Class A biosolids pose minimal risk associated with use on edible food crops as a result of prior treatment that eliminates pathogens.

The Part 503 rule specifies upper limits for nine trace elements (Arsenic, Cadmium, Copper, Lead, Mercury, Molybdenum, Nickel, Selenium and Zinc) in sewage sludge to be land applied. Four options exist for meeting pollutant limits: The ceiling concentration limits (CCL), exceptional quality (EQ), cumulative pollutant loading rate (CPLR), and annual pollutant loading rate (APLR). Subpart D of the 503 rule addresses alternatives for limiting vectors such as rodents, birds and other organisms that could potentially spread disease by transporting pathogens from the application site. The goal of vector attraction reduction efforts is to either reduce the attractiveness of the biosolids to vectors by reducing organic matter (e.g., digestion, alkaline addition) or prevent vectors from coming into contact with the biosolids (e.g., subsurface injection or tillage incorporation within a short period of time after land application).

When biosolids are applied on agricultural land, they are applied at an agronomic rate to meet nitrogen requirements of the crop (CFR 40, Part 503.11) and prevent nitrogen loss to the environment. Individual states are allowed to expand the federal Part 503 rules to address state specific needs and concerns. In Michigan, Part 24 Rules of MDEQ [117] expands the Part 503 definition to include phosphorus and potassium in total nutrient management plans for crops grown at biosolids land application sites [117]. Biosolids cannot be applied to agricultural land if the phosphorus Bray P1 soil test is greater than 150 ppm or 170 ppm based on the Mehlich 3 soil test.

Specified management practices for the land application of biosolids prohibit application of bulk sewage sludge if the site is likely to adversely affect a threatened of endanger species, cannot be applied to flooded, frozen or snow-covered ground or within 10 m from water bodies (CFR 40, Part 503.14). The state of Michigan expanded the general restrictions in Part 24 Rules to include a separation distance of 600 m from municipal wells, 30 m from domestic wells, homes and commercial buildings, and 15 m from surface waters. A minimum separation distance of 0.75 m between the soil surface and groundwater is required when biosolids are applied. There are many regional variations in required set-back distances and other management practices to protect the environment and public health [118,119,120,121,122].

Because biosolids are frequently used as a source of crop nutrients there are restrictions on the timing of biosolids application and delay time between land application and crop harvest. Michigan Part 24 Rules prohibits the harvest of food crops for periods ranging from 14 to 38 months depending upon the crop following application and the method of application. A landowner cannot harvest food crops, feed crops, and fiber crops for 30 days after biosolids are applied. Livestock are not permitted to graze on land for 30 days after a biosolids application.

In conclusion, current biosolid treatment technologies that have been required by Part 503 Rule, with existing regulations and guidelines governing the use of biosolid as a soil amendment, are sufficient to protect human health and the environment [20]. To date there is no documented scientific evidence that the Part 503 Rule has failed to protect public health.

BEST MANAGEMENT PRACTICES FOR LIVESTOCK MANURE TO PROTECT WATER QUALITY

Runoff from the farmstead, pastures and fields where manure has been stored, deposited or applied can transport pathogens, sediment, organic solids, and nutrients to surface waters. Tillage and manure management practices that quickly move manure into the soil and root zone will protect water quality by recycling valuable crop nutrients and stabilizing potential contaminants by filtration, and sorption.

Best management practices (BMPs) are proven practical and affordable soil and water conservation and management approaches. BMPs vary from region-to-region and field-to-field because of variability in site-specific conditions and for flexibility to address local concerns. BMPs are determined through a collaborative effort of local stakeholders including farmers, extension educators, agribusiness professionals, regulatory agencies and technical service providers, and include cover crops, conservation tillage, buffer strips and set-backs along with soil and manure testing to implement management practices that prevent the loss of nutrients and biological contaminants to the environment.

All states have Right-to-Farm laws. For example, in Michigan, the Farm Act (P.A. 1981, No. 93) was adopted and amended in 1987 (P.A. 1987, No. 240) to protect the environment and to protect crop and livestock producers from nuisance suits if they are in compliance with Federal, state and local laws, and follow recommended manure management practices [123]. P.A. 1981, No. 93 authorized the development of Generally Accepted Agricultural and Management Practices (GAAMPs) for farms and farm operations. GAAMPs for Manure Management and Utilization are scientifically based, updated annually, and establish base-level management options for runoff control and wastewater management, odor management, construction design and management for manure storage and treatment facilities, and manure land application [123].

When manure is applied to cropland it is important to identify the fields and high-risk areas in fields where runoff is likely to occur. High risk areas include sloping ground, fields adjacent to ditches and waterways, bottom land that tends to flood in wet weather, tile drained land, and drained land with surface inlets. Vegetative filters, buffer strips and grass waterways have long been used to separate cropped or manure-applied land from nearby waterways. Widely used soil conservation practices that stabilize soil and prevent overland flow, runoff and soil erosion will help prevent manure contaminants from reaching waterways [124].

Conservation tillage leaves crop residue on the soil surface and is an effective way to reduce runoff and erosion. Low-disturbance aeration tillage is compatible with no-till cropping and creates an absorptive surface in untilled ground that inhibits overland flow by fracturing the soil, increasing surface roughness, improving infiltration, and conserving crop residues [125]. Tillage is generally helpful in disrupting preferential flow paths. Tillage disrupts macropores, delays manure movement, and can greatly decrease bacteria concentration in effluent [126]. Problems are likely when high rates of manure are applied and when manure is applied on wet ground or when tiles lines are flowing. Efforts to minimize problems must include low application rates [126,127].

Soil conservation practices that stabilize soil will help keep manure in the proper place. High risk soils typically are fine-textured soils with night crawler burrows. Sandy loams are lower risk.

Overland flow of manure contaminants can contribute to localized ponding and preferential flow to subsurface drains. Cover crops protect the soil from wind and water erosion, recycle nutrients, and improve soil structure and fertility. Cover crops create an effective barrier to overland flow and manure contamination of waterways [125]. When manure is applied to a bare soil surface, nutrients and other contaminants accumulate at the surface and increase the chance of nutrient and bacterial transport in runoff water. When manure is applied to a vegetative surface, the plant biomass and organic matter filters bacteria and nutrients.

Cover crops inhibit overland flow, filter sediment and organic materials, assimilate nutrients, and extract water from the soil thereby increasing water holding capacity. Cover crops can influence the persistence of indicator organisms and pathogens in the root zone. Work by Gagliardi and Karns [128] has shown greater persistence of *E. coli* O157:H7 on rye roots (47–96 d) and alfalfa roots (92 d) than in bare soil (25–41 d), but its persistence on crimson clover and hairy vetch roots were similar to bare soil. When manure slurry is applied to a vegetative surface the near-surface zone of high biomass and organic matter can enhance adsorption, straining and filtering of pathogens. Lim *et al.* [129] showed complete coliform removal of up to 2×10^7 colony forming units (cfu) 100 mL^{-1} in passing a 6.1 m tall fescue filter strip. Coyne *et al.* [130] reported 43%–74% removal of coliforms up to 10^8 cfu 100^{-1} mL in passing a 9 m mixed Kentucky bluegrass and tall fescue filter strip.

Technologies have been developed to treat manure nutrients and some of these technologies may also promote removal and degradation of pathogens such as physical treatments (separation of the solids and liquids in manure slurry by settling, filtration, screening, or drying), chemical treatments (addition of coagulants, such as lime, alum, and organic polymers to manure), and biological treatment (composting and anaerobic digestion).

CONCLUSIONS AND RECOMMENDATIONS

A comprehensive management plan is needed to address problems of microbial water pollution in rural systems in a sustainable manner. There is a need to integrate the values and needs as well as the efforts of all stakeholders within a risk analysis framework. A key cornerstone of sustainable rural development is effective management of water and waste. Cooperation of farmers, township officials and individual homeowners is critical to achieve an effective plan. Awareness and education of responsible parties is a first step to an integrated plan and requires education of farmers, township officials, individual homeowners and water quality professionals regarding the benefits and risks associated with the use of biosolids and manure to help them make informed decisions about their choices. Interdisciplinary collaborations of scientists, engineers, and other professionals are essential. Risk assessment frameworks for human exposure to contaminants in rural systems need to be evaluated. These comprehensive frameworks need to include identification of direct and indirect exposure pathways, pathogen data compilation, and risk assessment model development for land application of animal manure compared to biosolids. Quantitative data is a key component in the development of an effective risk assessment framework.

Pollution source tracking can help identify origins of pollution (manure or biosolids). Microbial source tracking tools can be coupled with geographical information system (GIS) data to help identify nonpoint sources in rural areas and minimize the human risks. The associations between bacterial communities and nutrient and chemical concentrations need to be assessed in order to determine if specific microbial community structure could be associated with specific types of chemical inputs and land uses.

Another important issue is the identification of transport mechanisms of contaminants from manure and biosolids to groundwater and surface water. At this point simplistic safety guidelines, such as setback distances, are used for locating wells and waste application sites. More elaborate investigation based on physicochemical processes such as sorption, speciation, and biotic and abiotic transformations of contaminants in soil and water are needed. To achieve that, advanced source tracking laboratory techniques coupled with nested field sampling and process modeling is required.

Existing techniques for managing agricultural waste, such as land application in various crop systems, need to be re-evaluated based on their potential to remove and inactivate contaminants. Over the past decades, waste management focused on the effects of nutrients, especially N and P, on water quality as the most important environmental concern. Microbes originating from biosolid and manure applications are often low on the priority for regulation and best management practices. In many cases, untreated manure may be applied in the setback areas where biosolids land application is prohibited; relatively few regulations govern land application of manure. Federal and state requirements regarding manure pathogens content are needed in order to preserve a water quality in rural areas.

The critical issues that need further research and attention include risk assessment models and determination of exposure pathways, optimization and use of quantitative analytical methods, source tracking and transport mechanisms, and optimization of waste and water treatment. Investigation of alternative techniques for treatment of animal waste, such as anaerobic membrane bioreactors, as well as investigation of the feasibility of co-treatment of human and agricultural waste is of great interest.

ACKNOWLEDGMENTS

The authors would like to thank Michigan State University, East Lansing (U.S.A.) and Indian Institute of Technology, Delhi (India).

CONFLICTS OF INTEREST

The authors declare no conflict of interest.

REFERENCES

1. Nicholson, F.; Groves, S.; Chambers, B. Pathogen survival during livestock manure storage and following land application. In *ADAS Gleadthorpe Research Center, Meden Vale, Mansfield, Nottinghamshire NG20 9PF, UK*; National Academy Press: Washington, DC, USA, 2004. [Google Scholar]
2. Meslin, F.X. Global Aspects of Emerging and Potential Zoonoses: A WHO Perspective. In Proceedings of the 1st International Conference on Emerging Zoonoses, Jerusalem, Israel, 24–28 November 1996.
3. Slifko, T.R.; Smith, H.V.; Rose, J.B. Emerging parasite zoonoses associated with water and food. *Int. J. Parasitol.* 2000,*30*, 1379–1393. [Google Scholar] [CrossRef] [PubMed]
4. Sobsey, M.D.; Khatib, L.A.; Hill, V.R.; Alocilja, E.; Pillai, S. Pathogens in Animal Wastes and the Impacts of Waste Management Practices on Their Survival, Transport and Fate. In *White Paper, Midwest Plan Service*; Iowa State University: Ames, IA, USA, 2001. [Google Scholar]
5. Hubálek, Z. Emerging human infectious diseases: Anthroponoses, zoonoses, and sapronoses. *Emerg. Infect. Dis.* 2003,*9*, 403–404. [Google Scholar] [CrossRef] [PubMed]
6. Gannon, V.P. J.; Bolin, C.; Moe, C.L. Waterborne Zoonoses: Emerging Pathogens and Emerging Patterns of Infection. In *Waterborne Zoonoses: Identification, Causes and Control*; Cotruvo, J.A., Dufour, A., Rees, G., Bartram, J., Carr, R., Cliver, D.O., Craun, G.F., Fayer, R., Gannon, V.P.J., Eds.; WHO: London, UK, 2004. [Google Scholar]
7. Cliver, D.O.; Moe, C.L. Prospects of Waterborne Viral Zoonoses. In *Waterborne Zoonoses: Identification, Causes and Control*; Cotruvo, J.A., Dufour, A., Rees, G., Bartram, J., Carr, R., Cliver, D.O., Craun, G.F., Fayer, R., Gannon, V.P.J., Eds.; WHO: London, UK, 2004. [Google Scholar]
8. Palmer, S.; Brown, D.; Morgan, D. Early Qualitative Risk Assessment of the Emerging Zoonotic Potential of Animal Diseases. *Br. Med. J.* 2005, *331*, 1256–1260. [Google Scholar] [CrossRef]
9. Pourcher, A.M.; Morand, P.; Picard-Bonnaud, F.; Billaudel, S.; Monpoeho, S.; Federighi, M.; Moguedet, G. Decrease of enteric micro-organisms from rural sewage sludge during their composting in straw mixture. *J. Appl. Microbiol.* 2005,*99*, 528–539. [Google Scholar] [CrossRef] [PubMed]
10. Bofill-Mas, S.; Albinana-Gimenez, N.; Clemente-Casares, P.; Hundesa, A.; Rodriguez-Manzano, J.; Allard, A.; Calvo, M.; Girones, R. Quantitation and stabilityof human adenoviruses and polyomavirusJCPyV in wastewater matrices. *Appl. Environ. Microbiol.* 2006, *72*, 7894–7896. [Google Scholar] [CrossRef] [PubMed]
11. Pourcher, A.M.; Françoise, P.B.; Virginie, F.; Agnieszka, G.; Vasilica, S.; Gérard, M. Survival of fecal indicators and enteroviruses in soil after land-spreading of municipal sewage sludge. *Appl. Soil Ecol.* 2007, *35*, 473–479. [Google Scholar]

[CrossRef]

12. Monpoeho, S.; Maul, A.; Bonnin, C.; Patria, L.; Ranarijaona, S.; Billaudel, S.; Ferré, V. Clearance of Human-Pathogenic Viruses from Sludge: Study of Four Stabilization Processes by Real-Time Reverse Transcription-PCR and Cell Culture.*Appl. Environ. Microb.* 2004, *70*, 5434–5440. [Google Scholar] [CrossRef]

13. Monpoeho, S.; Maula, A.; Mignotte-Cadiergues, B.; Schwartzbrod, L.; Billaudel, S.; Ferre, V. Best Viral Elution Method Available for Quantification of Enteroviruses in Sludge by Both Cell Culture and Reverse Transcription-PCR. *Appl. Environ. Microbiol.* 2001, *67*, 2484–2488. [Google Scholar] [CrossRef] [PubMed]

14. Gobernaa, M.; Podmirsega, S.; Waldhubera, S.; Knappa, B.; Garcíab, C.; Insama, H. Pathogenic bacteria and mineral N in soils following the land spreading of biogas digestates and fresh manure. *Appl. Soil Ecol.* 2011, *49*, 18–25. [Google Scholar] [CrossRef]

15. Kudva, I.T.; Blanch, K.; Hovde, C.J. Analysis of Escherichia coli O157: H7 survival in ovine or bovine manure and manure slurry. *Appl. Environ. Microbiol.* 1998, *64*, 3166–3174. [Google Scholar] [PubMed]

16. Sinton, L.W.; Braithwaite, R.R.; Hall, C.H.; Mackenzie, M.L. Survival of indicator and pathogenic bacteria in bovine feces on pasture. *Appl. Environ. Microbiol.* 2007, *73*, 7917–7925. [Google Scholar] [CrossRef] [PubMed]

17. King, G.; Brooks, J.P.; Brown, S.; Gerba, C.; O'Connor, G.A.; Pepper, I.L. *Land Application of Organic Residuals: Public Health Threat or Environmental Benefit*; American Society for Microbiology: Washington, DC, USA, 2011. [Google Scholar]

18. Daniel, T.C.; Sharpley, A.N.; Lemunyon, J.L. Agricultural phosphorus and eutrophication: A symposium overview. *J. Environ. Qual.* 1998, *27*, 251–257. [Google Scholar] [CrossRef]

19. Scott, T.M.; Rose, J.B.; Jenkins, T.M.; Farrah, S.R.; Lukasik, J. Microbial source tracking: Current methodology and future directions. *Appl. Environ. Microbiol.* 2002, *68*, 5796–5803. [Google Scholar] [CrossRef] [PubMed]

20. Easterling, D.R.; Evans, J.L.; Groisman, P.Y.; Karl, T.R.; Kunkel, K.E.; Ambenje, P. Observed variability and trends in extreme climate events: A brief review. *Bull. Am. Met. Soc.* 2000, *81*, 417–425. [Google Scholar] [CrossRef]

21. Easterling, D.R.; Thomas, K.R.; Gallo, K.P. Observed climate variability and change of relevance to the biosphere. *J. Geophys. Res.* 2000, *105*, 20101–20114. [Google Scholar] [CrossRef]

22. Atherholt, T.B.; LeChevallier, M.W.; Norton, W.D.; Rosen, J.S. Effect of rainfall on Giardia and Crypto. *J. Am. Water Works Assoc.* 1998, *90*, 66–80. [Google Scholar]

23. Tate, K.W.; Atwill, E.R.; George, M.R.; McDougald, N.K.; Larsen, R.E. Crytposporidium parvum transport from cattle fecal deposits on California rangelands. *J. Range Manag.* 2000, *53*, 295–299. [Google Scholar] [CrossRef]

24. Kistemann, T.; Claben, T.; Koch, C.; Dangendorf, F.; Fischeder, R.; Gebel, J.; Vacata, V.; Exner, M. Microbial load of drinking water reservoir tributaries during extreme rainfall and runoff. *Appl. Environ. Microbiol.* 2002, *68*, 2188–2197. [Google Scholar] [CrossRef] [PubMed]
25. Gary, H.; Johnson, S.; Ponce, S. Cattle grazing impact on surface water quality in a Colorado front range stream. *J. Soil Water Conserv.* 1983, *38*, 124–128. [Google Scholar]
26. Niemi, R.M.; Niemi, J.S. Bacterial pollution of waters in pristine and agricultural lands. *J. Environ. Qual.* 1991, *20*, 620–627. [Google Scholar] [CrossRef]
27. Doran, J.W.; Linn, D.M. Bacteriological quality of runoff water from pastureland. *Appl. Environ. Microbiol.* 1979, *37*, 985–991. [Google Scholar] [PubMed]
28. Jawson, M.D.; Elliott, L.F.; Saxton, K.E.; Fortier, D.H. The effect of cattle grazing on indicator bacteria in runoff from a Pacific Northwest watershed. *J. Environ. Qual.* 1982, *11*, 621–627. [Google Scholar] [CrossRef]
29. Culley, J.L.B.; Phillips, P.A. Bacteriological quality of surface and subsurface runoff from manuredsandy clay loamsoil. *J. Environ. Qual.* 1982, *11*, 155–158. [Google Scholar] [CrossRef]
30. McMurry, S.W.; Coyne, M.S.; Perfect, E. Fecal coliform transport through intact soil blocks amended with poultry manure. *J. Environ. Qual.* 1998, *27*, 86–92. [Google Scholar] [CrossRef]
31. Howell, J.M.; Coyne, M.S.; Cornelius, P.L. Effect of sediment particle size and temperature on fecal bacteria mortality rates and the fecal coliform/fecal streptococci ratio. *J. Environ. Qual.* 1996, *25*, 1216–1220. [Google Scholar] [CrossRef]
32. Fleming, R.J.; Bradshaw, S.H. *Contamination of Subsurface Drainage Systems during Manure Spreading*; ASAE Paper No. 92-2618; American Society of Agricultural Engineers: St. Joseph, MI, USA, 1992. [Google Scholar]
33. Shipitalo, M.J.; Gibbs, F. Potential of Earthworm Burrows to Transmit Injected Animal Waste to Tile Drains. *Soil Sci. Soc. Am. J.* 2000, *64*, 2103–2109. [Google Scholar] [CrossRef]
34. Shipitalo, M.J.; Protz, R. Comparison of morphology and porosity of a soil under conventional and zero tillage. *Can. J. Soil Sci.* 1987, *67*, 445–456. [Google Scholar] [CrossRef]
35. Drees, L.R.; Karathanasis, A.D.; Wilding, L.P.; Blevins, R.L. Micromorphological characteristics of long-term no-till and conventionally tilled soils. *Soil Sci. Soc. Am. J.* 1994, *58*, 508–517. [Google Scholar] [CrossRef]
36. Pagliai, M.; Raglione, M.; Panini, T.; Maletta, M.; La Marca, M. The structure of two alluvial soils in italy after 10 years of conventional and minimum tillage. *Soil Tillage Res.* 1995, *34*, 209–223. [Google Scholar] [CrossRef]
37. Geohring, L.D.; Van Es, H.M. Soil Hydrology and Liquid Manure Applications. In. In Proceedings of the Liquid Manure Application Systems Conference,

Rochester, NY, USA, 1994; pp. 166–174.

38. Evans, M.R.; Owens, J.D. Factors affecting the concentration of fecal bacteria in land drainage water. *J. Gen. Micro.*1972, *71*, 477–485. [Google Scholar] [CrossRef]

39. Dean, D.M.; Foran, M.E. The effects of farm liquid waste applications on tile drainage. *J. Soil Water Conser.* 1992, *47*, 368–369. [Google Scholar]

40. McLellan, J.E.; Fleming, R.J.; Bradshaw, S.H. *Reducing Manure Output To Streams from Subsurface Drainage Systems*; ASAE Paper No. 93–2010; American Society of Agricultural Engineers: St. Joseph, MI, USA, 1993. [Google Scholar]

41. Jamieson, R.C.; Gordon, R.J.; Sharples, K.E.; Stratton, G.W.; Madani, A. Movement and Persistence of Fecal Bacteria in Agricultural Soils and Subsurface Drainage Systems: A Review. *Can. Biosyst. Engin.* 2002, *44*, 1–9. [Google Scholar]

42. Kon, T.; Weir, S.C.; Howell, E.T.; Lee, H.; Trevors, J.T. Repetitive element (REP)-polymerase chain reaction (PCR) analysis of Escherichia coli isolates from recreational waters of southeastern Lake Huron. *Can. J. Microbial.* 2009, *55*, 269–276. [Google Scholar] [CrossRef]

43. McLellan, S.L.; Salmore, A.K. Evidence for localized bacterial loading as the cause of chronic beach closings in a freshwater marina. *Water Res.* 2003, *37*, 2700–2708. [Google Scholar] [CrossRef] [PubMed]

44. Rose, J.B.; Verhougstrate, M. Investigation of Water Quality and Sources Associated with Buck Creek Watershed. Michigan State University, East Lansing, MI, USA, 18 September 2008. [Google Scholar]

45. Sengelov, G.; Agerso, Y.; Halling-Sorensen, B.; Baloda, S.B.; Andersen, J.S.; Jensen, L.B. Bacterial antibiotic resistance levels in Danish farmland as a result of treatment with pig manure slurry. *Environ. Intern.* 2003, *28*, 587–595. [Google Scholar] [CrossRef]

46. Pei, R.; Kim, S.-C.; Carlson, K.; Pruden, A. Effect of River Landscape of the sediment concentrations of antibiotics and corresponding antibiotic resistance genes (ARG). *Water Res.* 2006, *40*, 2427–2435. [Google Scholar] [CrossRef] [PubMed]

47. Pruden, A.; Pei, R.; Storteboom, H.; Carlson, K. Antibiotic Resistance Genes as Emerging Contaminants: Studies in the Northern Colorado. *Environ. Sci. Tech.* 2006, *40*, 7445–7450. [Google Scholar] [CrossRef]

48. Batt, A.L.; Bruce, I.B.; Aga, D.S. Evaluating the vulnerability of surface waters to antibiotic contamination from varying wastewater treatment plant discharges. *Environ. Pollut.* 2006, *142*, 295–302. [Google Scholar] [CrossRef] [PubMed]

49. Chee-Sanford, J.C.; Aminov, R.I.; Krapac, I.J.; Garrigues-Jeanjean, N.; Mackie, R.I. Occurrence and Diversity of Tetracycline Resistance Genes in Lagoons and Groundwater Underlying Two Swine Production Facilities. *Appl. Environ. Microbiol.* 2001, *67*, 1494–1502. [Google Scholar] [CrossRef] [PubMed]

50. Sapkota, A.R.; Curriero, F.C.; Gibson, K.E.; Schwab, K.J. Antibiotic-resistant enterococci and fecal indicators in surface water and groundwater impacted by a concentration swine feeding operation. *Environ. Health Perspect.* 2007, *115*, 1040–1045. [Google Scholar] [CrossRef] [PubMed]

51. Song, W.; Huang, M.; Rumbeiha, W.; Li, H. Determination of amprolium, carbadox, monensin, and tylosin in surface water by liquid chromatography/tandem mass spectrometry. *Rapid Commun. Mass. Spectrom.* 2007, *21*, 1944–1950. [Google Scholar] [CrossRef] [PubMed]

52. Witte, W. Medical consequences of antibiotic use in agriculture. *Science* 1998, *279*, 996–997. [Google Scholar] [CrossRef] [PubMed]

53. Smith, K.E.; Besser, J.M.; Hedberg, C.W.; Leano, F.T.; Bender, J.B.; Wicklund, J.H.; Johnson, B.P.; Moore, K.A.; Osterholm, M.T. Quinolone-resistant campylobacter jejuni Infections in Minnesota, 1992–1998. *N. Engl. J. Med.* 1999,*340*, 1525–1532. [Google Scholar] [CrossRef] [PubMed]

54. Munir, M.; Xagoraraki, I. Levels of Antibiotic Resistance Genes in Manure, Biosolids, and Fertilized Soil. *J. Environ. Qual.* 2011, *40*, 248–255. [Google Scholar] [CrossRef] [PubMed]

55. Gerba, C.P.; Pepper, I.L.; Whitehead, L.F., III. A risk assessment of emerging pathogens of concern in the land application of biosolids. *Water Sci. Technol.* 2002, *46*, 225–230. [Google Scholar] [PubMed]

56. Brooks, J.P.; Tanner, B.D.; Gerba, C.P.; Haas, C.N.; Pepper, I.L. Estimation of bioaerosol risk of infection to residents adjacent to a land applied biosolids site using an empirically derived transport model. *J. Appl. Microbiol.* 2005, *98*, 397–405. [Google Scholar] [CrossRef]

57. Tanner, B.D.; Brooks, J.P.; Gerba, C.P.; Haas, C.N.; Josephson, K.L.; Pepper, I.L. Estimated occupational risk from bioaerosols generated during land application of class B biosolids. *J. Environ. Qual.* 2008, *37*, 2311–2321. [Google Scholar] [CrossRef]

58. Gerba, C.P.; Castro-del, C.; Brooks, J.O.; Pepper, I.L. Exposure and risk assessment of *salmonella* in recycled residuals.*Water Sci. Technol.* 2008, *57*, 1061–1065. [Google Scholar] [CrossRef] [PubMed]

59. Kumar, A.; Wong, K.; Xagoraraki, I. Effect of Detection Methods on Risk Estimates of Exposure to Biosolids-Associated Human Enteric Viruses. *Risk Anal.* 2012, *32*, 916–929. [Google Scholar] [CrossRef] [PubMed]

60. Brooks, J.P.; McLaughlin, M.R.; Gerba, C.P.; Pepper, I.L. Land application of manure and class B biosolids: An occupational and public quantitative microbial risk assessment. *J. Environ. Qual.* 2012, *41*, 2009–2023. [Google Scholar] [CrossRef] [PubMed]

61. Dungan, R.S. Estimation of Infectious Risks in Residential Populations Exposed to Airborne Pathogens During Center Pivot Irrigation of Dairy Wastewaters. *Environ. Sci. Technol.* 2014, *48*, 5033–5042. [Google Scholar] [CrossRef]

62. Jahne, M.A.; Rogers, S.W.; Holsen, T.M.; Grimberg, S.J. Quantitative microbial

risk assessment of bioaerosols from a manure application site. *Aerobiologia* 2014. [Google Scholar] [CrossRef]

63. Low, S.Y.; Paez-Rubio, T.; Baertsch, C.; Kucharski, M.; Peccia, J. Off-site Exposure to Respirable Aerosols Produced during the Disk-incorporation of Class B Biosolids. *J. Environ. Eng.* 2007, *133*, 987–994. [Google Scholar] [CrossRef]
64. Eisenberg, J.N. *Application of a Dynamic Model to Assess Microbial Health Risks Associated with Beneficial Uses of Biosolids*; Water Environment Research Foundation: Alexandria, VA, USA, 2006; pp. 1–124. [Google Scholar]
65. Galada, H.G.; Gurian, P.L.; Joe, A.; Kumar, A.; Olson, B.; Olson, M.; Richter, E.; Teng, J.; Zhang, H.; Xagoraraki, I.; *et al. Site Specific Risk Assessment Tool for Land Applied Biosolids*; Water Environment Research Foundation: Alexandria, VA, USA, 2012. [Google Scholar]
66. Atwill, E.R.; Hou, L.; Karle, B.M.; Harter, T.; Tate, K.W.; Dahlgren, R.A. Transport of *cryptosporidium parvum* oocysts through vegetated buffer strips and estimated filtration efficiency. *Appl. Environ. Microbiol.* 2002, *68*, 5517–5527. [Google Scholar] [CrossRef] [PubMed]
67. Bradford, S.A.; Tadassa, Y.F.; Pachepsky, Y. Transport of *Giardia* and manure suspensions in saturated porous media.*J. Environ. Qual.* 2006, *35*, 749–757. [Google Scholar] [CrossRef]
68. Muirhead, R.W.; Collins, R.P.; Bremer, P.J. Numbers and transported state of *Escherichia coli* in runoff direct from fresh cowpats under simulated rainfall. *J. Appl. Microbiol.* 2006, *42*, 83–87. [Google Scholar] [CrossRef]
69. Chetochine, A.S.; Brusseau, M.L.; Gerba, C.P.; Pepper, I.L. Leaching of phage from class B biosolids and potential transport through soil. *Appl. Environ. Microbiol.* 2006, *72*, 665–671. [Google Scholar] [CrossRef] [PubMed]
70. US EPA. *The Standards for the Use or Disposal of Sewage Sludge*; Final 40 CFR Part 503 Rules; EPA 822/Z-93/001; US EPA: Washington, DC, USA, 1993. [Google Scholar]
71. Bibby, K.; Viau, E.; Peccia, J. Viral metagenome analysis to guide human pathogen monitoring in environmental samples. *Lett. Appl. Microbiol.* 2011, *52*, 386–392. [Google Scholar] [CrossRef] [PubMed]
72. Michigan Department of Environmental Quality (MDEQ). *Administrative Rules, Part 24, Land Application of Biosolids*; Michigan Department of Environmental Quality, Surface Water Quality Division: Lansing, MI, USA, 1999. [Google Scholar]
73. Jacobs, L.W.; McCreary, D.S. *Applying Biosolids to Land in Michigan*; Extension Bulletin E-2780; Michigan State University: East Lansing, MI, USA, 2001. [Google Scholar]
74. Evanylo, G.K. *Agricultural Land Application of Biosolids in Virginia*; Publication 452–302; College of Agriculture and Life Sciences, Virginia Polytechnic Institute and State University: Blacksburg, VA, USA, 2009. [Google Scholar]
75. Eash, N.S.; McClurkan, J.; Burns, R.T. *Best Management Practices (BMP's) for*

Land Application of Biosolids; University of Tennessee: Knoxiville, TN, USA, 1997. [Google Scholar]

76. Barbarick, K.A.; Ippolito, J.A. Nutrient assessment of a dry land wheat agroecosystem after 12 years of biosolids applications. *Agron. J.* 2007, *99*. [Google Scholar] [CrossRef]
77. Arnold, K.; Dunn, J.; Carpenter, J.D. *Best Management Practices for Biosolids Land Application*; Extension Bulletin WQ426; University of Missouri Extension: Columbia, MO, USA, 1994. [Google Scholar]
78. Michigan Commission of Agriculture and Rural Development (MDARD). *Generally Accepted Agricultural and Management Practices for Manure Management and Utilization*; MDARD: Lansing, MI, USA, 2014. [Google Scholar]
79. Harrigan, T.M.; Northcott, W.; Rector, N.; Bolinger, D. *Keeping Land-Applied Manure in the Root Zone: Part 1: Sediment and Contaminant Runoff*; Extension Bulletin WO-1036; Michigan State University: East Lansing, MI, USA, 2007. [Google Scholar]
80. Harrigan, T.M.; Mutch, D.R.; Snapp, S.S. Slurry-Enriched Seeding of Biosuppressive Covers. *Appl. Eng. Agric.* 2006, *22*, 827–834. [Google Scholar] [CrossRef]
81. Harrigan, T.M. Manure on Tile-Drained Cropland. In *Michigan Dairy Review*; Michigan State University: East Lansing, MI, USA, 2005; Volume 10, pp. 10–12. [Google Scholar]
82. Harrigan, T.M.; Northcott, W.; Rector, N.; Bolinger, D. *Keeping Land-Applied Manure in the Root Zone: Part 2: Tile-drained Land*; Extension Bulletin WO-1037; Michigan State University: East Lansing, MI, USA, 2007. [Google Scholar]
83. Gagliardi, J.V.; Karns, J.S. Persistence of *Escherichia coli* O157:H7 in soil and on plant roots. *Environ. Microbiol.* 2002, *4*, 89–96. [Google Scholar] [CrossRef] [PubMed]
84. Lim, T.T.; Edwards, D.R.; Workman, S.R.; Larson, B.T.; Dunn, L. Vegetated filter strip removal of cattle manure constituents in runoff. *Trans. ASAE* 1998, *41*, 1375–1381. [Google Scholar] [CrossRef]
85. Craun, M.F.; Craun, G.F.; Calderon, R.L.; Beach, M.J. Waterborne Outbreaks in the United States. *J. Water Health* 2006,*4*, 19–30. [Google Scholar]
86. Craun, G.F. Waterborne disease outbreaks in the United States of America: Causes and prevention. *World Health Stat. Q.* 1992, *45*, 192–199. [Google Scholar]
87. Craun, G.F.; Calderon, R.L.; Nwachuku, N. Causes of waterborne outbreaks reported in the United States. In *Drinking Water and Infectious Disease: Establishing the Links*; Hunter, P.R., Waite, M., Ronchi, E., Eds.; CRC Press: London, UK, 2003; pp. 1991–1998. [Google Scholar]
88. Lee, S.H.; Levy, D.A.; Craun, G.F.; Beach, M.J.; Calderon, R.L. Surveillance for waterborne-disease outbreaks—United States, 1999–2000. *Morb. Mortal. Wkly. Rep. Surveill. Summ.* 2002, *51*, 1–47. [Google Scholar]

89. Liang, J.L.; Dziuban, E.J.; Craun, G.C.; Hill, V.; Moore, M.R.; Gelting, R.J.; Calderon, R.L.; Beach, M.J.; Roy, S.L. Surveillance for waterborne disease and outbreaks associated with drinking water and water not intended for drinking-US, 2003–2004. *MMWR Surveill. Summ.* 2006, *55*, 31–65. [Google Scholar] [PubMed]

90. US EPA. *Prepublication of the Ground Water Rule Federal Register Notice*; EPA-HQ-OW-2002-0061; FRL-RIN 2040-AA97; US EPA: Washington, DC, USA, 2006. [Google Scholar]

91. Hrudey, S.E.; Hrudey, E.J. *Safe Drinking Water: Lessons from Recent Outbreaks in Affluent Nations*; IWA-publishing: London, UK, 2004. [Google Scholar]

92. Curriero, F.C.; Patz, J.A.; Rose, J.B.; Lele, S. The association between extreme precipitation and waterborne disease outbreaks in the United States, 1948–1994. *Am. J. Pub. Health* 2001, *91*, 1194–1201. [Google Scholar]

93. Hrudey, S.E.; Payment, P.; Huck, P.M.; Gillham, R.W.; Hrudey, E.J. A fatal waterborne disease epidemic in Walkerton, Ontario: Comparison with other waterborne outbreaks in the developed world. *Water Sci. Technol.* 2003, *47*, 7–14. [Google Scholar] [PubMed]

94. National Research Council (NRC): Committee on Toxicants and Pathogens in Biosolids Applied to Land. *Biosolids Applied to Land: Advancing Standards and Practices*; The National Academies Press: Washington, DC, USA, 2002; pp. 1–12. [Google Scholar]

95. Hoxie, N.J.; Davis, J.P.; Vergeront, J.M.; Nashold, R.D.; Blair, K.A. Cryptosporidiosis—Associated mortality following a massive waterborne outbreak in Milwaukee, Wisconsin. *Am. J. Health* 1997, *87*, 2032–2035. [Google Scholar] [CrossRef]

96. Craun, G.F.; Calderon, R.L.; Craun, M.F. Waterborne outbreaks caused by zoonotic pathogens in the USA. In*Waterborne Zoonoses*; IWA Publishing: London, UK, 2004; pp. 120–135. [Google Scholar]

97. Eisenberg, J.N.S.; Moore, K.; Soller, J.A.; Eisenberg, D.; Colford, J.M., Jr. Microbial risk assessment framework for exposure to amended sludge projects. *Environ. Health Perspect.* 2008, *116*, 727–733. [Google Scholar] [CrossRef] [PubMed]

98. Lapen, D.R.; Topp, E.; Edwards, M.; Sabourin, L.; Curnoe, W.; Gottschall, N.; Bolton, P.; Rahman, S.; Coelho, B.B.; Payne, M.; *et al.* Effect of liquid municipal biosolids application method on tile and groundwater quality. *J. Environ. Qual.* 2008, *37*, 925–936. [Google Scholar] [CrossRef] [PubMed]

99. Gottschall, N.; Edwards, M.; Topp, E.; Bolton, P.; Payne, M.; Curnoe, W.E.; Coelho, B.B.; Lapen, D.R. Nitrogen, phosphorous, and bacteria tile and groundwater quality following direct injection of dewatered municipal biosolids into soil. *J. Environ. Qual.* 2009, *38*, 1066–1075. [Google Scholar] [CrossRef]

100. Venglovsky, J.; Sasakova, N.; Vargova, M.; Pacajova, Z.; Placha, I.; Petrovsky, M.; Harichova, D. Evolution of temperature and chemical parameters during composting of the pig slurry solid fraction amended with natural zeolite. *Bioresour.*

Tech. 2005, *96*, 181–189. [Google Scholar] [CrossRef]

101. Venglovsky, J.; Martinez, J.; Placha, I. Hygienic and ecological risks connected with utilization of animal manures and biosolids in agriculture. *Livest. Sci.* 2006, *102*, 197–203. [Google Scholar] [CrossRef]
102. Thurston-Enriquez, J.; Gilley, J.; Eghball, B. Microbial quality of runoff following land application of cattle manure and swine slurry. *J. Water Health* 2005, *3*, 157–171. [Google Scholar] [PubMed]
103. Brooks, J.P.; Maxwell, S.L.; Rensing, C.; Gerba, C.P.; Pepper, I.L. Occurrence of antibiotic-resistant bacteria and endotoxin associated with the land application of biosolids. *Can. J. Microbiol.* 2007, *53*, 616–622. [Google Scholar] [PubMed]
104. North East Biosolids and Residuals Association (NEBRA). *A National Biosolids Regulation, Quality, End Use, and Disposal Survey*; NEBRA: Tamworth, NH, USA, 2007. [Google Scholar]
105. U.S. Environmental Protection Agency. *Emerging Technologies for Biosolid Management*; EPA 832-R-06–005; U.S. Environmental Protection Agency: Washington, DC, USA, 2006. [Google Scholar]
106. Pepper, I.L.; Brooks, J.P.; Gerba, C.P. Pathogens in biosolids. *Adv. Agron.* 2006, *90*, 1–41. [Google Scholar]
107. Lyberatos, G.; Sklivaniotis, M.; Angelakis, A.N. Management of Biosolids in EUREAU Countries. *Fresenius Environ. Bull.* 2011, *20*, 2489–2495. [Google Scholar]
108. Singh, R.P.; Agrawal, M. Potential benefits and risks of land application of sewage sludge. *Waste Manag.* 2008, *28*, 347–358. [Google Scholar] [CrossRef] [PubMed]
109. Lu, Q.; He, Z.L.; Stoffella, P.J. Land application of biosolids in the USA: A review. *Appl. Environ. Soil Sci.* 2012. [Google Scholar] [CrossRef]
110. United Nations Industrial Development Organization (UNIDO); International Fertilizer Development Center (IFDC).*Fertilizer Manual*, 3rd ed.; Kluwer Academic Press: Norwell, MA, USA, 1998. [Google Scholar]
111. Stehouwer, R.C.; Wolf, A.M.; Doty, W.T. Chemical monitoring of sewage sludge in Pennsylvania: Variability and application uncertainty. *J. Environ. Qual.* 2000, *29*, 1686–1695. [Google Scholar] [CrossRef]
112. Shepherd, M.A.; Withers, P.J. Phosphorus leaching from liquid digested sewage sludge applied to sandy soils. *J. Agric. Sci.* 2001, *136*, 433–441. [Google Scholar]
113. He, Z.L.; Alva, A.K.; Calvert, D.V.; Li, Y.C.; Stoffella, P.J.; Banks, D.J. Nutrient availability and changes in microbial biomass of organic amendments during field incubation. *Compost. Sci. Util.* 2000, *8*, 293–302. [Google Scholar] [CrossRef]
114. Hue, N.V. Sewage sludge. In *Soil Amendments and Environmental Quality*; Lewis Publishers: Boca Raton, FL, USA, 1995; pp. 199–247. [Google Scholar]
115. Pepper, I.L.; Brooks, J.P.; Sinclair, R.G.; Gurian, P.L.; Gerba, C.P. Pathogens and indicators in United States Class B biosolids: National and historic distributions. *J. Environ. Qual.* 2010, *39*, 2185–2190. [Google Scholar] [CrossRef] [PubMed]

116. US EPA. *Cyanobacteria and Cyanotoxins: Information for Drinking Water Systems*; Fact Sheet EPA-810F11001; US EPA: Washington, DC, USA, 2012. [Google Scholar]

117. Viau, E.; Peccia, J. Survey of Wastewater Indicators and Human Pathogen Genomes in Biosolids Produced by Class A and Class B Stabilization Treatments. *Appl. Environ. Microbiol.* 2009, *75*, 164–174. [Google Scholar] [CrossRef] [PubMed]

118. Wong, K.; Onan, B.; Xagoraraki, I. Quantification of enteric viruses, indicators and salmonella in Class B anaerobic digested biosolids by culture and molecular methods. *Appl. Environ. Microb.* 2010, *76*, 6441–6448. [Google Scholar] [CrossRef]

119. Guzman, C.; Jofre, J.; Montemayor, M.; Lucena, F. Occurrence and levels of indicators and selected pathogens in different sludges and biosolids. *J. Appl. Microbiol.* 2007, *103*, 2420–2429. [Google Scholar] [CrossRef] [PubMed]

120. Animal Health Institute. 2002. Available online: http://www.ahi.org (accessed on 29 June 2014).

121. Burkholder, J.; Libra, B.; Weyer, P.; Heathcote, S.; Kolpin, D.; Thorne, P.S.; Wichman, M. Impacts of waste from concentrated animal feeding operations on water quality. *Environ. Health Perspect.* 2007, *115*, 308–312. [Google Scholar] [CrossRef] [PubMed]

122. Wilson, S.C. Hogwash! Why Industrial Animal Agriculture Is Not beyond the Scope of Clean Air Act Regulation. *Pace Environ. L. Rev.* 2007, *24*, 439. [Google Scholar]

123. US EPA. *Protocol for Developing Pathogen TMDLs*; EPA 841-R-00–002; United States Environmental Protection Agency: Washington, DC, USA, 2001. [Google Scholar]

124. Jongbloed, A.W.; Lenis, N.P. Environmental concerns about animal manure. *J. Anim. Sci.* 1998, *76*, 2641–2648. [Google Scholar] [PubMed]

125. Hanselman, T.A.; Graetz, D.A.; Wilkie, A.C. Manure-borne estrogens as potential contaminants: A review. *Environ. Sci. Technol.* 2003, *37*, 5471–5478. [Google Scholar] [CrossRef] [PubMed]

126. Johnson, A.C.; Williams, R.J.; Matthiessen, P. The total potential steroid hormone contribution of farm animals to freshwaters: The United Kingdom as a case study. *Sci. Total Environ.* 2006, *362*, 166–178. [Google Scholar] [CrossRef] [PubMed]

127. Gerba, C.P.; Smith, E.J. Sources of pathogenic microorganisms and their fate during land application of wastes. *J. Environ. Qual.* 2005, *34*, 42–48. [Google Scholar] [PubMed]

128. Guan, T.Y.; Holley, R.A. Pathogen survival in swine manure environments and transmission of human enteric illness: A review. *J. Environ. Qual.* 2003, *32*, 383–392. [Google Scholar] [CrossRef] [PubMed]

129. Hutchinson, M.L.; Walters, L.D.; Avery, S.M.; Munro, F.; Moore, A. Analyses of livestock production, waste storage, and pathogen levels and prevalences in

farm manures. *Appl. Environ. Microb.* 2005, *71*, 1231–1236. [Google Scholar] [CrossRef]

130. Coyne, M.S.; Gilfillen, R.A.; Rhodes, R.W.; Blevins, R.L. Soil and fecal coliform trapping by grass filter strips during simulated rain. *J. Soil Water Conserv.* 1995, *50*, 405–408. [Google Scholar]

16

CHAPTER

COMBINED SODIUM HYDROXIDE AND AMMONIUM HYDROXIDE PRETREATMENT OF POST-BIOGAS DIGESTION DAIRY MANURE FIBER FOR COST EFFECTIVE CELLULOSIC BIOETHANOL PRODUCTION

Sasikumar Elumalai, Aicardo Roa- Espinosa, John L Markley and Troy M Runge

Keywords: Biogas, Digestion, Manure fiber, Sodium hydroxide, Ammonium hydroxide, Alkaline pretreatment, Enzyme saccharification

BACKGROUND

Global consumption of non-renewable fossil fuels in the transportation sector has increased vigorously during the last three decades with simultaneous increment in price of fuels [1, 2]. For economic and environmental reasons, it is critical to find replacements for fossil fuels. Renewable, second-generation cellulosic biofuels offer the potential to improve energy security and reduce the deleterious environmental impact of first generation biofuels [3, 4, 5, 6]. However, challenges remain in converting lignocellulosic biomass into sustainable biofuels in a cost- and energy-effective manner at large-scale [7, 8, 9]. The most commonly investigated lignocellulosic feedstocks for potential ethanol production are agricultural (crop residues) and forestry wastes (mill residues). Both of these feedstocks are natural composites consisting of three main biopolymers; cellulose, hemicellulose, and lignin [10, 11, 12, 13, 14, 15]. Several studies [16, 17, 18, 19, 20] have demonstrated the potential of manure fibers (either pre- or post-biogas digestion) as a lignocellulose feedstock for the production of

Citation: Sasikumar Elumalai, Aicardo Roa- Espinosa, John L Markley and Troy M Runge. Combined sodium hydroxide and ammonium hydroxide pretreatment of post-biogas digestion dairy manure fiber for cost effective cellulosic bioethanol production. DOI: 10.1186/2043-7129-2-12.

biofuels and value-added chemicals. The composition of manure fibers depends on the animal feed and the conditions of anaerobic digestion if carried out. The dry fibers typically have a high content of both carbohydrate (40−43%) and lignin (20−25%). Manure fiber is plentiful. In average, dairy cattle produce about 12.0 gal of manure per 1,000 lb live weight per day with 14.4 lb total solids. For example, the United States alone produces 110 million tonnes (d.b.) of manure annually. This manure supply could generate about 60 tonnes of biogas along with 60 million dry tonnes of undigested fibers that could produce an additional 7.6 billion liters (1.7 billion gallons) of ethanol [21, 22]. Traditionally, most manure has been spread on fields, but digestion for biogas production is becoming more common in almost all countries [23, 24, 25, 26]. The undigested manure byproduct of biogas production primarily is applied as a nutrient to farmland, but a small part is utilized for animal bedding [27], manure composts for organic fertilizer [28, 29, 30], and even the manufacture of particleboard [31]. Because manure fibers are known to be highly recalcitrant to enzymatic digestion, efforts on the conversion of manure fibers into biofuels have been limited [20]. Anaerobic digestion of manure for the production of biogas consumes hemicellulose and nearly all-available soluble sugars and leave cellulose and lignin untouched [20, 32]. In addition, undesired components associated with nitrogenous extractives and ash increase the cost of biofuels production [33, 34, 35]. The recalcitrant nature of biomass is attributed to tight lignin wrapping, which prevents the accessibility of the biomass carbohydrate fractions (cellulose/hemicellulose) to enzymes, hemicellulose sheathing, cellulose crystallinity, and degree of polymerization [9, 36]. Lignin not only hinders enzyme accessibility to cellulose but also provides the non-productive and/or irreversible binding of enzymes [37]. Therefore, a pretreatment step mechanical and/or chemical is necessary to modify the lignocellulose complex matrix structure in such a way as to disrupt lignin, dissolve hemicelluloses, and break down the cellulose crystallinity in order to enhance substrate accessibility to enzymes and in turn, release more fermentable sugars [38, 39].

In general, nitrogenous matter in anaerobically digested fiber increases its alkalinity (to pH 8.5−9.0). Therefore, alkaline pretreatments are expected to require less chemical than acidic pretreatments [20]. Alkaline biomass pretreatment methods using either sodium hydroxide [40, 41] or aqueous ammonia [42] have been studied in recent years and shown to have high efficiency and low cost [14, 38]. Sodium hydroxide treatment effectively depolymerizes and removes the most labile biomass components, such as hemicelluloses and lignin, causes swelling that increases enzyme accessible surface area (for solvation and saponification reactions) and reduces the degree of polymerization and crystallinity of cellulose [40]. Aqueous ammonia reacts selectively with lignin by cleaving C-O-C bonds in lignin and ether and ester bonds in lignin-carbohydrate complexes, but carbohydrate removal and/or degradation is limited. In addition, these treatments cause significant morphological changes in the lignocellulose to improve enzyme accessibility [43, 44]. However, aqueous ammonia may not be effective for the pretreatment of substrates having relatively higher lignin (wood feedstocks) [38]. According to reviews, maximum delignification (~64%) with enzymatic saccharification (~65%) could be achieved for anaerobically digested manure fiber by using dil. NaOH under elevated temperature [20, 45]. Also, the addition of supplementary reagents to the alkali pretreatment chemicals, such as oxidizing agents

[46] or lime [47], has been shown to further improve delignification and subsequent enzymatic digestion of lignocellulose substrates [48]. As pretreatment protocols for post-biogas digestion (PBD) manure fiber, which contain high residual lignin, aimed at improving enzymatic digestion for ethanol fermentation, we explored the use of dilute sodium hydroxide and/or aqueous ammonia. The addition of NaOH to NH_4OH shifts the equilibrium to form gaseous NH_3, a reversible reaction that could be used to facilitate its recovery and reuse so as to improve the cost-effectiveness of this process (Figure 1) [49]. The fibers were then enzymatically saccharified to convert glucan to glucose to determine the effectiveness of the pretreatments. The pretreated fibers were also examined by scanning electron microscopy and x-ray diffraction measurements to determine cellulose crystallinity.

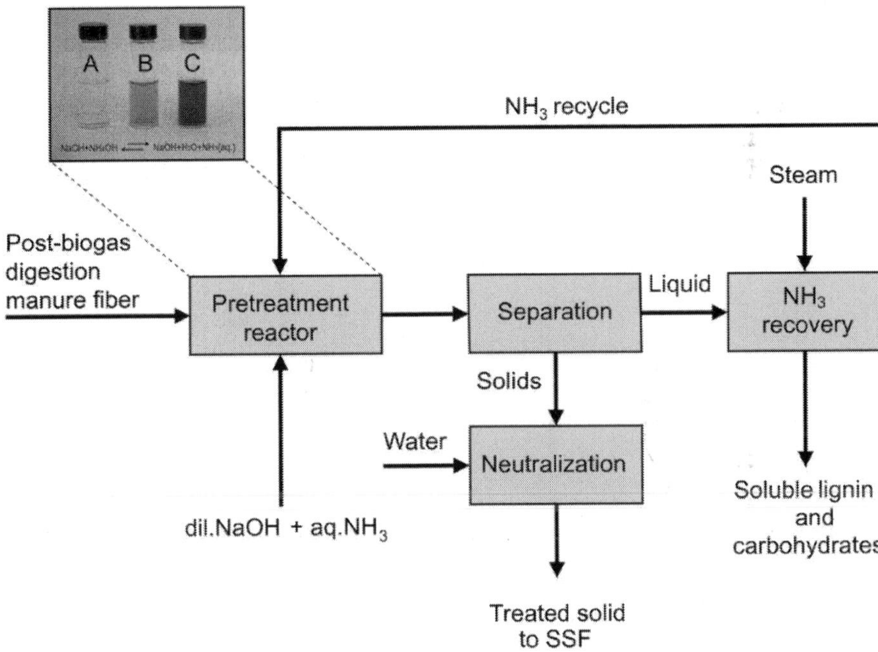

Figure 1: Schematic representation of the pretreatment of post-biogas digestion manure fibers for cellulosic bioethanol production by NaOH and NH_4OH. The inset shows (A) reference blank, (B) 0.25% conc. NH_4OH (5 mL), and (C)0.25% conc. NaOH (2.5 mL) plus 0.25% conc. NH_4OH (2.5 mL) [49].

RESULTS AND DISCUSSION

Characterization of post-biogas digestion manure fiber

As determined by mechanical sieve analysis, the post-biogas digestion (PBD) manure fiber that had undergone alkaline pretreatment displayed a fiber size range from 2.4 mm to < 75 μm with a calculated number average diameter (D_N) of 0.041 mm and volume surface mean diameter (D_S) of 0.415 mm [50]. Particles smaller than 18-mesh (1.0 mm) accounted for 93% of PDB fibers (dry basis), as compared to about 73% of pre-biogas digestion manure fibers. The effects anaerobic bacterial digestion lead to smaller particle sizes [20]. In general, particle size plays a significant role in the effectiveness of the pretreatment and fermentation steps [51, 52], with smaller fiber size being advantageous for bioconversion. The PBD manure fiber contained 41.1% carbohydrate by weight, of which 23.6% was glucose and 17.5% other sugars (xylose, galactose, arabinose and mannose) (Table 1). The carbohydrate content of PBD manure fiber is 25–28% lower than those of other commonly used substrates for cellulosic ethanol production (corn stover, switch grass, sugarcane bagasse, and wheat straw) [15]. More accessible carbohydrate sugars are digested in the animal and during anaerobic digestion.

Table 1: Compositional analyses result of post-biogas digestion manure fiber (anaerobically bacterial digested)

Component	[a] Post-biogas digestion manure fiber, dry basis
Carbohydrate	
Glucose	23.6 ± 0.3%
Xylose, Galactose, Arabinose, and Mannose	17.5 ± 0.3%
Lignin	
Acid insoluble lignin	27.6 ± 0.1%
Acid soluble lignin	1.8 ± 0.0%
Lignin monomer	
[b] S/G ratio	0.3
S/H ratio	1.9
H/G ratio	0.1
Uronic acid	2.6 ± 0.1%
Ash	13.7 ± 0.1%
Extractives	11.7 ± 0.1%
Fiber elements	
C	39.1 ± 0.4%
N	3.3 ± 0.2%
P	1.1 ± 0.0%
K	1.4 ± 0.0%
Ca	4.8 ± 0.6%
Mg	0.8 ± 0.1%
S	1.0 ± 0.0%
Cl	0.4 ± 0.1%
Fe	0.3 ± 0.1%
Zn	0.1 ± 0.0%

[a] Data are average of two replicates. Numbers with ± values represent standard errors; [b] Syringyl (S), guaiacyl (G), and *p*-hydroxyphenyl (H) lignin.

The low carbohydrate content of PBD manure fibers leads to carbohydrate/lignin ratios 60-65% lower than in other agricultural biomass. PBD manure fibers contained 27.6%

acid insoluble lignin (ash free and Klason) and 1.8% acid soluble lignin. Analysis of lignin monomers yielded 19:71:10 syringyl (S):guaiacyl (G):*p*-hydroxyphenyl (H) on dry basis, ratios consistent with corn stover lignin [42, 53]. Fiber ash, a non-reactive and undesired component of manure fiber for biofuels production, which negatively affects ethanol yields particularly from thermochemical ethanol production [54], accounted for 13.7% (dry basis). It has been reported that manure fiber also contains ~12% solvent extractives, composed mainly of nitrogenous materials, nonstructural sugars, inorganics, waxes, oils, and other compounds [55]. The predominant component, nitrogenous material, comes from indigestible forage proteins and ammonia and other nitrogen compounds in urine and manure. This nitrogen could be a potential nutrient source for microbial growth in ethanol fermentation [17, 35]; however, biomass extractives interfere with analytical measurements [56, 57] and thus were not considered in our study.

Alkaline pretreatment of PBD manure fiber

We investigated the pretreatment of PBD manure fiber by dilute sodium hydroxide and ammonium hydroxide (SHAH). We studied the effects of different pretreatment parameters, including alkali loading, temperature, and residence time, on the recovery and subsequent enzymatic digestion of PBD manure fiber (Table 2). We used low-severity protocols to evaluate pretreatment under conditions that minimized the cost of chemicals and the energy needed to heat the samples: 100°C for 5 min at concentrations of 2−10% (by dry fiber weight) of NaOH and NH_3[41, 58]. These pretreatment protocols led to substantial decreases in fiber residual lignin (16−40% delignification) and improved carbohydrate recovery (80-67%) and higher carbohydrate concentration (2-10%). The pretreatment yields were calculated based on a comparison between the weight of contents present in the sample before (initial) pretreatment and the weight of contents present in the solids remaining after pretreatment. These low severity pretreatment conditions achieved higher fiber delignification than higher severity conditions [20, 47, 59]. We also investigated the effects of higher temperature (120°C at15 psi) and longer residence time (60 min) at different concentrations of NaOH and ammonia (Table 2). The more stringent conditions led to modest increases in total sugars and reductions in total lignin. The results were consistent with the literature [20] and observed approximately 2-3-fold increase on pretreatment recovery.

We compared the effects of SHAH pretreatment protocols with those using NaOH and aqueous NH_3 singly under conditions of 121°C at 15 psi and 60 min (Table 2). In terms of solids recovered, total sugars, and total lignin, treatment with NaOH alone was equivalent to SHAH pretreatment; for both, increased alkali loading led to lower solids recovery, higher total sugars, and lower total lignin. Treatment with aqueous NH_3 alone led to higher solids recovery, slightly lower total sugars and higher total lignin. Increasing concentrations of aqueous ammonia had little effect on glucose, total sugars, or total lignin. A linear model relating modified severity parameter that combines the effects of temperature, time, and alkali concentration to the percentage removal of lignin was used for the determination of fiber delignification during alkaline pretreatments. The model was developed by plotting log (M_0) vs percent delignification, as given in Figure 2. The modified severity parameter model was validated by plotting the experimental vs model predicted values of fiber delignification (Additional file 1: Figure S1) and observed $R^2>0.94$ for all pretreatment methods; indicating good predictive ability of the model.

Table 2: Pretreatment recovery and solid residues composition of post-biogas digestion manure fiber after treatment using NaOH and NH_3 at different conditions

Alkali loading%, gm/gm dry solids		Temperature (°C)	Residence time (min)	[a] Modified CS factor, log M_0	Solids recovery%	Residue composition%, dry basis		
Dilute NaOH	Aqueous NH_3					Glucose	[b] Total sugars	[c] Total lignin (AIL + ASL)
2.0	2.0	100	5	0.09 (n = 0.89)	80.6 ± 0.1	24.2 ± 0.2	41.1 ± 0.6	31.1 ± 0.1
4.0	4.0	100	5	0.35	76.6 ± 0.0	24.7 ± 0.4	42.0 ± 0.4	31.0 ± 0.4
6.0	6.0	100	5	0.51	72.0 ± 0.5	25.0 ± 0.3	41.4 ± 0.5	29.0 ± 0.4
8.0	8.0	100	5	0.62	68.9 ± 0.1	27.3 ± 0.7	44.4 ± 1.2	28.6 ± 0.2
10.0	10.0	100	5	0.71	64.6 ± 0.3	28.2 ± 1.5	43.2 ± 1.2	27.6 ± 0.5
2.0	2.0	121/15 psi	60	0.26 (n = 3.1)	78.1 ± 0.2	25.3 ± 0.2	42.1 ± 0.2	31.5 ± 0.1
4.0	4.0	121/15 psi	60	1.20	71.2 ± 0.6	26.6 ± 0.8	42.9 ± 1.0	28.8 ± 0.3
6.0	6.0	121/15 psi	60	1.74	63.0 ± 0.2	30.0 ± 0.7	46.9 ± 1.0	26.5 ± 0.1
8.0	8.0	121/15 psi	60	2.14	57.0 ± 0.6	32.5 ± 0.0	52.3 ± 0.0	22.8 ± 0.0
10.0	10.0	121/15 psi	60	2.43	51.2 ± 0.0	32.8 ± 0.0	53.5 ± 0.0	19.5 ± 0.1
2.0	-	121/15 psi	60	0.21 (n = 2.2)	79.0 ± 0.1	25.5 ± 0.0	42.4 ± 0.0	33.2 ± 0.4
4.0	-	121/15 psi	60	0.89	72.6 ± 0.8	27.0 ± 1.1	44.2 ± 1.2	30.5 ± 0.2
6.0	-	121/15 psi	60	1.28	64.2 ± 0.1	31.1 ± 0.5	48.5 ± 1.3	27.3 ± 0.1

8.0	-	121/15 psi	60	1.56	58.5 ± 0.1	32.2 ± 0.0	48.9 ± 0.0	23.9 ± 1.3
10.0	-	121/15 psi	60	1.77	52.0 ± 0.3	31.8 ± 0.0	50.0 ± 0.0	20.1 ± 0.1
-	2.0	121/15 psi	60	0.21 (n = 2.2)	83.1 ± 0.4	22.9 ± 1.2	41.2 ± 1.8	33.1 ± 0.3
-	4.0	121/15 psi	60	0.89	79.5 ± 1.2	22.7 ± 0.1	42.2 ± 0.3	33.2 ± 0.2
-	6.0	121/15 psi	60	1.28	79.3 ± 1.6	22.2 ± 0.3	42.0 ± 0.0	32.6 ± 0.7
-	8.0	121/15 psi	60	1.56	77.2 ± 0.4	23.2 ± 0.4	43.1 ± 0.8	32.4 ± 0.2
-	10.0	121/15 psi	60	1.77	76.8 ± 0.8	23.2 ± 0.4	43.1 ± 0.8	31.2 ± 0.8
R^2					0.95	0.90	0.79	0.87
RMSE					2.26	1.19	1.84	1.57
Tukey's HSD ($P < 0.05$)					43.34	17.84	25.5	16.07

[a]Modified combined severity factor; [b]Total sugars including glucose, xylose, galactose, arabinose and mannose; [c] Total lignin including acid insoluble lignin (AIL) and acid soluble lignin (ASL) after ash correction. Date are average of two replicates. Numbers with ± values represent standard errors.

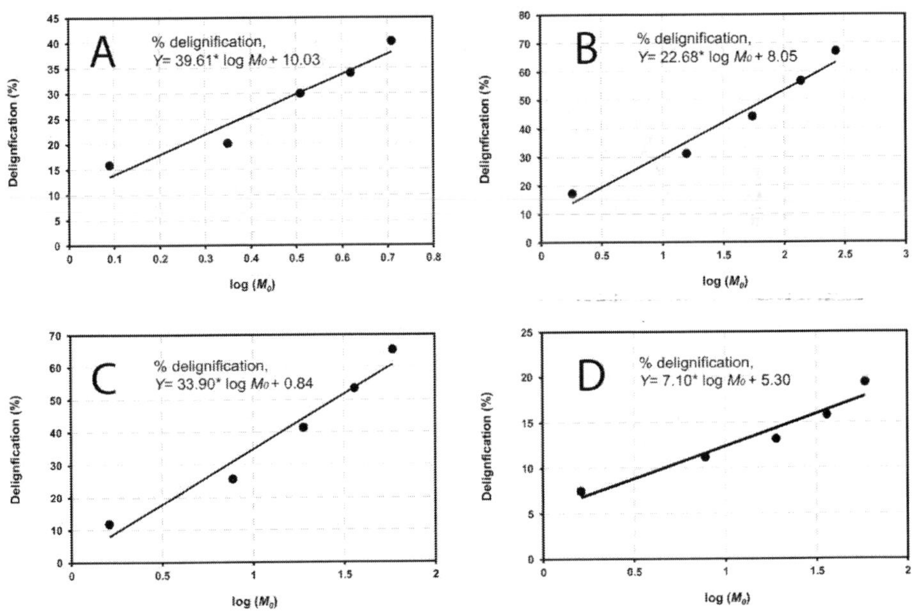

Figure 2: Plot of percent delignification of manure fibers vs modified severity parameter (log M_0) for alkaline pretreatment treated at. (A) 100°C using combined NH_3 and NaOH, (B) 121°C/15 psi using combined NH_3 and NaOH,(C) 121°C/15 psi using NaOH, and (D) 121°C/15 psi using NH_3.

Effects of alkaline pretreatment of PBD manure fiber on subsequent enzyme saccharification

We carried out enzymatic digestion of pretreated PDB manure fiber to determine the release of fermentable sugar for ethanol fermentation (Figure 3). The typical enzymatic hydrolysis profile showed rapid saccharification over 6 h followed by leveling off thereafter (Figure 3A).

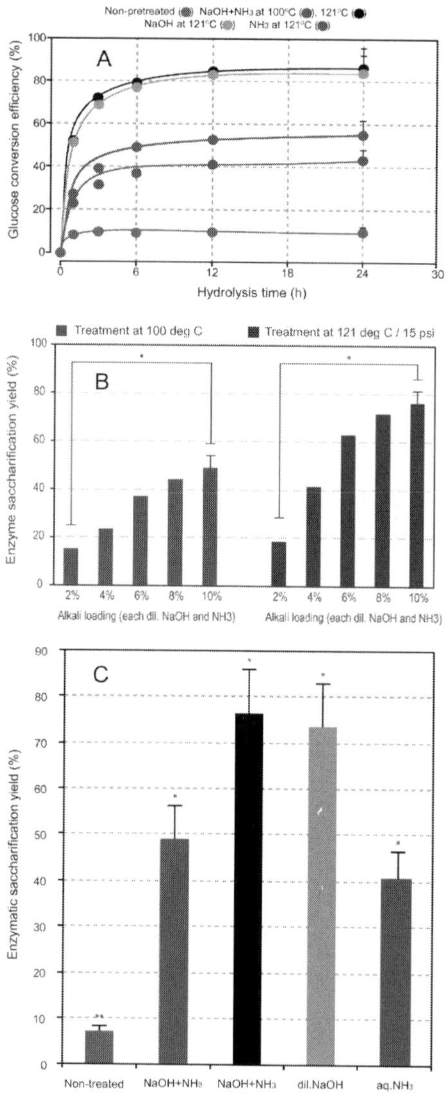

Figure 3: Enzymatic digestibility of manure fibers both non-pretreated and pretreated at 10% alkali loading under different conditions. (A) Glucose conversion efficiency, (B) total saccharification yield of manure fibers treated under different conditions after 24 h, and (C) total sac-

charification yield of manure fibers after 24 h. Error bars represent root mean square error. (**) Non-significant and (*) significantly different at 95% confidence level, p value<0.001.

Enzymatic hydrolysis results of SHAH pretreatment at 100°C (Figure 3B) showed saccharification yields of 15-49% with 18-55% glucose conversion after 24 h with a corresponding increase in concentration of 5-3% points difference between each concentration increment. Approximately 1.5 − 2.0-fold increase in saccharification yield was achieved with 121°C at 15 psi and 60 min pretreatment (Figure 3B), indicating that removal of residual lignin (relatively 8-66% higher delignification) and other substrate features had significant impact on the improved enzyme accessibility for fiber digestion [60]. SHAH pretreatment achieved 3 − 8% and 1.5 − 2.5-fold higher saccharification than separate NaOH and NH_3pretreatments, respectively (Figure 3C). Also, it was approximately 3-fold higher saccharification than Teater et al. [20].

Effects of pretreatment of PBD manure fiber on surface structure and cellulose crystallinity

We used scanning electron microscopy (SEM) to determine the effects of pretreatment on the surface features of the fibers. PBD manure fibers that were not pretreated (Figure 4A) or pretreatment with NaOH alone (Figure 4B) or aqueous NH3 alone (Figure 4C) exhibited rigid and highly ordered surface structure. By contrast the SEM image of fibers that underwent SHAH pretreatment exhibited sponge-like structures and an apparent increase in fiber porosity (Figure 4D) [39, 40].

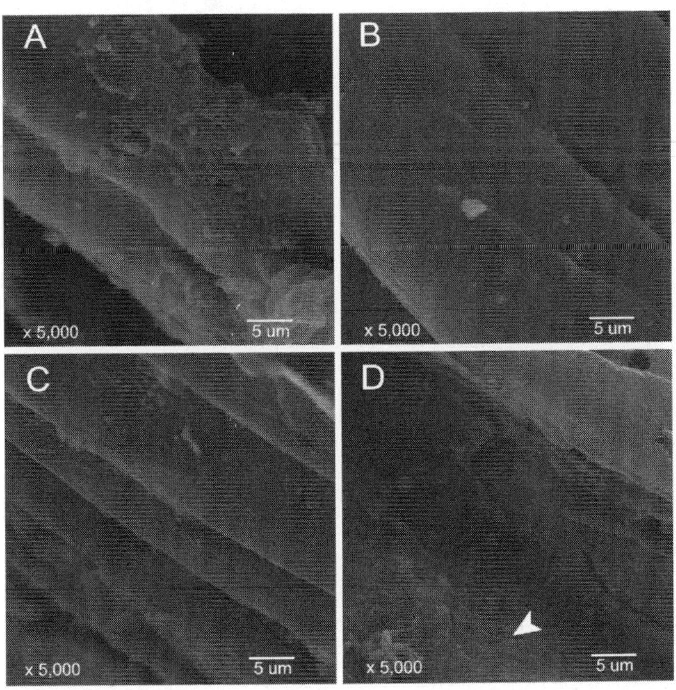

Figure 4: Scanning electron microscope images of manure fibers before and after treat-

ment at 121°C and 10% alkali loading level. (A) non-pretreated (control), (B) pretreated by NH_3 alone, (C) pretreated by NaOH alone, and (D)pretreated by combined NH_3 and NaOH.

We used X-ray powder diffraction pattern to determine the effects of pretreatment on cellulose crystallinity. The results (shown in Figure 5A) showed that the three treatment protocols at 121°C and 15 psi reduced crystallinity in the order: 10% NH_3 < 10% NaOH < 10% NaOH + NH_3. SHAH pretreatment greatly alters the crystalline structure by the competitive reaction of both alkalis resulting in the formation of different allomorphs that have different unit cell dimensions, chain packaging, and hydrogen bonding relationships [61, 62]. Studies recognized that alkaline pretreatment causes swelling of cellulose, leading to the decrease of degree of polymerization and crystallinity, and increases the surface area that facilitates more substrate exposed to cellulase attack [48]. However, the poor negative correlation between the cellulose crystallinity index and enzymatic digestion of fibers under different conditions ($R^2 < 0.1$) (Figure 5B−D), might be due to the dissolution of amorphous materials (xylan and lignin) and/or interference of other soluble materials [63, 64].

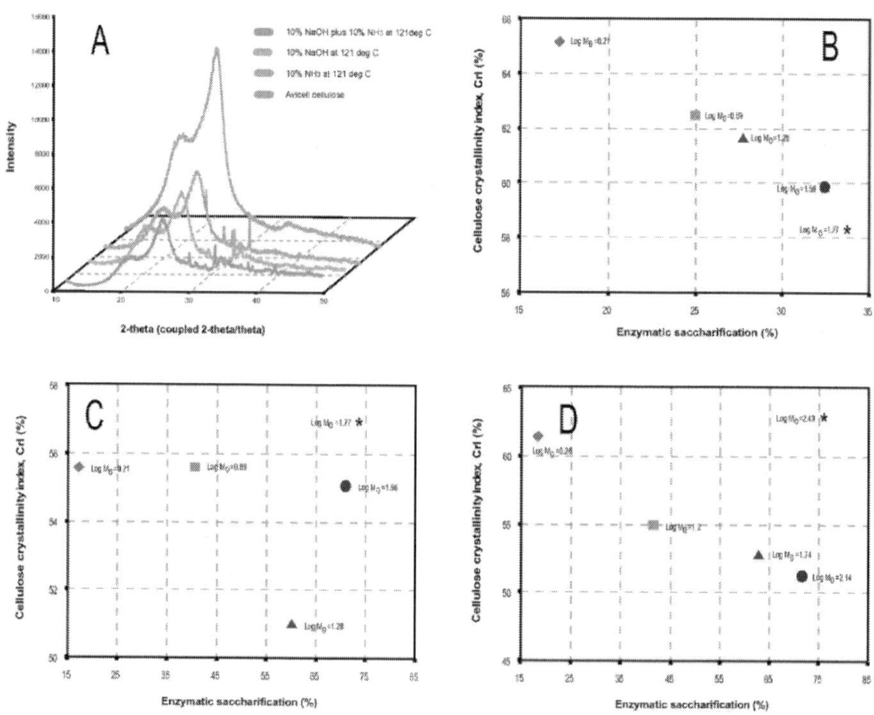

Figure 5: Cellulose crystallinity index (CrI) of alkaline-treated manure fibers and its relationship to the enzymatic digestibility (24 h) . (A) powder X-ray diffraction spectrum of fibers after treatment at different conditions, (B)correlation between enzymatic digestibility and CrI of separate NH_3 treated fibers, (C) correlation between enzymatic digestibility and CrI of separate NaOH treated fibers, and (D) correlation between

enzymatic digestibility and CrI of both NaOH and NH_3 treated fibers, under different conditions. Data are averages of two replicates.

Effect of pretreatment conditions on fiber delignification and enzymatic saccharification

We carried out a detailed study of the relationship between pretreatment parameters and fiber composition following enzymatic digestion by using a central composite design experiment with 3 dependent factors and 3 different levels (Table 3): 3 alkali loadings (x_1) of 2.0, 4.0 and 6.0% of each NaOH and NH_3, 3 treatment temperatures (x_2) of 80, 100, and 121°C, and three residence times (x_3) of 5, 30, and 60 min. The experimental parameters were selected on the basis of a previous SHAH pretreatment study on PBD manure fiber. The results showed a dependence on the pretreatment conditions of sugars released, mainly hemicellulose which is more vulnerable to chemical attack, and lignin (Table 4). The model identified that, within the studied range of experiments, chemical loading had the most significant effect on both sugar dissolution (regression coefficient, $\beta_3 = -5.4$) and delignification ($\beta_3 = 10.5$).

Table 3: Central composite experimental design and the corresponding pretreatment responses with subsequent enzyme saccharification of post-biogas digestion manure fibers

Independent variables			Dependent variables, Y		
[a] % NaOH and NH_3 loading, gm/gm dry fiber, x_1	Temperature (°C), x_2	Residence time (min), x_3	Total sugar recovery%	Delignification% dry basis	Enzyme saccharification%
8.0	100	30	87.3 ± 0.0	22.9 ± 0.3	24.5 ± 0.8
12.0	80	5	87.4 ± 0.0	25.7 ± 1.3	19.9 ± 4.3
8.0	100	30	87.3 ± 0.0	22.9 ± 0.3	24.5 ± 0.8
12.0	121	5	87.5 ± 0.4	21.6 ± 1.1	23.9 ± 1.2
8.0	100	60	86.0 ± 0.4	26.3 ± 0.8	34.0 ± 1.2
4.0	121	5	94.6 ± 0.1	5.9 ± 0.9	2.5 ± 0.4
12.0	80	60	82.7 ± 0.6	32.8 ± 0.5	44.5 ± 1.0
8.0	100	5	91.5 ± 0.1	17.1 ± 0.2	16.2 ± 0.1
8.0	100	30	87.3 ± 0.0	22.9 ± 0.3	24.5 ± 0.8
12.0	100	30	79.4 ± 0.1	37.1 ± 0.3	36.5 ± 3.8
4.0	80	60	93.4 ± 0.1	12.2 ± 0.1	10.9 ± 0.1
8.0	121	30	80.2 ± 0.8	30.3 ± 1.0	49.0 ± 2.8
12.0	121	60	73.3 ± 0.4	44.3 ± 0.8	59.0 ± 0.5
8.0	100	30	87.3 ± 0.0	22.9 ± 0.3	24.5 ± 0.8
4.0	100	30	93.1 ± 0.1	10.0 ± 1.3	13.7 ± 0.4
4.0	121	60	88.3 ± 0.1	17.5 ± 1.3	27.5 ± 0.2
4.0	80	5	95.2 ± 0.1	11.1 ± 1.1	3.1 ± 0.2
8.0	80	30	89.5 ± 0.1	21.1 ± 0.0	24.8 ± 0.5
8.0	100	30	87.3 ± 0.0	22.9 ± 0.3	24.5 ± 0.8

[a] Percentage of each NaOH and NH_3 added to the dry manure solids. Data are average of two replicates. Numbers with ± values represent standard errors.

Increasing alkali loadings from 2.0 to 6.0% led to a decrease in total sugar recovery by 20% and a decrease in residual lignin by 40%. The correlation coefficient values for the models ($R^2 \geq 0.95$) indicate that a large fraction of the variation in responses results from differences in the independent variables. Although enhanced removal of residual lignin is expected to improve subsequent enzymatic digestibility, the simultaneous loss of residual carbohydrate should decrease the yield of sugars through enzymatic hydrolysis [46].

Table 4: Statistical analysis of the effect of pretreatment parameters on manure fiber pretreatment recovery and following enzymatic saccharification

	Total sugar recovery		Delignification		Enzymatic saccharification	
[a]R^2	0.96		0.97		0.95	
Prob. > F	< 0.0001*		< 0.0001*		< 0.0001*	
Terms	Estimate	[b]Prob. $> \lvert t \rvert$	Estimate	[b]Prob. $> \lvert t \rvert$	Estimate	[b]Prob. $> \lvert t \rvert$
Temp (°C)	−2.43	0.0011*	1.66	0.061	5.68	0.0035*
Time (min)	−3.26	0.0001*	5.17	< 0.0001*	11.0	< 0.0001*
Chemical loading (% on dry solids)	−5.42	< 0.0001*	10.4	< 0.0001*	12.6	< 0.0001*
Temp x Temp	−1.50	0.1613	1.43	0.360	7.13	0.0345*
Time x Time	2.41	0.0372*	−2.61	0.113	−4.65	0.139
Chemical loading x Chemical loading	−0.079	0.938	−0.76	0.620	−4.64	0.140
Temp x Time	−1.73	0.015*	3.26	0.0046*	3.47	0.0681
Temp x Chemical loading	−0.44	0.463	0.91	0.322	0.33	0.848
Time x Chemical loading	−1.37	0.0414*	2.14	0.0363*	3.36	0.0755

[a] Correlation coefficient; [b] p-value of the two-tailed test; * Numbers with asterisk indicate that the term has a significant effect at 95% confidence level.

Following enzymatic saccharification, chemical loading ($\beta_3=12.6$) had more significant effect than residence time ($\beta_2=11.0$) or treatment temperature ($\beta_1=5.9$). Three linear effects and one quadratic effect were observed with subsequent enzymatic digestibility of the treated fibers. An increase in chemical loading from 2.0 to 6.0% led to a 57% saccharification yield with 64% glucose conversion. This may be due to the enhanced removal of enzyme barriers, including residual lignin (~30%) and hemicelluloses (~22%), and surface modification during pretreatment which improves enzyme accessibility [9]. Linear terms of delignification and enzymatic saccharification correlated positively with the treatment parameters, indicating that these have the greatest effect on substrate deconstruction. The second most important parameter affecting the overall process was the residence time [65]. This suggests that longer treatment times reduce fiber recalcitrance that limits sugar degradation [66] and improve energy utilization.

CONCLUSIONS

In the present study, enzyme recalcitrant post-biogas digestion (PBD) manure fibers were subjected to an innovative pretreatment method involving combined alkalis (dilute sodium hydroxide and aqueous ammonia). The effects of pretreatment conditions were studied, including alkali loading on fibers, treatment temperature, and residence time. The results show that the dual alkali treatment improves fiber delignification (maximum 67.1%) and subsequent enzymatic digestion (maximum of 76.3%) of PBD manure fibers. Furthermore, the pretreatment alters the surface structural characteristics of the fiber apparently making them more prone to enzyme attack for enhanced sugar release. A positive factor in the economic viability of PBD manure fiber for cellulosic bioethanol production, is the high availability and relatively low cost of the feedstock. On the other hand, the costs of chemical consumption need to be taken into account, although these could be mitigated in part by recovery and reuse of the gas phase ammonia formed during the alkali reaction. In addition, it may be possible to improve the efficiency of the process by combined maceration (mechanical milling) and alkaline pretreatment with both NaOH and NH_3, and studies to evaluate this approach are ongoing.

METHODS

Manure samples

Post-biogas digestion manure fibers were collected from Maple Leaf Dairy Farm, Cleveland, Wisconsin. The cattle feed was a mixture of alfalfa, corn silage and other proteins according to the National Research Council nutrient requirements of dairy cattle. The anaerobic digestion was running with a hydraulic retention time of 14–15 days at 35 − 40°C. The slurry containing undigested solids were separated by a 2.0 mm screen screw press. The collected fibers contained 60 − 65% moisture; they were air-dried and ground with a laboratory hammer mill (Christy & Norris Ltd., England, Model No. 1024XC) and then sieved. The fiber fraction within 40–50 mesh was used for the analysis.

Combined alkaline pretreatment

Sodium hydroxide (50% wt. Fisher Catalog No. SS254-4) and ammonium hydroxide (30% wt. Fisher Catalog No. 125) were used pretreatment. Weighed quantities of fiber in 50 ml Oak Ridge thermal resistant tubes (Fisher Catalog No. 05-563-10G) were treated with NaOH and/or NH_3 (at different loadings of each alkali 2 − 10% w/w) at room temperature for 2 h, followed by heating at 100 or 121°C for 1 h. The solid-to-liquid ratio was maintained at 1:7. After pretreatment, the supernatant was collected following centrifugation at 3,900 rpm (Eppendorf 510R) for 20 min, and the solid residues obtained were thoroughly washed with water until the pH reached neutrality. Finally, the solid residues were dried in a freeze-dryer (VirTis freezemobile 35ES) and stored at −80°C (New Brunswick U-700 freezer).

Enzymatic digestibility

Enzyme saccharification of fibers, both non-treated (control) and chemically treated, was carried out according to the standard NREL procedure (LAP 42629). Saccharification was conducted in 50 mL Falcon tubes at 2.0% (w/v) substrate consistency level using sodium acetate buffer pH 4.8. Tetracycline antibiotic was added at 0.02% (w/v) to prevent microbial contamination, Enzymes used in this study, Cellic CTec2 (cellulase complex containing cellulose and β-glucosidase) and Cellic HTec2 (hemicellulases including xylanase), were generously provided by Novozymes (Franklinton, NC). The Cellic CTec2 and Cellic HTec2 loadings on substrates were 5% and 2% w/w (gm enzyme/gm dry fiber), respectively. Substrates were pre-incubated at 50°C in sodium acetate buffer for 24 h prior to the addition of enzymes. Hydrolysis was conducted at 50°C in a shaker (New Brunswick Scientific Excella E24) at 200 rpm for 24 h. Samples were collected intermittently and analyzed for sugar concentration using High Performance Liquid Chromatography (HPLC) System (Agilent Technologies 1200 series). The HPLC was equipped with Bio-rad deashing micro-guard column (Cat No. 125–0118, Bio-Rad, CA) and Agilent Hi-Plex H (7.7 × 300 mm, 8 μm) analytical column operated at 60°C with 5 mM H_2SO_4 mobile phase at the flow rate of 0.7 mL/min. A refractive index detector (Agilent Technologies) was operated at 55°C. The mobile phase was filtered through a 0.22 μm nylon membrane (Millipore Corporation, MA) and degassed. The released glucose and other sugars (xylose, galactose, arabinose and mannose) at each time interval were used to calculate the glucose conversion and saccharification efficiency of the substrate as a percentage to the potential sugars available in the substrates. Each data point was the average of two replicates.

Analytical methods

Manure fiber moisture, extractives and ash contents were determined according to National Renewable Energy Laboratory (NREL) analytical procedures LAP 012, LAP 010 and LAP 005, respectively. Similarly, carbohydrate analysis of non-treated and chemically pretreated manure fiber was carried out according to NREL procedure LAP 009. Samples (0.3 g) were weighed (W_1) in a 5 ml centrifuge tube and hydrolyzed with 3 ml 72% H_2SO_4 (v/v) for 60 min. The hydrolyzate was diluted to 4% acid concentration (v/v) and autoclaved for 60 min at 121°C at 15 psi. The hydrolysis solution was vacuum filtered through the previously weighed filtering crucible. The filtrate was collected (F_1) and analyzed for carbohydrate and acid soluble lignin determination. Carbohydrate content, including glucose, xylose, galactose, arabinose, and mannose sugars, were analyzed on an HPLC System (Agilent Technologies 1200 series) equipped with a Bio-Rad deashing micro-guard column (Cat No. 125–0118, Bio-Rad, CA) and an Agilent Hi-Plex H (7.7 × 300 mm, 8 μm) analytical column with a mobile phase of 5 mM H_2SO_4 operated at a flow rate of 0.7 mL/min at 60°C. The mobile phase was filtered through a 0.22 μm nylon membrane (Millipore Corporation, MA) and degassed, and peaks were detected by a refractive index detector (Agilent Technologies) operated at 55°C.

Acid insoluble lignin was calculated gravimetrically as acid-insoluble residue after correction for ash content. The lignin collected during filtration was washed with water and dried overnight in an oven at 105°C. The weight of the crucible with lignin was recorded (W_2), and the sample was ashed in muffle furnace for 4 h at 575°C. Finally, the weight of the crucible with ash content was recorded (W_3). Acid insoluble lignin (AIL) content of the manure fiber was calculated by the following equation:

$$AIL\,(\%) = \frac{(W_2 - W_3)}{W_1 \times m_i} \times 100$$

where m_i is the initial moisture content of the manure sample. The filtrate F_1 was measured for the acid soluble lignin at 208 nm using UV/Vis spectroscopy (Agilent Cary 60) with 4% (v/v) sulfuric acid as reference blank.

Lignin monomer composition

Manure fiber lignin composition was determined by derivatization followed by reductive cleavage method [67].

Colorimetric assay of uronic acid

A *m*-hydroxydiphenyl colorimetric assay was followed for the determination of uronic acid content [39, 68]. All chemicals were purchased from Sigma Aldrich (St. Louis, MO) and used as such. 200 µL of the acid hydrolyzate filtrate (F_1) was added to 1.2 mL H_2SO_4-tetraborate solution (476.8 mg sodium tetraborate dissolved in 500 mL 18 M H_2SO_4). Followed by, heating in a boiling water bath for 5 min and ice cooled. 20 µL of 0.15% m-hydroxydiphenyl reagent (22.5 mg 3-phenylphenol dissolved in 15 mL 0.5% w/v NaOH) was added to the reaction mixture and vortexed immediately until to get a dark pink color. Finally, the reagent mixture was read after the original color development at 520 nm using UV/Vis spectrophotometer (Agilent Cary 60). The uronic acid content was calculated from the OD value using the glucouronic acid/galacturonic acid calibration curve.

Scanning electron microscopy

The manure residues were collected after pretreatment at different conditions and washed with distilled water and vacuum dried. The dry samples both non-treated and chemically treated were coated with gold in a SeeVac Auto conductavac IV sputter coater and scanned by scanning electron microscope (Hitachi S-570 LaB_6, Tokyo, Japan) at accelerating voltage of 10.0 kV (12.7 stub size).

Crystallinity index measurement

Cellulose crystallinity index (CrI) of both treated and non-treated manure fiber was measured by powder X-ray diffraction (PXRD) method using a Bruker D2 Phaser instrument (Bruker AXS Inc., Madison, WI). Dried samples (~0.5 g) were ground to a powder < 100 µm size and pressed into 40-mm diameter pellets. The pellets were measured in Bragg-Brentano geometry using a LynxEye detector with 4° opening. Ni-filtered copper radiation was generated at 30 kV/10 mA, and the pellets were scanned from 5° to 50° by 0.02° steps at 1 s each. The divergence slit was 0.6 mm, and the primary and secondary soller slits were 2.5° and 4°, respectively. TOPAS software version 4.2 was used to calculate the CrI of samples from the ratio of the area of all crystalline peaks to the total area by the deconvolution method [69].

Elemental analyses of manure fiber

Elemental analysis of the manure fiber was carried out by using a wavelength dispersive X-ray fluorescence (WDXRF) spectrophotometer S8 Tiger (Bruker AXS Inc., Madison, WI). About 10 g of an air-dried, non-pretreated PBD manure fiber sample was ground with inert binding material (amyl acetate, 5% collodion) (Bruker AXS GmbH, Germany) at a 5:1 ratio to assist grinding performance, increase pellet stability, and reduce material rewelding in vessel. This was followed by compression pressing of the powder (30 KN/m^2) for 15 s in a 40-cm (dia) aluminum cup. The XRF spectrophotometer was equipped with 2 collimators (0.23° and 0.46°) and a set of 6 analyzer crystals (XS-GE-C, XS-CEM, XS-55, PET, LiF200 and LiF220). The measurement method used 27 kV/150 mA excitation for light elements and 60 kV/67 mA excitation for heavy elements using a Rhodium tube. The elemental composition was calculated by using QUANT-EXPRESS calibration (Bruker AXS GmbH, Germany).

Severity parameter and statistical data analysis

The severity parameter (R_0), a factor intended to quantify the energy intensity or severity of a pretreatment strategy, was initially defined by Overend and Chornet (1987) to relate temperature and time for steam explosion pretreated based on the assumption that the pretreatment effect follows first-order kinetics and obeys the Arrhenius equation [70]. Chum et al., (1990) later developed a modified severity parameter to use for sulfuric acid pretreatment that relates concentration with an introduced arbitrary constant, temperature and residence time as follows [71]:

$$M_0 = t \times C^n \times exp\left[\frac{(T_r - T_b)}{14.75}\right]$$

Where M_0 is the modified severity parameter; t is the residence time (min); C is the concentration of chemical (wt.%); T_r is the reaction temperature; T_b is the base temperature; n is the arbitrary constant. Silverstein et al. (2007) used the above equation for sodium hydroxide pretreatment by replacing the acid concentration with the alkali concentration [41]. Similarly, modified severity parameters of the alkaline pretreatment using NaOH and/or NH_3 were calculated from the above equation at different n-values obtained by data training while keeping positive log (M_0). The more detailed relationship between the pretreatment parameters and pretreatment recovery following enzymatic digestion (dependent variables) of the PBD manure fiber was explored using less runs by using central composite design experiments [64, 72]. Statistical software SAS JMP Pro version 10 was used for the design of experiments and to analyze the experimental data obtained. 3×3 central composite design experiments comprised a total of 19 runs with different combinations of pretreatment conditions derived by altering the three independent variables, alkali loading both NaOH and NH_3 ($2 - 10\%$ of each loading on dry manure fiber), treatment temperature (80, 100 and 120°C), and residence time (5, 30 and 60 min). The parameter levels were selected based on the previous study. The experimental data obtained were fitted to the following second order polynomial regression equation to identify the key variables:

$$Y = \beta_0 + \sum_{i=1}^{n} \beta_i x_i + \sum_{i=1}^{n} \beta_{ii} x_i^2 + \sum_{i=1}^{n} \sum_{j=1}^{n} \beta_{ij} x_i x_j$$

Where *Y* is the measured response (carbohydrate recovery, delignification efficiency and subsequent enzymatic digestibility); *i*, *j* are the linear and quadratic coefficients respectively; β_0 is the regression coefficient; *x* is the independent variable (alkali loading, temperature and residence time). The quality of model fit was expressed by the coefficient of determination, R^2 value. Model terms were evaluated based on the probability, p value with 95% confidence level.

DECLARATIONS

Acknowledgements

We gratefully acknowledge the United States Department of Agriculture-National Institute of Food and Agriculture for its financial support (USDA BRDI Grant number 2012-10006-19423). The authors thank GLBRC UW-Madison (Ruwan Ranatunga) for generously provided a Corn Stover-2010 sample for the comparison study. The authors thank Dharshana Padmakshan and Steven D. Karlen (GLBRC UW-Madison) for providing assistance with lignin monomer detection of manure fiber.

Competing Interests

The authors declare that they have no competing interests.

Authors' contributions

SE and TR designed the research with SE performing the experiments; SE, AR, TR, and JM analyzed the data and wrote the paper. All authors read and approved the final manuscript.

REFERENCES

1. Energy Perspectives: Fossil fuels dominate U.S. energy consumption. Book Energy Prespectives: Fossil fuels dominate U.S. energy consumption. 2012, Washington, DC: Annual Energy Review-U.S. Energy Information and Administration
2. International Energy Agency: Energy Statistics of OECD countries. Book Energy Statistics of OECD countries. 2013, Moscow: International Energy Agency Statistics
3. Simmons BA, Loque D, Blanch HW: Next-generation biomass feedstocks for biofuel production. Genome Biol. 2008, 9: 242.241-242.246.
4. Lynd LR, Cushman JH, Nichols RJ, Wyman CE: Fuel Ethanol from Cellulosic Biomass. Science. 1991, 251: 1318-1323. 10.1126/science.251.4999.1318.

5. Wyman CE: What is (and is not) vital to advancing cellulosic ethanol. Trends Biotechnol. 2007, 25: 153-157. 10.1016/j.tibtech.2007.02.009.
6. Slade R, Bauen A, Shah N: The greenhouse gas emissions performance of cellulosic ethanol supply chains in Europe. Biotechnol Biofuels. 2009, 2: 1-19. 10.1186/1754-6834-2-1.
7. Himmel ME, Ding SY, Johnson DK, Adney WS, Nimlos MR, Brady JW, Foust TD: Biomass recalcitrance: Engineering plants and enzymes for biofuels production. Science. 2007, 315: 804-807. 10.1126/science.1137016.
8. Pan XJ, Xie D, Gilkes N, Gregg DJ, Saddler JN: Strategies to enhance the enzymatic hydrolysis of pretreated softwood with high residual lignin content. Appl Biochem Biotechnol. 2005, 121: 1069-1079.
9. Chandra RP, Bura R, Mabee WE, Berlin A, Pan X, Saddler JN: Substrate pretreatment: The key to effective enzymatic hydrolysis of lignocellulosics?. Adv Biochem Eng Biot. 2007, 108: 67-93.
10. Hu G, Heitmann JA, Rojas OJ: Feedstock Pretreatment Strategies for Producing Ethanol from Wood, Bark, and Forest Residues. Bioresources. 2008, 3: 270-294.
11. Betancur GJV, Pereira N: Sugar cane bagasse as feedstock for second generation ethanol production. Part II: Hemicellulose hydrolysate fermentability. Electron J Biotechnol. 2010, 13: 1-11.
12. Silva NLC, Betancur GJV, Vasquez MP, Gomes ED, Pereira N: Ethanol Production from Residual Wood Chips of Cellulose Industry: Acid Pretreatment Investigation, Hemicellulosic Hydrolysate Fermentation, and Remaining Solid Fraction Fermentation by SSF Process. Appl Biochem Biotechnol. 2011, 163: 928-936. 10.1007/s12010-010-9096-8.
13. Vasquez MP, da Silva JNC, de Souza MB, Pereira N: Enzymatic hydrolysis optimization to ethanol production by simultaneous saccharification and fermentation. Appl Biochem Biotechnol. 2007, 137: 141-153.
14. Sun Y, Cheng JY: Hydrolysis of lignocellulosic materials for ethanol production: a review. Bioresource Technol. 2002, 83: 1-11. 10.1016/S0960-8524(01)00212-7.
15. Sarkar N, Ghosh SK, Bannerjee S, Aikat K: Bioethanol production from agricultural wastes: An overview. Renew Energ. 2012, 37: 19-27. 10.1016/j.renene.2011.06.045.
16. Yue ZB, Teater C, Liu Y, MacLellan J, Liao W: A Sustainable Pathway of Cellulosic Ethanol Production Integrating Anaerobic Digestion With Biorefining. Biotechnol Bioeng. 2010, 105: 1031-1039.
17. Liao W, Liu Y, Liu CB, Wen ZY, Chen SL: Acid hydrolysis of fibers from dairy manure. Bioresource Technol. 2006, 97: 1687-1695. 10.1016/j.biortech.2005.07.028.
18. Wen ZY, Liao W, Chen SL: Hydrolysis of animal manure lignocellulosics for reducing sugar production. Bioresource Technol. 2004, 91: 31-39. 10.1016/S0960-8524(03)00166-4.
19. Chen S, Wen Z, Liao W, Liu C, Kincaid RL, Harrison JH, Elliott DC, Brown MD,

Stevens DJ: Studies into using manure in a biorefinery concept. Appl Biochem Biotech. 2005, 121: 999-1015.

20. Teater C, Yue ZB, MacLellan J, Liu Y, Liao W: Assessing solid digestate from anaerobic digestion as feedstock for ethanol production. Bioresource Technol. 2011, 102: 1856-1862. 10.1016/j.biortech.2010.09.099.

21. Gollehon N, Caswell M, Ribaudo M, Kellogg B, Lander C: Confined animal production and manure nutrients. J Agr Resour Econ. 2000, 25: 726-726.

22. United States Department of Agriculture: Farms, Land in Farms, and Livestock Operations. Book Farms, Land in Farms, and Livestock Operations. 2009, Washington, DC: USDA National Agricultural Statistics Service

23. Yadav A, Gupta R, Garg VK: Organic manure production from cow dung and biogas plant slurry by vermicomposting under field conditions. Int J Recycling Organic Waste in Agric. 2013, 2: 21-10.1186/2251-7715-2-21.

24. Yadvika S, Sreekrishnan TR, Kohli S, Rana V: Enhancement of biogas production from solid substrates using different techniques—a review. Bioresour Technol. 2004, 95: 1-10. 10.1016/j.biortech.2004.02.010.

25. Alvarez R, Villca S, Lidén G: Biogas production from llama and cow manure at high altitude. Biomass Bioenergy. 2006, 30: 66-75. 10.1016/j.biombioe.2005.10.001.

26. Iyagba ET, Mangibo IA, Mohammad YS: The study of cow dung as co-substrate with rice husk in biogas production. Scientific Res & Essay. 2009, 4: 861-866.

27. Petersen SO, Sommer SG, Beline F, Burton C, Dach J, Dourmad JY, Leip A, Misselbrook T, Nicholson F, Poulsen HD, et al: Recycling of livestock manure in a whole-farm perspective. Livest Sci. 2007, 112: 180-191. 10.1016/j.livsci.2007.09.001.

28. Miller DN, Berry ED: Cattle feedlot soil moisture and manure content: 1. Impacts on greenhouse gases, odor compounds, nitrogen losses, and dust. J Environ Qual. 2005, 34: 644-655. 10.2134/jeq2005.0644.

29. Johnson GA, Davis JG, Qian YL, Doesken KC: Topdressing turf with composted manure improves soil quality and protects water quality. Soil Sci Soc Am J. 2006, 70: 2114-2121. 10.2136/sssaj2005.0287.

30. Olson BM, Papworth LW: Soil chemical changes following manure application on irrigated alfalfa and rainfed timothy in southern Alberta. Can J Soil Sci. 2006, 86: 119-132. 10.4141/S05-024.

31. Spelter H, Winandy J, Zauche T: Anaerobically Digested Bovine Biofiber as a Source of Fiber for Particleboard Manufacturing: An Economic Analysis. Bioresources. 2008, 3: 1256-1266.

32. Tambone F, Genevini P, D'Imporzano G, Adani F: Assessing amendment properties of digestate by studying the organic matter composition and the degree of biological stability during the anaerobic digestion of the organic fraction of MSW. Bioresource Technol. 2009, 100: 3140-3142. 10.1016/j.biortech.2009.02.012.

33. Kissinger WF, Koelsch RK, Erickson GE, Klopfenstein TJ: Characters of manure

harvested from beef cattle feedlots. Appl Eng Agric. 2007, 23: 357-365.

34. Larney FJ, Ellert BH, Olson AF: Carbon, ash and organic matter relationships for feedlot manures and composts. Can J Soil Sci. 2005, 85: 261-264. 10.4141/S04-060.

35. Reinertsen SA, Elliott LF, Cochran VL, Campbell GS: Role of Available Carbon and Nitrogen in Determining the Rate of Wheat Straw Decomposition. Soil Biol Biochem. 1984, 16: 459-464. 10.1016/0038-0717(84)90052-X.

36. Hsu TA: Pretreament of biomass. Handbook on Bioethanol: Production and Utilization. Edited by: Wyman C. 1996, Washington, DC: Taylor & Francis, 179-195.

37. Rahikainen JL, Martin-Sampedro R, Heikkinen H, Rovio S, Marjamaa K, Tamminen T, Rojas OJ, Kruus K: Inhibitory effect of lignin during cellulose bioconversion: The effect of lignin chemistry on non-productive enzyme adsorption. Bioresource Technol. 2013, 133: 270-278.

38. Kumar P, Barrett DM, Delwiche MJ, Stroeve P: Methods for Pretreatment of Lignocellulosic Biomass for Efficient Hydrolysis and Biofuel Production. Ind Eng Chem Res. 2009, 48: 3713-3729. 10.1021/ie801542g.

39. Elumalai S, Tobimatsu Y, Grabber JH, Pan XJ, Ralph J: Epigallocatechin gallate incorporation into lignin enhances the alkaline delignification and enzymatic saccharification of cell walls. Biotechnol Biofuels. 2012, 5: 59-71. 10.1186/1754-6834-5-59.

40. Chen Y, Stevens MA, Zhu YM, Holmes J, Xu H: Understanding of alkaline pretreatment parameters for corn stover enzymatic saccharification. Biotechnol Biofuels. 2013, 6: 8-17. 10.1186/1754-6834-6-8.

41. Silverstein RA, Chen Y, Sharma-Shivappa RR, Boyette MD, Osborne J: A comparison of chemical pretreatment methods for improving saccharification of cotton stalks. Bioresour Technol. 2007, 98: 3000-3011. 10.1016/j.biortech.2006.10.022.

42. Kim TH, Kim JS, Sunwoo C, Lee YY: Pretreatment of corn stover by aqueous ammonia. Bioresource Technol. 2003, 90: 39-47. 10.1016/S0960-8524(03)00097-X.

43. Kim TH, Lee YY: Pretreatment and fractionation of corn stover by ammonia recycle percolation process. Bioresource Technol. 2005, 96: 2007-2013. 10.1016/j.biortech.2005.01.015.

44. Kim TH, Lee YY: Pretreatment of corn stover by soaking in aqueous ammonia. Appl Biochem Biotech. 2005, 121: 1119-1131.

45. Yue ZB, Teater C, MacLellan J, Liu Y, Liao W: Development of a new bioethanol feedstock - Anaerobically digested fiber from confined dairy operations using different digestion configurations. Biomass Bioenerg. 2011, 35: 1946-1953. 10.1016/j.biombioe.2011.01.035.

46. Gupta R, Lee YY: Investigation of biomass degradation mechanism in pretreatment of switchgrass by aqueous ammonia and sodium hydroxide. Bioresource Technol.

2010, 101: 8185-8191. 10.1016/j.biortech.2010.05.039.

47. Xu JL, Cheng JJ: Pretreatment of switchgrass for sugar production with the combination of sodium hydroxide and lime. Bioresource Technol. 2011, 102: 3861-3868. 10.1016/j.biortech.2010.12.038.
48. Chang VS, Holtzapple MT: Fundamental factors affecting biomass enzymatic reactivity. Appl Biochem Biotech. 2000, 84–6: 5-37.
49. Parsons TR, Maita Y, Lalli CM: A manual of chemical and biological methods for seawater analysis. 1984, Paris: Pergamon Press
50. McCabe WL, Smith JC, Harriott P: Operations involving particulate solids. Unit Operations of Chemical Engineering. Edited by: McCabe WL, Smith JC, Harriot P. 1993, Singapore: McGraw Hill, 925-5
51. Humbird R, Davis R, Tao L, Kinchin C, Hsu D, Aden A, Schoen P, Lukas J, Olthof B, Worley M, Sexton D, Dudgeon D: Dilute-Acid Pretreatment and Enzymatic Hydrolysis of Corn Stover. Process Design and Economics for Biochemical Conversion of Lignocellulosic Biomass to Ethanol. 2011, National Renewable Energy Laboratory /TP-5100-47764, Golden, CO, 275-3000. 303
52. Bhatia L, Johri S, Ahmad R: An economic and ecological perspective of ethanol production from renewable agro waste: a review. AMB Express. 2012, 2: 1-19. 10.1186/2191-0855-2-1.
53. Fox SC, McDonald AG: Chemical and Thermal Characterization of Three Industrial Lignins and Their Corresponding Lignin Esters. Bioresources. 2010, 5: 990-1009.
54. Daystar JS, Venditti RA, Gonzalez R, Jameel H, Jett M, Reeb CW: Impacts of Feedstock Composition on Alcohol Yields and Greenhouse Gas Emissions from the NREL Thermochemical Ethanol Conversion Process. Bioresources. 2013, 8: 5261-5278.
55. Kylliainen O, Holmbom B: Chemical composition of components in spruce bark waters. Pap Puu-Pap Tim. 2004, 86: 289-292.
56. Sluiter JB, Ruiz RO, Scarlata CJ, Sluiter AD, Templeton DW: Compositional Analysis of Lignocellulosic Feedstocks. 1. Review and Description of Methods. J Agr Food Chem. 2010, 58: 9043-9053. 10.1021/jf1008023.
57. Thammasouk K, Tandjo D, Penner MH: Influence of extractives on the analysis of herbaceous biomass. J Agr Food Chem. 1997, 45: 437-443. 10.1021/jf960401r.
58. Chen Y, Stevens MA, Zhu Y, Holmes J, Xu H: Understanding of alkaline pretreatment parameters for corn stover enzymatic saccharification. Biotechnol Biofuels. 2013, 6: 1-10. 10.1186/1754-6834-6-1.
59. Chaturvedi V, Verma P: An overview of key pretreatment processes employed for bioconversion of lignocellulosic biomass into biofuels and value added products. 3 Biotech. 2013, 3: 415-431.
60. Yu ZY, Jameel H, Chang HM, Park S: The effect of delignification of forest biomass on enzymatic hydrolysis. Bioresource Technol. 2011, 102: 9083-9089. 10.1016/j.biortech.2011.07.001.

61. Mittal A, Katahira R, Himmel ME, Johnson DK: Effects of alkaline or liquid-ammonia treatment on crystalline cellulose: changes in crystalline structure and effects on enzymatic digestibility. Biotechnol Biofuels. 2011, 4: 41-55. 10.1186/1754-6834-4-41.

62. Krassig HA: Methods of activation. Cellulose. Edited by: Krassig HA. 1993, Netherlands: Taylor & Francis, 215-258. 11

63. Sathitsuksanoh N, Zhu ZG, Wi S, Zhang YHP: Cellulose Solvent-Based Biomass Pretreatment Breaks Highly Ordered Hydrogen Bonds in Cellulose Fibers of Switchgrass. Biotechnol Bioeng. 2011, 108: 521-529. 10.1002/bit.22964.

64. Kim SB, Lee SJ, Lee JH, Jung YR, Thapa LP, Kim JS, Um Y, Park C, Kim SW: Pretreatment of rice straw with combined process using dilute sulfuric acid and aqueous ammonia. Biotechnol Biofuels. 2013, 6: 109-119. 10.1186/1754-6834-6-109.

65. Cheng YS, Zheng Y, Yu CW, Dooley TM, Jenkins BM, VanderGheynst JS: Evaluation of High Solids Alkaline Pretreatment of Rice Straw. Appl Biochem Biotech. 2010, 162: 1768-1784. 10.1007/s12010-010-8958-4.

66. Miron J, Benghedalia D: Effect of Hydrolyzing and Oxidizing-Agents on the Composition and Degradation of Wheat Straw Monosaccharides. Eur J Appl Microbiol. 1982, 15: 83-87. 10.1007/BF00499511.

67. Lu FC, Ralph J: The DFRC method for lignin analysis. Part 3. NMR studies. J Wood Chem Technol. 1998, 18: 219-233. 10.1080/02773819809349578.

68. Blumenkr N, Asboehan G: New Method for Quantitative-Determination of Uronic Acids. Anal Biochem. 1973, 54: 484-489. 10.1016/0003-2697(73)90377-1.

69. Park S, Baker JO, Himmel ME, Parilla PA, Johnson DK: Cellulose crystallinity index: measurement techniques and their impact on interpreting cellulase performance. Biotechnol Biofuels. 2010, 3: 10-19. 10.1186/1754-6834-3-10.

70. Overend RP, Chornet E: Fractionation of Lignocellulosics by Steam-Aqueous Pretreatments. Philos T R Soc A. 1987, 321: 523-536. 10.1098/rsta.1987.0029.

71. Chum HL, Johnson DK, Black SK, Overend RP: Pretreatment Catalyst Effects and the Combined Severity Parameter. Appl Biochem Biotech. 1990, 24–5: 1-14.

72. Sasikumar E, Viruthagiri T: Optimization of Process Conditions Using Response Surface Methodology (RSM) for Ethanol Production from Pretreated Sugarcane Bagasse: Kinetics and Modeling. Bioenerg Res. 2008, 1: 239-247. 10.1007/s12155-008-9018-6.

INDEX

N

O

P

R

S

V

W